Walter Peschka

Liquid Hydrogen

Fuel of the Future

Translated by
Edmund A. Wilhelm and Ulrike Wilhelm

Springer-Verlag Wien New York

Prof. Dr. Ing. habil. Walter Peschka

Deutsche Forschungsanstalt für Luft- und Raumfahrt e.V.
Stuttgart, Bundesrepublik Deutschland

Softcover reprint of the hardcover 1st edition 1992
Printed on acid-free paper

With 147 Figures

Cover-design: T. Erben, Wien

ISBN-13: 978-3-7091-9128-6 e-ISBN-13: 978-3-7091-9126-2
DOI: 10.1007/978-3-7091-9126-2

Je crois que l'eau sera un jour employée comme combustible, que l'hydrogène et l'oxygène, utilisés isolément ou simultanément, fourniront une source de chaleur et de lumière inépuisables et d'une intensité que la houille ne saurait avoir. Un jour, les soutes de steamers et le tenders des locomotives, au lieu de carbon, seront chargés de ces deux gaz comprimés, qui brûleront dans les foyers avec une énorme puissance calorifique.
... L'eau est le carbon de l'avenir.

Jules Verne, L'Ile Mystérieuse, 1870, Tome II, Chap. XI.

I believe that water will one day be employed as fuel, that hydrogen and oxygen which constitute it, used singly or together, will furnish an inexhaustible source of heat and light, of an intensity of which coal is not capable. Some day the coal bunkers of steamships and the tenders of locomotives will, instead of coal, be stored with these two condensed gases, which will burn in the furnaces with enormous heat energy.
... Water is the coal of the future.

Jules Verne, Mysterious Island, 1870, Part II, Chap. XI.

Translation of the Preface to the German Edition

This book is based on published material, oral presentations and lecture courses, as well as the author's personal research in the specific field of space technology and in the general areas of energy storage and transfer, and cryogenics.

The science and technology of liquid hydrogen—once essential prerequisites for the rapid development of space technology—are now also proving to be more and more important for the energy production of the future. Hydrogen as an energy carrier can generally mediate the existing disparity between nuclear energy and regenerative energy, both of which are indispensable for the future.

Hydrogen, as a secondary energy carrier, can be produced from these primary energy sources with minimal environmental impact and without the detrimental, long-term pollution effects of current fossil fuel technology. Hydrogen, therefore, represents the ultimate in energy technology.

The initial, large-scale application of hydrogen as a secondary energy was as a high-energy rocket propellant.

The procedures for its large scale liquefaction, storage and employment were generally developed in the U.S. Currently in Europe similar activities are being conducted only in France. The effort in West Germany involves testing hydrogen–oxygen and hydrogen–fluorine rocket engines, studying also the physical and technical characteristics of slush hydrogen—mixture of the solid and liquid phase—and is concentrating currently on R&D applications of liquid hydrogen as an alternate fuel. Similar activities are also being conducted in Japan and Canada.

The objectives of this book are: to provide an overview of the state of the art of liquid hydrogen energy technology, indicate the enormous potential for future applications of this unique energy carrier, and stimulate the development of further R&D work.

In order to focus the scope of this presentation only information dealing with liquid hydrogen and its capability as a secondary energy carrier is developed in detail. Appropriate attention is given to current research initiatives, magnetocaloric refrigeration for example, as well as to aviation and ground transportation applications, both of which are of considerable importance for the energy technology of the future.

The expert will surely miss some topics from this already widely diversified field which could not be properly included in this book. At this point the standard reference, "Technology and Uses of Liquid Hydrogen" (R.B. Scott, ed., 1964), must be noted.[1] This book, written in the pioneer days of liquid hydrogen technology, provides the reader with a sense of the commitment and dedication with which, in a short time, liquid hydrogen technology evolved from laboratory applications to large-scale industrial use. We can only hope for the same kind of dedication in accomplishing future work that has long been recognized as necessary in the field of energy technology.

Complete coverage of the technical literature was not possible within this framework and was not attempted. Selection was based on ease of access as well as providing for the reader, in so far as possible, a practical introduction to this unique field. Publications prior to the latter part of 1983 were reviewed while material published during the time this book was in preparation were only partially reviewed.

Before concluding I would like to express my gratitude to all the friends and colleagues here and abroad who, through their work, contributed to the inspiration and preparation of this book.

This book is dedicated to those who, through their unselfish effort and determination, have already contributed to the establishment of the foundation for an alternate energy technology via the "coal of the future" for the benefit of future generations.

Sindelfingen, May 1984 W. Peschka

[1] Also compare Sloop, J. F.: Liquid Hydrogen as a Propulsion Fuel (p. 100).

Preface

This English version largely corresponds to the German version from the year 1984. In this version the author has sought to include the essential developments which have been made since 1984 concerning liquid hydrogen as an energy carrier. In the same way the reference to literature was also expanded as much as possible.
In Chap. 6, "Liquid Hydrogen as Fuel" the developments of the last six years which were marked by increased activity by renowned automobile producers were primarily taken into consideration in Sect. 6.2 "Ground Transport".
In order to provide a more extensive description in Sect. 5.2, Chap. 7 of the German version "LH_2 in Nuclear Research and Nuclear Technology" was left out.
The author expresses his gratitude to the publisher Springer-Verlag, Vienna as well as to Mr. and Mrs. E. Wilhelm who provided the essential contributions for the publication of the English version.
Above all, the author sincerely expresses his deep gratitude to Dr. Fred J. Edeskuty and Walter F. Stewart from the Los Alamos National Laboratory who provided valuable advice, important supplements and improvements for the English manuscript.

Sindelfingen, October 1991 W. Peschka

Translators Preface

At the outset of our translation effort we realized that we were not sufficiently familiar with many of the technical terms in the field of liquid hydrogen. In order to become better acquainted with the terminology we decided to do some research of our own. This research touched on many of the topic areas covered in the author's book including physical chemistry, thermodynamics, cryogenics and energy systems. In the course of this research it became apparent that there is a definite niche to be filled by Professor Peschka's book "Liquid Hydrogen as an Energy Carrier".

We were able to find lots of technical literature on liquid hydrogen. There are numerous technical papers on various methods for the production of liquid hydrogen as well as its potential uses. Most of this literature dealt with a specific aspect of liquid hydrogen and therefore quickly went into detail that could be followed only by the most resolute of readers. We were not able to find a single reference which provided us with a general, easily read, overview of liquid hydrogen. Among other things, the author's book fills this void.

We also noted during our research that, although the U.S. has the most experience dealing with liquid hydrogen, most of the documentation on liquid hydrogen was originally written in a language other than English. This did not help us in our efforts. Although we wanted to use technical terms currently, we did not want to fall into the same traps as previous translators. Therefore, in certain cases experts in the field of liquid hydrogen may take exception to our choice of terminology. However, we feel confident that we have achieved the author's primary intent. We have provided an English version that is easily read and consistent in its use of terminology.

Mr. & Mrs. E. Wilhelm — Weil der Stadt

Contents

Introduction 1

Hydrogen is the ninth most abundant chemical element on Earth. The entire universe consists of more than 90% hydrogen, in the stars in a plasma state and interstellar matter mostly in an atomic state. In the solar system, aside from the sun, interplanetary gas as well as the outer planets (Jupiter, Saturn, Uranus and Neptune) consist primarily of hydrogen.

On Earth, especially in the atmosphere, free hydrogen can be found only in negligible amounts. It emanates from volcanos mixed with other gases and it can be found in underground inclusions of potash salt beds.

Hydrogen can be set free through various pyrolitic processes as for instance in the formation of fossil fuels in the geological past. In a combined state, hydrogen exists in water as well as in almost all organic compounds. As a chemical component of water it is available in sufficiently large quantities on the Earth's surface, whereas in its gaseous form it cannot be held within the atmosphere by the Earth's gravity.

The production of hydrogen from water requires the input of a considerable amount of energy. Primary energy is used in the production of the energy carrier, hydrogen, which aside from its availability, is distinguished by its compatibility with other energy carriers such as electricity or synthetic natural gas (SNG).

From an ecological point of view hydrogen is unequalled for environmental safety. In the production of hydrogen from water the same amount of oxygen is produced, which is later used in a practical, almost pollution-free combustion. In contrast, the combustion of fossil energy sources affects the balance of the oxygen and carbon dioxide in the atmosphere. This hydrogen cycle, more specifically this water cycle, has a considerably higher capacity with regard to energy transfer than the currently used carbon cycle.

Hydrogen, described by Cavendish[1] in 1766 as "inflammable air" and called "hydrogenium" by Lavoisier, was liquefied for the first time by Dewar in 1898. For more than half a century liquid hydrogen was produced only in small amounts and its use was basically limited to laboratory tests. Initial applications were principally for the generation of low temperatures and

[1] Phil. Trans., 1766, p. 141, p. 145.

the measure of the specific heat of solids down to approximately 13.8 K, the lowest possible temperature attainable with liquid hydrogen. In addition to these tests, primarily carried out by Simon in Oxford with the objective of proving Nernst's theorem, work on the free energy of chemical bonding in the temperature range from liquid hydrogen to ambient temperature was later carried out, especially at the University of California by W. F. Giauque from 1920 to 1930.

For a long time, the unusual behavior of the specific heat of hydrogen, especially at low temperatures, could not be explained. Due to the theoretical work in quantum-theory by E. Schrödinger and W. Heisenberg, the latter was able to prove that the two forms of the hydrogen molecule, ortho-hydrogen with parallel nuclear spins and para-hydrogen with anti-parallel nucelar spins were responsible. Ortho-hydrogen is about three times more abundant than para-hydrogen at ambient temperature, whereas at low temperature the latter exists almost exclusively. The large difference in the internal energy between ortho- and para-hydrogen has considerable consequences for practical applications. If hydrogen in its "normal composition" (75% ortho and 25% para) is liquefied, the liquid hydrogen initially has these ortho-para-composition, however the thermodynamic equilibrium at 20 K corresponds to a composition of 99.8% para-hydrogen. Liquid hydrogen of "normal composition" slowly changes to para-hydrogen with considerable heat emission. This is especially important in the long-term storage of hydrogen because the adjustment of the equilibrium must be accomplished during liquefaction with the help of appropriate catalysts to avoid additional excessive evaporation.

Following pioneer work by Herrick L. Johnston at the Cryogenic Laboratory of the Ohio State University from 1940–1950, the National Bureau of Standards (now National Institute for Standards and Technology, NIST) established a laboratory in Boulder, Colorado in 1952 (commissioned by the U.S. Atomic Energy Commission) to produce larger amounts of liquid hydrogen and perform intensive testing with the goal of technical applications. Included among these applications were primarily the liquid hydrogen bubble chambers for nuclear research, the first of which was built at the Lawrence Radiation Laboratory (LRL) in 1959, as well as nuclear and chemical rocket propulsion.

Nuclear rocket propulsion systems were tested intensively from 1960 to 1970, but they never advanced past the experimental stage. However, the use of liquid hydrogen together with liquid oxygen as a rocket fuel has been developed intensively since 1959. The first successful launch of a hydrogen–oxygen rocket from Cape Kennedy on the 27th of November 1963—the upper stage of a NASA Centaur rocket with an Atlas booster—led to the NASA Apollo program and the Space Shuttle. These projects resulted in a large-scale technical application of liquid hydrogen within a short time.

In addition, the plant constructed and operated by Linde's Eismaschinen AG in 1958 for the production of deuterium by means of the fractional distillation of liquid hydrogen should be noted. The first significant technical handling of large amounts of liquid hydrogen in Europe took place in this plant and in a similar plant constructed by Sulzer in Switzerland.
In addition to these recent applications of liquid hydrogen we can foresee its future applications as an energy carrier. Current experience indicates, to a large degree, that a conversion of the energy supply to non-fossil carriers of primary energy will be necessary around the beginning of the next century in order to guarantee satisfaction of the energy requirements of the future as well as for ecological reasons.
Generally, nuclear energy and solar energy, in all their forms, are considered as primary energy carriers. But the use of hydrogen as a secondary energy carrier, due to its compatibility with existing energy carriers like electricity and synthetic energy carriers derived from coal, must be kept open as an option[2].
As a result of its storage and transportation characteristics liquid hydrogen offers possible advantages in the supply of energy in general and as an alternative fuel. It is unequalled for environmental compatibility in aviation as well as in ground transportation. Although these applications will surely not be the first to be implemented on a large-scale, they offer the possibility of demonstrating the use and safe handling of hydrogen as an energy carrier on an adequate scale and they can make important contributions toward substantially more complex applications in the future.

[2] "The Forever Fuel—The Story of Hydrogen" by P. Hoffmann, Boulder, Colorado, Westview-Press (1981), is recommended to the reader who is interested in getting a quick overview.

Hydrogen Production 2

Today, hydrogen serves, primarily, as a raw material in chemical processes. It is used only on a small, although increasing, scale as an energy carrier, as in rocket fuel for example.
To be used as a future energy carrier an increase in the level of current production by at least two orders of magnitude is required. Therefore, large-scale production procedures are an essential prerequisite for the use of hydrogen as an energy carrier in the future.

2.1 Hydrogen from Fossil Raw Materials

At the present time hydrogen is generally produced on a large-scale by steam reforming, with petroleum or natural gas as the initial material and also by the partial oxidation of crude oil [1–3, 12, 13]. With the steam reforming of hydrocarbons a reaction takes place in the presence of an appropriate catalyst in the pressure range from atmospheric to approximately 30 bar at temperatures between 700 and 900 °C as follows

$$C_nH_m + nH_2O \rightarrow nCO + (0.5m + n)H_2.$$

Partial oxidation takes place at a pressure between 20 and 60 bar and a temperature in the range from 1100 to 1500 °C in an exothermic reaction as follows

$$C_nH_m + 0.5nO_2 \rightarrow nCO + 0.5mH_2.$$

Through "shifting", CO is oxidized in a catalytic exothermic reaction

$$CO + H_2O \rightarrow CO_2 + H_2.$$

The steam-iron process [4] is of some importance, in which, at high temperatures between 950 and 980 °C and pressures between 2 and 3 bar, iron initially oxidizes and hydrogen is released

$$2Fe + 3H_2O \rightarrow Fe_2O_3 + 3H_2$$
$$3Fe + 4H_2O \rightarrow Fe_3O_4 + 4H_2.$$

After the oxidation process, the iron oxide produced is reduced in a cyclic operation by an exothermic reaction with a reduction gas, generally CO.

$$Fe_2O_3 + CO \rightarrow 2FeO + CO_2$$
$$FeO + CO \rightarrow Fe + CO_2$$
$$Fe_3O_4 + 4CO \rightarrow 3Fe + 4CO_2$$

A recently developed endothermic procedure [5] uses the catalytic cracking of hydrocarbons in a temperature range between 700 and 980 °C, as for example

$$CH_4 \rightarrow C + 2H_2.$$

The heat required for the process can be regeneratively supplied by a parallel combustion of the carbon with air.
The water–gas process [6] with coke or coal is the oldest method used for the production of CO and hydrogen. The endothermic reaction takes place at temperatures around 1000 °C in accordance with the simplified reaction equation

$$C + H_2O \rightarrow CO + H_2.$$

During intermittent operation the heat required for the reaction is supplied from the combustion of carbon with air. With the use of pure oxygen instead of air, continuous exothermic operation is possible by adding the oxygen to the steam.
Several other procedures have been developed that are primarily for the production of synthetic gas and hydrogen using coke or coal with air or pure oxygen [6, 7, 12–20].
Furthermore, it must be mentioned that coal gas, through "dry distillation" of coal—carbonization, i.e., heating up to 1200 to 1300 °C under vacuum—contains a high percentage of hydrogen (see Table 2).
Coal gas is also an accepted heating gas, in "Siemens–Martin Furnaces", in industrial power plants for example, and it was formerly used as "town gas" for domestic heating and cooking.

Table 1. Advanced coal gasification processes

	Pressure	Characteristics	
Koppers process	1 bar	autotherm, high O_2 requirement	[7]
Winkler process	1 bar	autotherm, high O_2 requirement fluidized bed	[8, 9]
Lurgi process	30 bar	autotherm, low O_2 requirement solid bed	[10, 11]

Table 2. Average composition of coal gas in Vol.%

Hydrogen	53%
Methane	23%
Nitrogen	12%
Carbon monoxide	2.5%
Ethane	2%

Since the heat requirement of coal gasification is a considerable cost factor, there exists considerable interest in the use of cost effective nuclear process heat produced by gas-cooled high-temperature reactors (see [21–26]).
The main problem with hydrogen production by nuclear coal gasification is the transmission of the considerable endothermic reaction heat to the mixture of the coal and the vaporization medium. Since nuclear heat is supplied in the form of hot helium, which has to be returned to the reactor with practically no loss and free from impurities, heat exchangers are necessary. The transfer of heat in these exchangers creates considerable technical problems due to the corrosion, erosion, and pollution of the heating surfaces.
All methods of hydrogen production from fossil raw materials essentially involve water decomposition through the oxidation of carbon to CO and CO_2. The amount of CO_2 released into the atmosphere is especially high with coal gasification. It is either 0.75 or 0.5 mol of CO_2 per mol H_2 (16.5 kg or 11 kg of CO_2 per kg of H_2 respectively) depending on whether the reaction heat, which has to be produced for transformation, is provided by coal combustion or non-fossil energy. Thus, environmental problems of the first degree (the atmospheric CO_2 content) can arise from hydrogen production via coal gasification as well as from the production of synthetic natural gas (SNG) and synthetic fuels based on hydrocarbons. Furthermore, in hydrogen production via coal gasification the considerable amount of sulfur dioxide as well as the ash produced must be eliminated in order to avoid environmental damage. In addition, the hydrogen must further undergo a more or less costly purification process, depending upon its intended use [48–52].

2.2 Electrolytic Hydrogen Production

For a long time hydrogen has been produced on a large scale through the decomposition of water via electrolysis [27, 28]. This is primarily important where cheap hydroelectric power is available (Norway, Canada) and where hydrogen is needed as a raw material, as for example in the production of artificial fertilizers by means of ammonia synthesis.

Hydrogen is also obtained as an important by-product in the chemical industry from chlorine production using chlorine alkali electrolysis. A hydrogen-pipeline network has long been in operation in the Ruhr industrial district to provide for its distribution and the satisfaction of energy requirements using the hydrogen produced by this process[3].

Because hydrogen obtained via electrolysis is about two to three times more expensive than hydrogen obtained from fossil raw materials, except in those areas having an overabundant supply of electricity, it is primarily used where there is a requirement for a high degree of purity.

The investment cost for the production plant essentially determines the price of the hydrogen. This cost, as well as the efficiency of the electrolyzer and the cost of the electricity, which reflects the operating costs, can be reduced by further development of existing procedures as well as through newly developed procedures of water electrolysis. The major development requirements involve the reduction of cell voltages, the increase of current density, the increased application of cheaper cell and electrode materials as well as the increase in operating temperature and pressure [30, 33].

From a thermodynamic standpoint the increase of the operating temperature is particularly advantageous, although essential material problems will have to be solved simultaneously. For water decomposition

$$H_2O \rightarrow H_2 + 1/2O_2 \tag{1}$$

without consideration of loss, the required electrical energy W for the reversible process is

$$W = \Delta G = \Delta J - T\Delta S. \tag{2}$$

ΔG is the change in the Gibbs function (free enthalpy) (237 kJ/mol at 25 °C and hydrogen at 1 bar pressure), J is the reaction enthalpy (286 kJ/mol), T is the absolute temperature, and ΔS is the reaction entropy (203 J/mol K).

For the required electrical energy the following applies

$$W = \Delta G = F \cdot n \cdot E \tag{3}$$

where F is the Faraday constant (96485 As/mol), $n = 2$, the number of interchanged electrons per elementary reaction (1), and E is the ideal (reversible) voltage of decomposition.

This results in

$$E = \frac{\Delta G}{n \cdot F} = 1.229\ \text{V}. \tag{4}$$

Due to irreversible processes in the reaction mechanism at the

[3] See C. Isting (Chem. Werke Hüls AG), pp. A2–6 of Eurocon: 74 Digest. The Hague, Netherlands: Royal Instn. Engrs. 1974.

electrodes—side-effect reactions as well as the influence of the reaction kinetics—and heat production to the resistance in the electroytes, the real decomposition voltage E_r is always higher than the ideal decomposition voltage E. This means that only the $E \cdot I$ (where I is the cell current) part of the supplied electrical power, $E_r \cdot I$, is used for hydrogen production, whereas the portion $(E_r - E) \cdot I$ is transformed into heat inside the cell. The voltage difference $E_r - E$ is described as "excess voltage" and should be kept as small as possible in the interest of high electrolyzer efficiency.
Since the change ΔG of reaction (1) is smaller than the reaction enthalpy change ΔJ by the amount $T \cdot \Delta S$, in the ideal case the cell operation requires a supply of electrical energy $W = \Delta G$ as well as a supply of heat $Q = T \cdot \Delta S$, which can be provided by the heat produced as a result of the excess voltage in the cell.
If the excess voltage $E_r - E$ exceeds the amount E'

$$E' = \frac{T \cdot \Delta S}{n \cdot F} = 0.253\ \text{V} \tag{5}$$

or in other words,

$$\text{if } E_r > E_{th} \quad (E_{th} = 1.482\ \text{V}), \tag{6}$$

then the heat production exceeds $T \cdot \Delta S$, the amount required for the reaction, and the cell emits heat. E_{th} is described as the thermoneutral cell voltage[4].
Electrolyzers currently in use are operated at cell voltages from 1.8 V to about 2.0 V, which corresponds with electrolyzer efficiencies between 60–70% [29, 30, 33].
With increasing temperature T the ideal decomposition voltage E decreases.

$$E = E_{th} - T \cdot \frac{\Delta S}{n \cdot F} \tag{7}$$

$$\frac{\Delta S}{n \cdot F} = 25\ \text{mV/K}. \tag{8}$$

This results in a decrease in the demand for electrical energy and an increase in the heat required (see Fig. 1). Since an increase in the temperature provides an additional improvement in the reaction kinetics at the electrodes and thus an increase in the current density, i.e., an improved cell utilization, this represents the most effective way to develop new, more efficient electrolyzers, aside from developing improved electrode materials.

---

[4] I.e., $E_{th} = E + \frac{T \cdot \Delta S}{n \cdot F} = \frac{\Delta J}{n \cdot F}$.

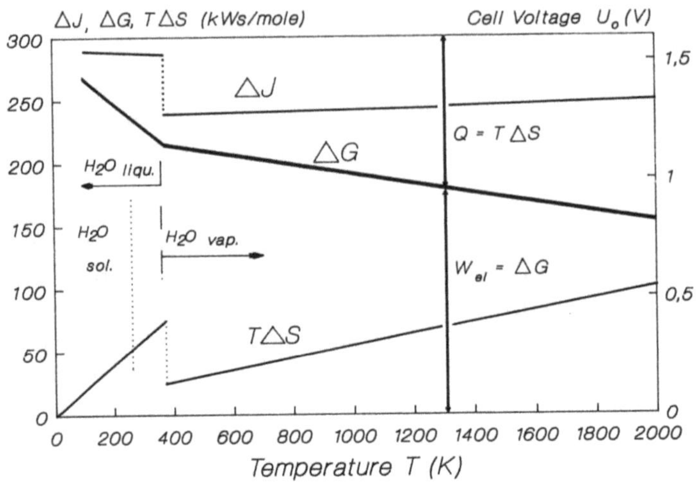

Fig. 1. Thermodynamics of water decomposition: change in Gibbs function (free enthalpy) ΔG, reaction enthalpy ΔJ and reaction entropy ΔS as a function of temperature T

Alkaline Low-Temperature Electrolysis

Conventional alkaline low-temperature electrolysis has been in use for more than 80 years. At operating temperatures between 60–80 °C, diluted potassium hydroxide (KOH) or diluted sodium hydroxide (NaOH) are normally used as electrolytes. The current, generally accepted filter-press bipolar cell has steel cathodes and nickel anodes or nickel-plated steel (which are separated by asbestos diaphragms) [27, 28, 31, 33].
The electrical efficiency, as a result of the net caloric value of the hydrogen (33 kWh/kg), is about 60–70%. With a maximum current density of 3000 A/m^2 applied to the electrode surface this corresponds to a cell voltage of about 1.8 V–2.0 V and an electrical energy consumption of more than 50 kWh/kg (4.5 kWh/Nm^3).
An improvement can be achieved through electrolysis under pressure, where pressures up to 30 bar are used (Zdanski–Lonza Procedure, Lurgi Procedure) [10].

Alkaline Medium-Temperature Electrolysis

The current low-temperature electrolysis procedure can be improved in efficiency (decrease in the excess electrode voltage and increase in current density) and in economy by shifting to higher operating temperatures and their associated higher operating pressures [33].

The goals of currently ongoing development projects in alkaline medium-temperature electrolyzers are operating temperatures in the range of 150–250 °C with corresponding pressures between 30 and 100 bar and cell voltages below 1.6 V (with a corresponding efficiency of 70–80%) with current densities in the range of 10^4 A/m^2. Durable diaphragms as well as longer lasting electrode materials are currently being developed in order to attain these goals [32].

Electrolysis with Solid-State Electrolytes

Here, instead of the diaphragm and liquid electrolyte, a solid-state ion conducting membrane (solid polymer electrolyte (SPE)) is used. Demineralized water, which is absorbed by the solid electrolyte, is used for electrolysis [31, 33, 34].

Low-temperature electrolyzers, which are currently used in smaller systems, use high polymer $[H_3O]^+$ ionic conductors made of perfluorated polysulfonate at operating temperatures up to 150 °C and operating pressures up to 200 bar. The high conductivity of the acidic electrolytes, as well as membranes less than 0.5 mm thick, decrease loss due to resistance and provide efficiencies of around 75–80%, even at high current densities (10^4 A/m^2).

Because of the constant corrosion associated with acidic solid electrolytes, precious metal electrodes and platinum catalysts have been required. Future large-scale applications of low-temperature, solid-state electrolysis depend largely on the development of more economical materials for electrodes and catalysts.

For thermodynamic reasons (see Fig. 1) the use of high operating temperatures, which are sought in high-temperature steam electrolysis, is advantageous because, with decreasing demand in electrical energy, a considerable part of the energy required for the decomposition of water can be supplied in the form of high-temperature heat.

The high-temperature cells currently under laboratory development use ceramic oxygen ionic conductors and operate in the temperature range between 800 and 1000 °C. At the present time technical application is not in sight because of the associated material problems. These problems in particular require a feasible technical solution in order to use such high temperatures in the electrolyte cell [35, 36].

2.3 Thermochemical Water Decomposition Procedure

From Fig. 1 it can be determined that above a temperature of about 1800–2000 K the required electrical energy for electrolytic water decomposition approaches zero, i.e., water decomposition is possible solely from the addition of heat via thermal dissociation [36]. At these temperatures the partial pressure of the hydrogen (about 10 mbar) is far too low for large-scale technical applications. The partial pressure of hydrogen first reaches the 1 bar range at temperatures between 3000 and 4000 K. The use of these temperatures is not technically feasible with current container materials. In addition, there is the problem of the separation of hydrogen and oxygen within this temperature range. A new starting point for the solution to these problems may result from either the further development of gaseous nuclear reactors, originally investigated for rocket propulsion (fissionable material in the plasma state) or through the transfer of such developments to a high-temperature solar technology which still remains to be developed.

Technically feasible procedures for water splitting via thermochemical processes are only applicable, according to the latest research developments, when the required temperatures do not exceed 800–1000 °C. Water splitting by multistage thermochemical closed cycle processes or hybrid processes are essentially feasible—where the latter includes steps which entail electrochemical reactions [37–40, 57–60]. These closed cycle processes consist of a sequence of chemical reaction stages with different chemical reactants in which water splitting is the end result and the other substances involved are regenerated, through an appropriate process, to their initial composition.

The main problem of achieving satisfactory overall energy efficiency with reasonable expenditure for the production process has not been solved despite intensive research. It cannot be determined if and when these thermochemical processes can be employed in large-scale applications.

2.4 Further Procedures for Hydrogen Production from Water

Here, the coupling of semiconductor photoelements (solar cells) with electrolyzers must be mentioned. This process will gain importance in the future with the reduction in cost for the solar generators needed to supply the electric current [41, 43]. As a result of a previously existing, practically ideal match between the current-voltage characteristics of photovoltaic solar generators and electrolyzers, a direct coupling, without the requirement for intermediate electronic switching circuits for power matching, is possible and should be attempted [42, 53–56]. Another method of hydrogen production via water electrolysis is the use of electricity from ocean thermal power plants [44]. Additional approaches for the production of hydrogen from water can be derived from the direct coupling of radiation energy with molecular processes. Among these are radiolysis [45], photo-biological processes based on natural photosynthesis, photochemical processes, as well as photoelectrochemical processes using semiconductors [46, 47]. All these procedures are currently at the basic research stage. Despite the currently attainable low efficiencies of less than 1%, with appropriate progress in the field of organic and nonorganic catalysts and photosensitizers their application in the future is not ruled out.

References

[1] Ullmanns Enzylkopädie der Technischen Chemie, Bd. 18, 3. Aufl., Wasserstoff.
[2] W. Balthasar, P. J. Hambleton: Industrial Scale, Production of Hydrogen from Natural Gas, Naphta and Coal. Int. J. Hydrogen Energy 5, No. 1 (1980).
[3] T. R. Roszkowski, N. W. Snyder: Present and Emerging Hydrogen Production Technology. In: Proc., 4th. World Hydr. Energy Conf. Vol. 1. pp. 47–68. Pasadena 1982.
[4] P. V. Clifton: The Schultz Steam Iron Process. Industr. Eng. Chem. *21*, 161–166 (1959).
[5] J. B. Pohlenz, L. O. Stine: New Process Promises Low Cost Hydrogen. Oil and Gas *23*, 127–132 (1962).
[6] H. W. Gratkowski: Kohlevergasungs-Verfahren. Ullmanns Enzyklopädie der Technischen Chemie, Bd. 10 (1958), Ergänzungsband (1970).
[7] K. H. Osthaus: Kohlevergasung nach dem Koppers-Totzek-Verfahren für den Kraftwerksbetrieb. VGB-Kraftwerkstechnik *54*, 221–224 (1974).
[8] J. F. Farnsworth, D. M. Mitsak, R. Wintrell: Production of Gas from Coal by the Koppers-Totzek-Process. In: Proc. Symp. Clean Fuels from Coal, pp. 143–162. Inst. of Gas Techn., Chicago, December 1973.
[9] I. N. Banchik: The Winkler Process for the Production of Low BTU-Gas from Coal. Proc. Symp. Clean Fuels from Coal, pp. 163–178. Inst. of Gas Techn., Chicago, December 1973.
[10] Anon.: Lurgi-Handbuch, 2. Aufl., Kap. 2.1. Frankfurt/Main, Lurgi Gesellschaften (1970).
[11] H. Bierbach, H. Jockel: Weiterentwicklung der Lurgi-Druckvergasung. Stahl und Eisen 7, 371–376 (1980).
[12] T. King, G. R. Hill: Erforschung und Entwicklung der Kohlevergasung in den U.S.A. GWF-Gas/Erdgas *115*, 201–205 (1974).
[13] K. A. Van Heek: Überblick über den internationalen Entwicklungsstand der Kohlevergasungsverfahren. Stahl und Eisen 7, 307–321 (1980).
[14] Ch. Brecht, H. W. Gratkowski, G. Hoffmann: Vergasung und Hydrierung der Kohle—Eine tabellarische Übersicht der in- und ausländischen Entwicklungen sowie der großtechnisch eingesetzten Verfahren. Gaswärme Intern. *29*, 367–387 (1980).
[15] R. J. Grace: Development of the BI-Gas Process. In: Proc., Symp. Clean Fuels from Coal, pp. 178–197. Inst. of Gas Techn., Chicago, December 1973.
[16] C. E. Fink: The CO_2-Acceptor Process. In: Proc. Symp. Clean Fuels from Coal, pp. 301–309. Inst. of Gas Techn., Chicago, December 1973.
[17] F. Schora Jr., B. S. Lee, J. Hübler: The Hygas Process. In: Proc. Symp. Clean Fuels from Coal, pp. 218–240. Inst. of Gas Techn., Chicago, December 1973.
[18] A. E. Cover, W. C. Schreiner, G. D. Skaperdas: Kellog's Coal Gasification Process. Chem. Eng. Progr. *69*, 31–36 (1973)
[19] A. J. Forney, J. P. McGee: The Synthane Process—Research Results and Prototype Plant Design. AGA Synthetic Pipeline Gas Symp., Chicago (1972).
[20] R. M. Nadkarni, C. Bliss, W. I. Watson: Underground Gasification of Coal. Chemtech. *6*, 230–237 (1974).

[21] N. Fichtner: Wirtschaftliche Aussichten von mit nuklearer Prozeßwärme erzeugtem technischen Wasserstoff. Studie im Auftr. d. BMFT, NT 125 (1971).
[22] KFA Jülich: Entwicklung von Verfahren zur Umwandlung fester fossiler Rohstoffe mit Wärme aus Hochtemperaturreaktoren. KFA Jülich GmbH, Halbj. Bericht 1973/I.
[23] R. Schulten: Kernernergie zur Prozeßwärmenutzung. Atomwirtschaft *55*, 491–494 (1972).
[24] R. Schulten: Nuclear Energy as a Primary Energy Source for Hydrogen Production. Int. J. Hydrogen Energy *5*, 281–292 (1980).
[25] R. N. Quade, L. Meyer: Nuclear Heat Source for Hydrogen Production. Int. J. Hydrogen Energy *4*, 101–110 (1979).
[26] K. H. Heek: Wasserdampfvergasung von Kohle unter Einkopplung nuklearer Prozeßwärme aus Hochtemperatur-Kerreaktoren. In: Kohlevergasung in der Energietechnik. In: Proc., VGB-Kraftwerkstechnik GmbH, Essen (1979).
[27] Anon.: Ullmanns Enzyklopädie der Technischen Chemie, Bd. 6. 1953.
[28] Anon.: Gmelins Handbuch der anorganischen Chemie, Wasserstoff. 8. Aufl., p. 227, 244, 257 (1958).
[29] N. C. Hallet: Cost and Systems Analysis of Liquid Hydrogen Production. NASA-CR-73.266, 1968.
[30] D. P. Gregory, J. O. M. Bockris: The Hydrogen Economy. New York, Plenum Press (1972).
[31] A. Gann: Hydrogen Production by Water Electrolysis. ESA TT-250, 1976, see also DLR-Mitt., Nr. 39–74, Über die Herstellung von Wasserstoff durch Wasserelektrolyse, (1974).
[32] R. Renaud, R. L. LeRoy: Separator Materials for Use in Alkaline Water Electrolysis. Int. J. Hydrogen Energy 7, 155–166 (1982).
[33] A. Menth, S. Stucki: Present State and Outlook of the Electrolytic H_2-Production Route. In: Proc., 2nd. World Hydrogen Energy Conf. Vol. 1, pp. 55–63 (1978).
[34] L. J. Nuttal, J. H. Russel: Solid Polymer Electrolyte Water Electrolysis. Int. J. Hydrogen Energy *5*, 75–84 (1980).
[35] W. Dönitz, W. Schmidberger: Concepts and Design for Scaling up High Temperature Water Vapour Electrolysis. Int. J. Hydrogen Energy 7, 321–330 (1982).
[36] W. Dönitz, R. Schmidberger, E. Steinheil, R. Streicher: Hydrogen Production by High Temperature Electrolysis of Water Vapour. Int. J. Hydrogen Energy *5*, 55–64 (1980).
[37] K. F. Knoche, H. Cremer, D. Breywisch, et al.: Experimental Theoretical Investigation of Thermochemical Hydrogen Production. Int. J. Hydrogen Energy *3*, 209–216 (1978).
[38] G. E. Beghi: Review of Thermochemical Hydrogen Production. Int. J. Hydrogen Energy *4*, 555–566 (1981).
[39] G. E. Beghi: A Decade of Research on Thermochemical Hydrogen at the Joint Research Centre, Ispra. Int. J. Hydrogen Energy *11*, 761–772 (1986).
[40] F. Deneuve, J. Roncato: Thermochemical of Hybride Cycles of Hydrogen Production—Techno-Economical Comparison with Water Electrolysis. Int. J. Hydrogen Energy *6*, 9–24 (1981).
[41] R. Dahlberg: Replacement of Fossil Fuels by Hydrogen. Int. J. Hydrogen Energy 7, 121–142 (1982).
[42] C. Carpetis: A Study of Water Electrolysis with Photovoltaic Solar Energy Conversion. Int. J. Hydrogen Energy 7, 287–310 (1982).
[43] W. J. D. Escher, J. A. Hanson: Ocean Based Solar-to-Hydrogen Energy Conversion Macro system. In: Hydrogen Energy, Part A (T. N. Veziroglu, ed.), pp. 209–229. New York, Plenum Press (1975).
[44] G. L. Dugger, H. L. Olsen, W. B. Shippen et al.: Tropical Ocean Thermal Power Plants and Potential Products. In: Proc. Solar Energy for Earth, Conf., Los Angeles, Calif., Am. Inst. for Aeron. and Astron., New York (1975). AIAA-paper 75-167, 16 p.

[45] Y. Ikezoe, S. Sato, S. Shimizu, H. Nakajima: Potential of Carbon Doixide Radiolysis for Hydrogen Production. Int. J. Hydrogen Energy 7, 539–544 (1982).
[46] E. Broda: Hydrogen Production Through Solar Radiation by Means of Water Photolysis in Membranes. Int. J. Hydrogen Energy *3*, 119–122 (1978).
[47] N. Getoff et al.: Wasserstoff als Energieträger, 173–322. Wien New York: Springer 1977.
[48] A. Kruis, W. Scholz: Tieftemperaturwäschen zur Gaszerlegung. Linde-Bericht aus Technik und Wissenschaft *17*, 15–23 (1964).
[49] N. C. Ubdegraff: Gas Prepurification for Low-Temperature Processing. Chem. Eng. Progr. *53*, 268–271 (1957).
[50] W. R. Wood, B. D. Storrs: Girbotol Purification Process. Oil Gas J. *37*, 47–48 (1938).
[51] E. Guccione: Cryogenic Washing Scrubs Hydrogen for Liquid-Fueled Rockets. Chem. Eng. *70*, 150–152 (1963).
[52] H. Tanz: High-Purity Hydrogen Production (Erzeugung von Hochreinem Wasserstoff). DECHEMA Monogr. *65*, 293–301 (1970).
[53] C. Carpetis: Break-Even and Optimization Conditions for Overal Energy Systems wherein Hydrogen Storage Facilities are used. Int. J. Hydrogen Energy *10*, 839–850 (1985).
[54] C. Carpetis: An Assessment of Electrolytic Hydrogen Production by Means of Photovoltaic Energy Conversion. Int. J. Hydrogen Energy 9, 969–991 (1984).
[55] C. Carpetis: Technoeconomic Comparisons of Leading Hydrogen Storage Options. In: Proc., 10th IECEC Conf. Vol. 4, pp. 1747–1756 (1983).
[56] C. Carpetis: On the Design of Solar Energy Systems with Production and Storage of Hydrogen. In: Proc., 4th Int. Solar Forum, Vol. 2, pp. 961–963 (1982).
[57] M. Oertel, W. Weirich et al.: The Lithium-Lithium Hydride Process for the Production of Hydrogen: Comparison of two concepts for 950 and 1300 °C HTR Helium Outlet Temperature, Int. J. Hydrogen Energy *12*, 211–218 (1987).
[58] H. Engels, J. E. Funk, et al.: Thermochemical Hydrogen Production, Int. J. Hydrogen Energy *12*, 291–296 (1987).
[59] J. E. Funk: Thermochemical Water Decomposition—Current Status. In: Recent Developments in Hydrogen Technology K. D. Williamson Jr., F. J. Edeskuty eds Vol. 1, pp. 1–18. Cleveland, CRC Press (1986).
[60] Ch. F. Blazek, E. J. Daniels, T. D. Donakowski, M. Novil: Economics of Hydrogen in the 80's and Beyond. In: Recent Developments in Hydrogen Technology K. D. Williamson Jr., F. J. Edeskuty eds Vol. 2, pp. 1–34. Cleveland, CRC Press (1986).

Hydrogen Liquefaction 3

Several procedures have been developed for hydrogen liquefaction, encompassing the range from laboratory liquefaction equipment to large-scale plants. The large-scale end of the spectrum and the developments associated with it were essentially influenced and promoted by space technology requirements. Very detailed literature exists concerning the different procedures for hydrogen liquefaction. This is the basis for the following description of hydrogen liquefaction procedures (see for example [1–4, 6, 7]).

3.1 Fundamental Principles of Hydrogen Liquefaction

For liquefaction, the gas, previously purified in an appropriate manner, has to be cooled down to an appropriately low temperature, i.e., its boiling point of 20.3 K. According to Fig. 2, the thermodynamically ideal liquefaction process consists of an isothermal increase of the pressure from p_1 up to the—very high—pressure p_2, which is chosen so that upon the following isentropic expansion, the gas completely condenses in the expander. Because of the required high pressure p_2 (more than 10^5 bar for hydrogen) this process is not technically feasible.
Technically sound processes use the methods depicted in Fig. 3, along a low-pressure isobar P_N or a high-pressure isobar P_H with internal heat exchange. In the low-pressure cycle the gas is continuously cooled from ambient temperature T_0 to the dew point TP (Fig. 3) while the sensible heat $\int c_p \, dT$, the condensation heat L, and the heat of conversion C_{op} of the ortho-para-conversion is continuously removed.
The high-pressure cycle proceeds along a supercritical isobar P_H above critical pressure and, after reaching entropy of the saturated liquid via isentropic expansion and external work, leads to condensed liquid. During this process, the cooling is accomplished, while avoiding phase changes, in a series of very uniform adjustments of the specific heat. This is advantageous with respect to the design of required heat exchanger.

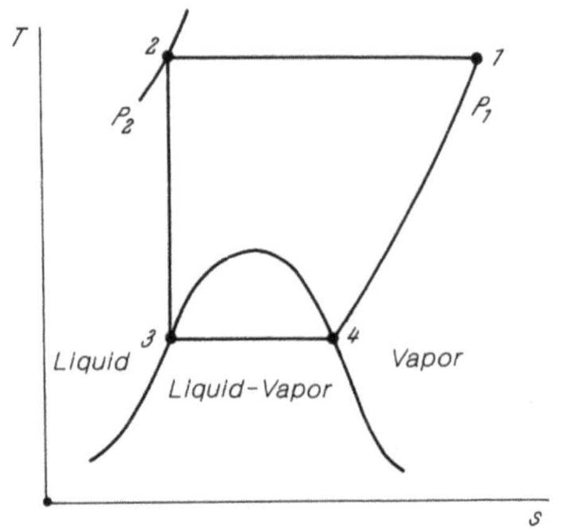

Fig. 2. Ideal thermodynamic liquefaction process

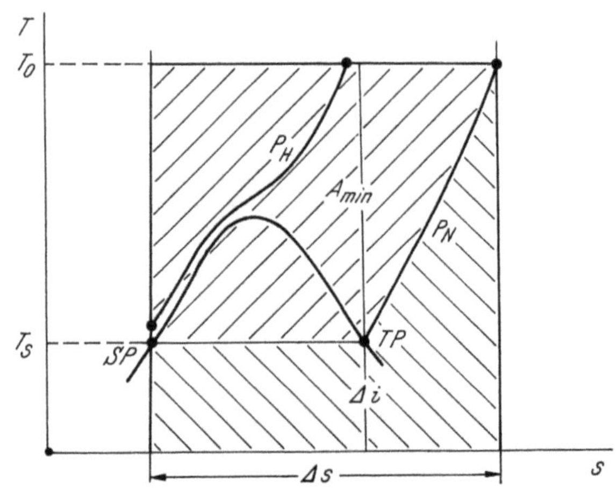

Fig. 3. Gas liquefaction along a low pressure or high pressure isobar (P_N or P_H). *SP* boiling point, T_P dew point, T_S boiling temperature, T_O ambient temperature, A_{min} minimum liquefaction work, Δi change of enthalpy, Δs change of entropy

Minimum Work of Liquefaction

Minimum work of liquefaction A_{min} is derived from the Carnot cycle (see Fig. 4). In order to remove the amount of heat Q at temperature T_2 from the system and release it at T_1, the work

$$A = A_{min} = (T_1 - T_2)(s_1 - s_2) \tag{9}$$

must be performed, whereas for the heat withdrawn at temperature T_2 the following applies:

$$Q = T_2(s_1 - s_2). \tag{10}$$

It follows that

$$A_{\min} = \frac{T_1 - T_2}{T_2} Q. \tag{11}$$

If heat is not isothermally released or removed, but instead is released along an isobar for example, then the cooling process can be divided into steps of differential Carnot cycles at the variable Temperature T (Fig. 5)

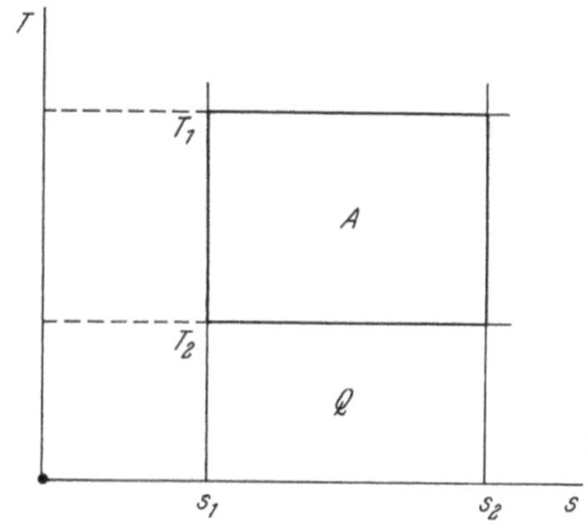

Fig. 4. Carnot Cycle for refrigeration. Q amount of heat removed at temperature T_2, A work input, $A + Q$ heat released at $T_1 > T_2$

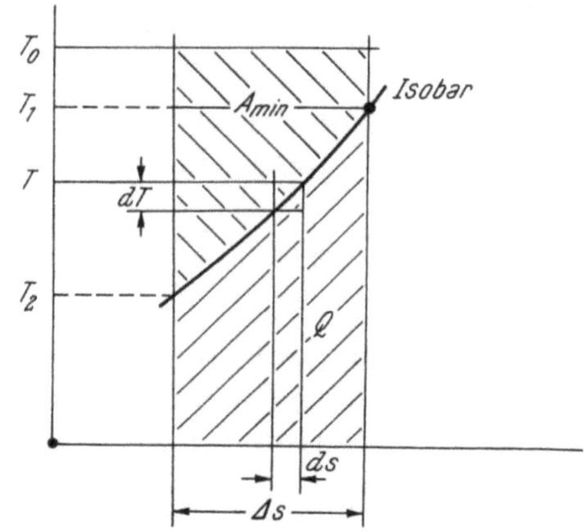

Fig. 5. Refrigeration Q and work $A_{\min}$ with non-isothermic heat removal along an isobar

and with heat release at Temperature T_0 the following applies

$$A_{\min} = \int_{T_1}^{T_2} \frac{T_0 - T}{T} \, dQ \tag{12}$$

because

$$dQ = di - \frac{dp}{\rho} \tag{13}$$

(p pressure, i enthalpy, ρ mass density).
Along the isobars the following applies

$$dp = 0, \quad dQ = di = c_p dT \tag{14}$$

and thus

$$A_{\min} = \int_{T_1}^{T_2} \frac{T_0 - T}{T} \, di = \int_{T_1}^{T_2} \frac{T_0 - T}{T} \, c_p \cdot dT. \tag{15}$$

Along the isobar, from Eq. (14) it follows that

$$ds = \frac{dQ}{T} = \frac{di}{T} \tag{16}$$

which results in

$$A_{\min} = T_0 \int_{T_1}^{T_2} ds - (i_2 - i_1) = T_0 \Delta s - \Delta i. \tag{17}$$

Thus, the lower hatched area of Fig. 3 represents the total heat removed from the system and the upper hatched area represents the minimum work of liquefaction $A_{\min}$. In the ideal case, i.e., without irreversible change in entropy due to component losses, every reversible process delivers the same result for minimum work of liquefaction in accordance with the first law of thermodynamics.
The application of the first law to the energy flow through a control surface, in the cycle in Fig. 2 for example, results in

$$\dot{q} + N = \sum_{\text{out}} \left(i + \frac{w^2}{2} \right) \cdot \dot{m} - \sum_{\text{in}} \left(i + \frac{w^2}{2} \right) \cdot \dot{m} \tag{18}$$

where $\dot{q} > 0$ and $N > 0$ represent the heat flux transferred to the system and the mechanical work done by the system respectively, and $\dot{m}$ represents the mass flow rate. Here the admissible neglecting of the kinetic energy changes per unit of mass $\frac{w^2}{2}$ as opposed to the enthalpy changes i results in a balance between the total energy transferred to the system and the enthalpy flux $\dot{m} \cdot i$

$$\dot{q} + N = \sum_{\text{out}} \dot{m} \cdot i - \sum_{\text{in}} \dot{m} \cdot i. \tag{19}$$

With Fig. 2 this results in

$$\dot{q} + N = \sum_{\text{out}} \dot{m} \cdot i - \sum_{\text{in}} \dot{m} \cdot i \quad \text{and} \quad i_3 = i_f \tag{20}$$

with

$$\dot{q} = -\dot{m} \cdot T_1(s_1 - s_2) \qquad \text{and} \quad s_2 = s_f \tag{21}$$

which results in[1]

$$\frac{N}{\dot{m}} = \frac{N}{\dot{m}_f} = T_1(s_1 - s_2) - (i_1 - i_f) \tag{22}$$

which corresponds to Eq. (17).
Furthermore, it should be pointed out that here $\dot{m} = \dot{m}_f$, i.e., the compressed gas mass flow rate $\dot{m}$ corresponds to the liquefied gas mass flow rate $\dot{m}_f$. Thus, here for the liquid yield φ applies

$$\varphi = \frac{\dot{m}_f}{\dot{m}} = 1. \tag{23}$$

Data for the minimum work of liquefaction for different gases is shown in Table 3.
It is evident from Table 3 that for helium the amount of work required for cooling the gas is greatly predominant, while with nitrogen it is small in comparison with the amount of work for condensation. With hydrogen the amounts are of the same order. Furthermore, for nitrogen and hydrogen,

Table 3. Minimum work of liquefaction of different gases

Gas	Boiling point (K)	Minimum work (kWh/kg)		
		Cooling	Condensation	Total
Methane	111.8	0.0769	0.2417	0.3186
Nitrogen	77.3	0.0547	0.1611	0.2158
Hydrogen[a]	20.4	1.5894	1.6777	3.27
Hydrogen[b]	20.3	2.2423	1.6777	3.92
Helium	4.2	1.9169	0.4133	2.3302

[a] Normal H_2, [b] at ortho-para equilibrium.

[1] i_f and s_f represent the enthalpy and entropy per unit of mass of the liquid respectively.

processes along a supercritical isobar have to be given priority due to the continuous removal of heat.

With hydrogen the heat of the ortho-para conversion must also be removed (Fig. 6), where, along the isobar $p = 1$ bar in conjunction with Eq. (15), the following results are achieved

$$A_{\min} = \int_{T_0}^{T_s} \frac{T_0 - T}{T} c_p \mathrm{d}T + \frac{T_0 - T_s}{T_s} \cdot L + \int_{C_1}^{C_2} \frac{T_0 - T_s}{T_s} C_{0 \to P} \cdot \mathrm{d}C$$

$$= (1.58 + 1.72 + 0.62)\mathrm{kWh/kg} = 3.92\,\mathrm{kWh/kg}, \qquad (24)$$

where c_p is the specific heat at constant pressure (Fig. 7), C the concentration of the ortho-hydrogen (para-concentration K_p in Fig. 6), C_{op} the temperature dependent ortho- to para-hydrogen conversion heat, T_s boiling

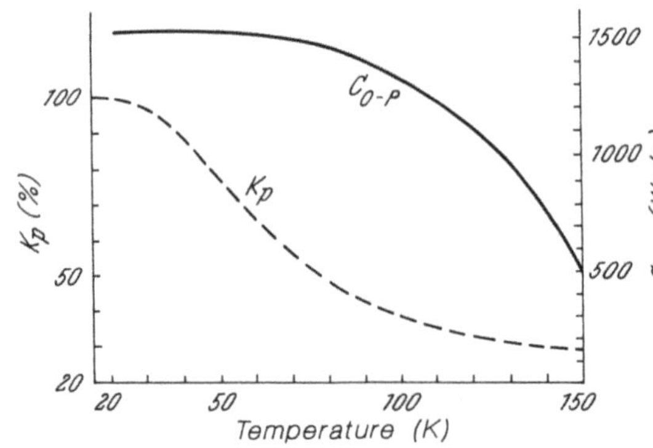

Fig. 6. Concentration K_p of para-hydrogen and conversion heat $C_{O\text{-}P}$ of the ortho-para conversion as a function of temperature (from [5]) (Ws = Wattsecond = Joule)

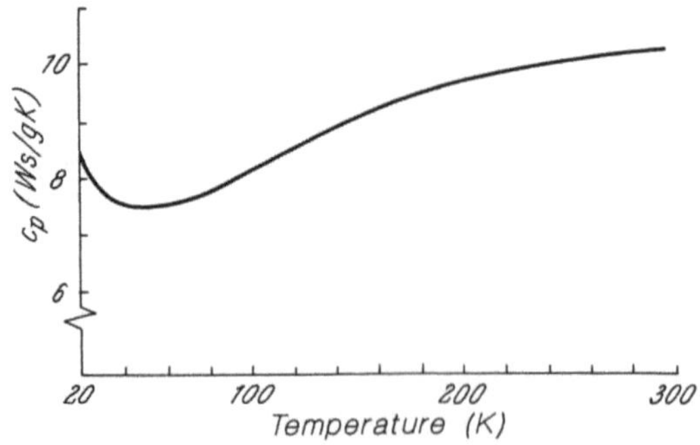

Fig. 7. Specific heat c_p of normal hydrogen at 1 bar (from [5]) (Ws = Joule)

temperature, L the heat of condensation. The total work of conversion A_k in dependence from the para-concentration is shown in Fig. 8.

3.1.1 *Refrigeration Processes as a Result of External Work*

Refrigeration processes use effects which lead to a decrease in the temperature of the working medium, either by external or internal work.
For hydrogen liquefaction, expansion engines are used for external work of the working medium and throttled expansion (Joule–Thomson expansion) for internal work. The working medium used is either the hydrogen itself or, in some cases, helium.
Furthermore, the adiabatic demagnetization of paramagnetic and ferromagnetic materials, where the electron gas is used as the working medium, have recently been recognized as refrigeration processes with external work and as having the potential for further application.
The adiabatic expansion of a gas by a work-producing device, such as a reciprocating engine or a turbine, from an isobar p_1 to isobar p_2 always leads to cooling of the gas which, in the ideal case, takes place isentropically from point 1 to point 2, with the expansion work $\Delta i = i_1 - i_2$ (see Fig. 9). The following isobaric heat absorption of the expanded gas $(2 \rightarrow 3)$ represents the desired cooling effect, where the gas absorbs the heat corresponding to the hatched area below the portion $(2 \rightarrow 3)$ of the isobar p_2. In the case of an ideal gas and where $T_3 = T_1$, this corresponds to the work of expansion, because $i_1 = i_3$.

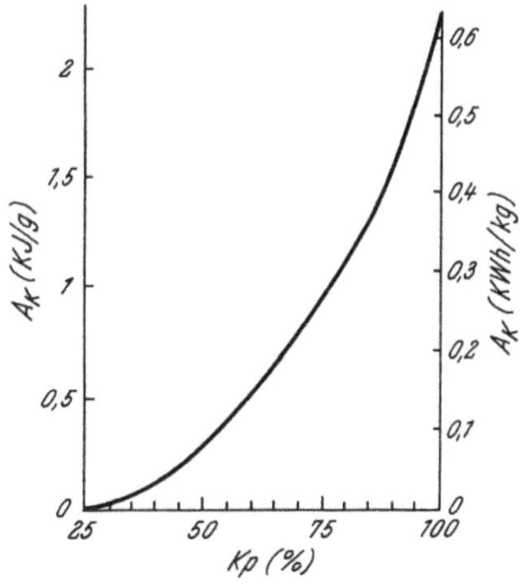

Fig. 8. Work of ortho-para conversion A_k as a function of the amount K_p of para-hydrogen (from [5])

Assuming an ideal gas—virtually valid for a real gas at temperatures above twice the critical temperature and pressures lower than about ten times more than the critical pressure—for the work of expansion A_{Ex} the following applies

$$A_{Ex} = i_1 - i_2 = c_p(T_1 - T_2). \tag{25}$$

Furthermore, for isentropic expansion of the ideal gas with constant specific heat and the specific heat ratio of $\kappa = c_p/c_v$, is

$$\frac{T_2}{T_1} = \left(\frac{p_2}{p_1}\right)^{(\kappa-1)/\kappa}, \tag{26}$$

$$c_p = \frac{\kappa R}{\kappa - 1}. \tag{27}$$

Substitution of (26) and (27) in (25) results in

$$A_{Ex} = \frac{\kappa}{\kappa - 1} R T_1 \left[1 - \left(\frac{p_2}{p_1}\right)^{(\kappa-1)/\kappa}\right]. \tag{28}$$

Data for the temperature drop as well as the expansion work for an ideal expansion engine are shown in Table 4.
The data shown in Table 4 cannot be achieved with actual expansion engines. Due to pressure losses at valves, incomplete expansion, heat transfer to the working medium, pressure losses during the intake and exhaust stroke as well as piston friction in piston engines, gap losses, impeller losses and the kinetic energy of the gas flowing from the turbine in expansion turbines, isentropic expansion of the working medium is not possible, i.e., the state of the gas after successful expansion corresponds with point 2′ on the isobar p_2, differing from point 2 in its higher temperature and higher entropy (see Fig. 9).

Table 4. Cooling by isentropic expansion starting at 293 K. Final pressure 1 bar as well as work of expansion per unit of mass for helium ($\kappa = 1.6$) and hydrogen ($\kappa = 1.4$)

Initial pressure (bar)	Final temperature (K)		Expansion work (J/g)	
	$\kappa = 1.4$	$\kappa = 1.6$	$\kappa = 1.4$	$\kappa = 1.6$
2	240	226	63	52
5	184	160	128	104
10	152	124	168	132
20	125	95	200	154
50	96	68	234	176
100	78	52	255	188

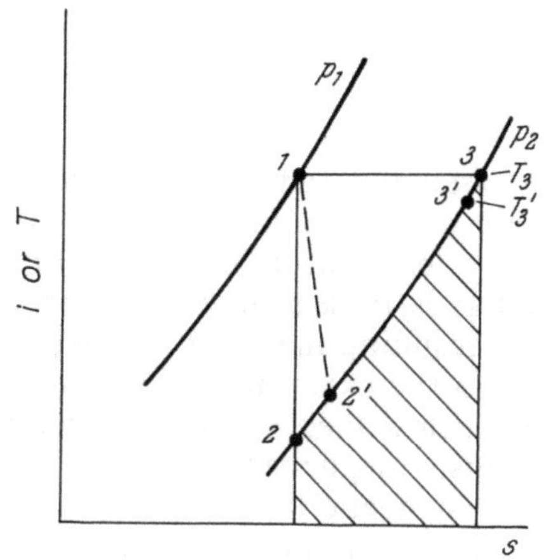

Fig. 9. Isentropic expansion and isobaric heat absorption in the i–s and T–s diagram

The adiabatic efficiency of the expander is defined by

$$\eta_{ad} = \frac{i_1 - i'_2}{i_1 - i_2}. \tag{29}$$

For piston expanders $\eta_{ad} = 0.7$–0.8, whereas turbine expanders in general demonstrate better results [1, 3].

In a continuous closed cycle, the return of the working medium to its initial state is necessary (point 1 on the isobar p_1). Due to the temperature difference required to maintain the heat flux in the heat exchangers, the temperature range on the isobar p_1 available for cooling is limited by $T'_3 < T_3$.

To return to the initial state 1, a compression of the working medium $(3 \to 1)$ is required, which can be accomplished isothermally only in the ideal case.

The deviation from the ideal thermodynamic behavior of a compressor can be characterized by the isothermal efficiency

$$\eta_{is} = \frac{A_{is}}{A_r}. \tag{30}$$

For good reciprocating compressors $\eta_{is} = 0.8$–0.85.

A_{is} is the isothermal compression work and A_r the real compression work, where the mechanical losses such as friction, etc., are not taken into consideration.

Details on expanders and compressors will not be further developed here, but only the extensive literature will be noted (see [1, 3–5]).

Magnetocaloric Refrigeration

General

The application of an external magnetic field to magnetic materials leads to a change in the state of order and thus to a change in the magnetic entropy. This results in a temperature change in the whole system. [8,9] With paramagnetic materials, these changes in magnetic entropy are caused by an alignment of the resulting magnetic moments of nonspinsaturated electron shells parallel to the external field, whereas the same occurs in ferromagnetic materials below their Curie point temperature through the alignment of spontaneously magnetized domains which exist without the presence of an external magnetic field. Moreover, the application of an external magnetic field results in an alignment of the magnetic moments of the elementary magnets—consisting of angular magnetic moments and magnetic spin moments of nonsaturated electron shells—in accordance with the Maxwell–Boltzmann energy distribution. Therefore, at equilibrium the following applies for the probability w_i of energy states u_i which can be attained by the particles,

$$w_i \sim \frac{e^{-u_i/kT}}{\sum_i e^{-u_i/kT}} \tag{31}$$

which results in a preference for the lower energy states [8,9].
Due to their magnetic spin moment ($\mu_B = 9.27 \times 10^{-21}$ Oe·cm^3, Bohr's magneton) electrons, for example, assume both energy states $u_{1,2} = \pm \mu_B H$ in a magnetic field H, which correspond to parallel and antiparallel alignment to the magnetic field H.
According to Eq. (31), after applying an external field H, a preference for the lower energy state occurs, which corresponds with the alignment of the magnetic moments parallel to the external field. At constant temperature (isothermal magnetization), heat is given off. With increasing magnetic field strength and decreasing temperature, the alignment of the electron spins increases along with the preference for a lower energy state and thus the degree of order of the system. This increase in the degree of order of the system corresponds to a decrease in the magnetic entropy s_m, which as a whole represents the decrease in entropy caused by the magnetic field.
In the ideal case, using isentropic demagnetization, a final temperature is attained which is below the initial temperature and corresponds with the decrease in magnetic entropy s_m.
However, these idealized states are valid in this form only when the spin system was completely disordered before the magnetization and when the change in magnetic entropy represents the major portion of the total entropy change that occurred [10–12].

Complete disorder of the spin system in the absence of an external magnetic field can be attained only with decreasing interaction within the spin system. Especially at low temperatures a partial alignment of the spin system and thus a "zero-field splitting" of the energy states can occur even with a decreasing external magnetic field, through electrostatic interaction with ions in solids, interaction of the magnetic moments of the electrons with one another or through interaction of the nuclear magnetic moments.
In order to achieve a maximum change in the total entropy, the entropy fraction of the lattice vibrations has to be small relative to the change in magnetic entropy, otherwise the low temperature which is attained immediately after removal of the magnetic field in the electron spin system would be sufficient only for a minimal cooling of the lattice. In addition, with the application of a moderate magnetic field (< 1 Tesla) large changes in the magnetic entropy using isothermal magnetization can be achieved for paramagnetic materials with high magnetic entropy values (maximum magnetic moments due to nonsaturated electron shells) only below a few degree Kelvin.
For these reasons, the application of the magnetocaloric refrigeration process has been limited so far to the temperature range below about 5 K. Thus, its application in hydrogen liquefaction is excluded because of the relatively high boiling point of hydrogen (20.4 K).
Figure 10 shows this par example for gadolinium phosphate (effective magnetic moment per atom $\mu_{eff} = 7.9\,\mu_B$). From this it can be seen that considerable changes in the magnetic entropy and thus effective cooling can be attained only below approximately 5 K [14–16].
The lowest attainable temperature using isentropic demagnetization of paramagnetic materials (approximately 10^{-3} K) is essentially limited via the previously mentioned zero-field splitting. Lower temperatures ($< 10^{-6}$ K) are attained through an analogous application to nuclear magnetic moments [12, 13].
It has been demonstrated that ferromagnetic materials can produce large changes in magnetic entropy in the range of their Curie point temperature [12, 15–17]. This is caused by the balance between the ordering influence of the exchange interaction within the spontaneous magnetized domains and the thermal agitation which counteracts this ordering effect. Below the Curie point temperature the exchange interaction is predominant whereas above this temperature—where no ferromagnetism exists—the thermal agitation is predominant. Therefore, the application of a strong magnetic field near the Curie point temperatures causes a larger change in magnetization and thus in the magnetic entropy, than at higher temperatures, where only paramagnetic magnetization occurs, or at low temperatures. There no large change in magnetization and in the entropy of the spin system is possible due to spontaneous magnetization and magnetic saturation. Prerequisites for this to occur are minimum hysteresis

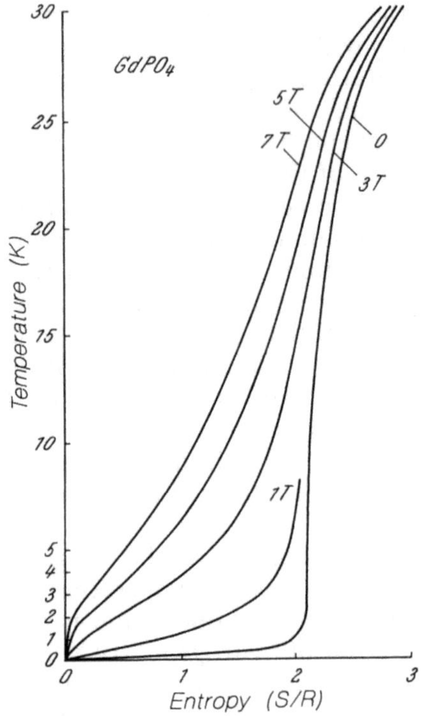

Fig. 10. Entropy S of gadolinium phosphate as a function of the magnetic field (in Tesla) (from [22]), R gas constant

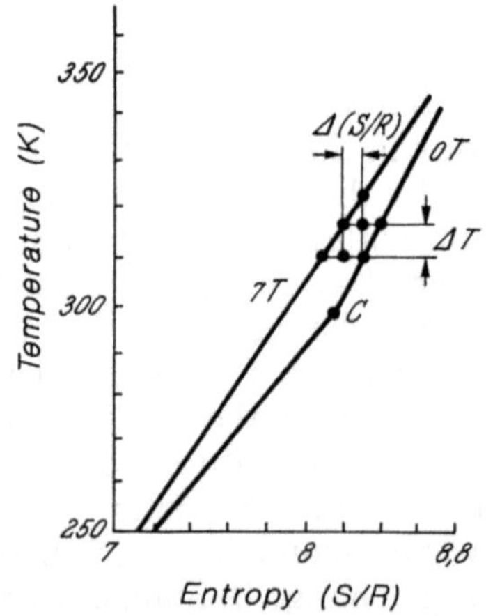

Fig. 11. The change in temperature and entropy for gadolinium as a result of magnetization in the vicinity of the Curie Point temperature C (293 K) (from [15])

losses which require the use of "magnetically soft" materials which have a hysteresis loop as narrow as possible and high magnetic saturation.
There are several ferromagnetic materials with Curie point temperatures in the range from 20 K to ambient temperature [15, 16]. In particular these include the "rare earth metals" as well as their alloys and compounds. Essentially, this principally also permits the use of the magnetocaloric refrigeration process for the liquefaction of hydrogen.
Figure 11 shows possible temperature changes as a result of a magnetic field of 7 *T* for gadolinium in the vicinity of its Curie point temperature (293 K) as well as the change in entropy attainable with isothermal magnetization.

Thermodynamic Relationships

In the following discussion expressions for entropy, specific heat and magnetization of a substance in an external magnetic field H are necessary (see Ref. [15, 16]).
With a change of the magnetization M (magnetic dipole moment per unit of mass) in an external field H, a magnetizable substance is supplied with energy $H\mathrm{d}M$. When additional heat $\mathrm{d}q$, but no external work is supplied, in accordance with the first law of thermodynamics the following applies for the internal energy u

$$\mathrm{d}u = \mathrm{d}q + H\mathrm{d}M \tag{32}$$

or after transposition

$$\mathrm{d}q = \mathrm{d}u - H\mathrm{d}M. \tag{33}$$

A comparison with a form of the first law containing mechanical work $p\mathrm{d}v$

$$\mathrm{d}q = \mathrm{d}u + p\mathrm{d}v \tag{34}$$

shows that (34) can be derived from (33), by substituting pressure p for the field intensity H and the volume v for the magnetization M. In the same manner the following applies for enthalpy i

$$\mathrm{d}q = \mathrm{d}i + M\mathrm{d}H. \tag{35}$$

After the introduction of the entropy $s = s(T, H)$

$$\frac{\mathrm{d}q}{T} = \mathrm{d}s = \left(\frac{\partial s}{\partial T}\right)_H \mathrm{d}T + \left(\frac{\partial s}{\partial H}\right)_T \mathrm{d}H \tag{36}$$

the following is obtained

$$T\mathrm{d}s = \mathrm{d}u - H\mathrm{d}M = \mathrm{d}i + M\mathrm{d}H. \tag{37}$$

The following is obtained from (35) and (36) for the specific heat with constant field H

$$c_H = T\left(\frac{\partial s}{\partial T}\right)_H. \tag{38}$$

In (38) $\left(\frac{\partial s}{\partial T}\right)_H$ can be replaced by magnetization M.
Through the comparison of coefficients from (35) and (36) the following is obtained

$$\left(\frac{\partial s}{\partial T}\right)_H = \frac{1}{T}\cdot\left(\frac{\partial i}{\partial T}\right)_H, \tag{39}$$

$$\left(\frac{\partial s}{\partial H}\right)_T = \frac{1}{T}\left[\left(\frac{\partial i}{\partial H}\right)_T + M\right]. \tag{40}$$

By forming the mixed derivative $\frac{\partial^2 s}{\partial T \partial H}$ from (39), and (40), and equating the same results in

$$\left(\frac{\partial i}{\partial H}\right)_T + M = T\left(\frac{\partial M}{\partial T}\right)_H. \tag{41}$$

Substituting (41) into (40) now results in

$$\left(\frac{\partial s}{\partial H}\right)_T = \left(\frac{\partial M}{\partial T}\right)_H. \tag{42}$$

Equation (42) represents the magnetic expression of Maxwell's thermodynamic relations.
From (38), (42) and (36) the following is obtained for entropy

$$\mathrm{d}s = \frac{c_H}{T}\mathrm{d}T + \left(\frac{\partial M}{\partial T}\right)_H \mathrm{d}H \tag{43}$$

Because ds is a total differential, applying integrability condition to (36) and (43) results in

$$\frac{\partial}{\partial H}\left(\frac{c_H}{T}\right)_T = \left(\frac{\partial^2 M}{\partial T^2}\right)_H \tag{44}$$

and after integration

$$c_H = c_0 + T\int_0^H \left(\frac{\partial^2 M}{\partial T^2}\right)\mathrm{d}H. \tag{45}$$

Here c_0 is the specific heat at zero magnetic field. Thus, for the calculation of c_H, information about the magnetization $M(H, T)$ is necessary.

The following applies in general to magnetization $M(H, T)$

$$M(H, T) = M_s \cdot B_J(x). \tag{46}$$

In addition the following is valid for the saturation magnetization M_s

$$M_s = n \cdot g \cdot \mu_B \cdot J \tag{47}$$

where n is the number of magnetic dipoles per unit of mass; μ_B is Bohr's magneton, $\mu_B = 0.917 \cdot 10^{-20}$ Oe·cm^3 $= 0.917 \cdot 10^{-23}$ A·m^2 and J is the total angular momentum quantum number of the electrons contributing to magnetization. Thus, for the total angular momentum $j = \frac{h}{2\pi}\sqrt{J(J+1)}$ applies, where h is Planck's constant.
For the Landé splitting factor q applies

$$g = 1 + \frac{J(J+1) + S(S+1) - L(L+1)}{2J(J+1)}. \tag{48}$$

It represents the breakdown of the total angular momentum of an electron shell into orbital angular momentum (L orbital quantum number) and spin angular momentum (S spin quantum number), or specifically the breakdown of the magnetic moments which are assigned to the angular momentum. The following applies for the orbital angular momentum

$$l = \frac{h}{2\pi}\sqrt{L(L+1)}, \text{ for the spin angular momentum } s = \frac{h}{2\pi}\sqrt{S(S+1)}.$$

For $g = 1$: total angular momentum is only orbital angular momentum, $g = 2$: total angular momentum is only spin angular momentum. This applies analogously to the magnetic moments assigned to the angular momentum.
For the Brillouin function $B_J(x)$ the following applies

$$B_J(x) = \frac{2J+1}{2J}\coth\frac{2J+1}{2J}x - \frac{1}{2J}\coth\frac{x}{2J}. \tag{49}$$

For paramagnetic materials the following applies

$$x = \frac{g \cdot \mu_B \cdot J \cdot H}{kT} \tag{50}$$

where $k = 1.380 \cdot 10^{-16}$ erg/K is the Boltzmann constant.
For ferromagnetic materials in the molecular field approximation [16] the following applies

$$x = g \cdot \mu_B \cdot J \cdot \left[H - \frac{D \cdot \rho}{G} M + \lambda \cdot \frac{\rho}{G} M \right] \Big/ kT. \tag{51}$$

Here D denotes the demagnetization factor, ρ density, G molecular mass, and λ molecular field constant. From (43), (45) and (49), c_H and s can be calculated for paramagnetic materials with the help of (50).
In general, from (42) the magnetic entropy s_m can be obtained as follows

$$s_m = \int_0^H \left(\frac{\partial M}{\partial T}\right)_H \cdot dH = n \cdot k \cdot \ln \frac{\sin h \dfrac{2J+1}{2J}}{\sin h \dfrac{x}{2J}} - n \cdot k \cdot x \cdot B_J(x), \tag{53}$$

whereas the field dependent specific heat can be obtained from (38) and (45). Further calculations of thermodynamic values for ferromagnetic materials were conveniently performed by means of a computer program.
The following approximations are obtained for paramagnetic materials at small field density and sufficiently high temperature $T (x \ll 1)$:
With $x \to 0$ the following is obtained from (46) and (49) for the magnetization after expansion of

$$\coth x = \frac{1}{x} + \frac{x}{3} + \cdots$$

$$M = n \cdot \frac{g^2 \cdot \mu_B^2 \cdot J(J+1)}{3k} \cdot \frac{H}{T} = \chi \cdot H. \tag{54}$$

This is Curie's Law [8] with the temperature dependent susceptibility

$$\chi = \frac{C}{T} \tag{55}$$

and the Curie constant

$$C = \frac{J(J+1) n \cdot g^2 \cdot \mu_B^2}{3k}. \tag{56}$$

From (53) the following is obtained for the magnetic entropy s_m

$$\lim_{x \to 0} s_m = n \cdot k \cdot \ln(2J+1) - n \cdot k \cdot \frac{J(J+1) g^2 \cdot \mu_B^2 H^2}{3 k^2 T^2}$$

$$= n \cdot k \cdot \ln(2J+1) - C \cdot \frac{H^2}{T^2}. \tag{57}$$

From Eqs. (49) and (50) in general the following can be obtained for the pure spin magnetism with two energy levels $J = \frac{1}{2}$, $g = 2$, which was given as an example on page 26

$$M = n \cdot \mu_B \cdot \operatorname{tgh} \frac{\mu_B H}{kT} \tag{58}$$

or in the extreme case $\frac{H}{T} \to 0$ Curie's Law

$$M = \frac{n\mu_B^2}{kT} H = \chi \cdot H. \tag{59}$$

With (57) the decrease of temperature with isentropic demagnetization of a paramagnetic substance can now be estimated. For the total entropy s_{tot} the following applies

$$s_{tot} = s_g + s_E + s_m. \tag{60}$$

The lattice entropy s_g using a Debye approximation is

$$s_g(T) = 77.9 \cdot n \cdot k \left(\frac{T}{\theta_D}\right)^3 = A \cdot \left(\frac{T}{\theta_D}\right)^3 \tag{61}$$

with the typical Debye temperature for the affected material θ_D. The entropy s_E of the conduction electrons can be ignored here because it is negligible with regard to the lattice entropy.
Therefore, from (61) and (57) the following can be used as an approximation of total entropy

$$s_{tot} = s_g(T) + s_m = A \cdot \left(\frac{T}{\theta_D}\right)^3 + n \cdot k \cdot \ln(2J+1) - \frac{C \cdot H^2}{T^2}. \tag{62}$$

With isentropic demagnetization s_{tot} remains constant whereas the temperature drops[2] to T_u. Therefore the following applies

$$s_{tot} = s_g(T) + s_m(T) = s_g(T_u) + s_m(T_u). \tag{63}$$

Using (61) and (57) the temperature $T_u < T$ which results after isentropic demagnetization can be calculated while ignoring the zero-field splitting. Considering that with isothermal magnetization $s_m(T)$ can at best approach zero, i.e.,

$$\frac{C \cdot H^2}{T^2} = n \cdot k \cdot \ln(2J+1), \tag{64}$$

---

[2] From Eq. (43) the temperature change ΔT with $ds = 0$ is obtained from the integration of

$$\Delta T = -T \cdot \int_{H_1}^{H_2} \frac{1}{c_H} \cdot \left(\frac{\partial M}{\partial T}\right)_H \cdot dH.$$

the following is obtained

$$T_u \geqslant T \cdot \left[1 - \frac{n \cdot k \cdot \ln(2J+1)}{s_g(T)} \right]^{1/3} . \tag{65}$$

With the approximation $x \ll 1$ (see (54)), (64) can most easily be fulfilled through a large value of the Curie constant C, whereas in order to obtain a minimum T_u the lattice entropy $s_g(T)$ should be as small as possible (see (65)), and thus the most favorable conditions for achieving a minimum final temperature T_u are established.
Using Eqs. (49), (53) and (61) to estimate for ferromagnetic materials shows, that under optimum conditions, isentropic decreases in temperature of approximately 1 K per Tesla (1 Tesla = 1 Vs/m^2) can be achieved in the vicinity of the Curie point temperature.

3.1.2 Refrigeration Processes as a Result of Internal Work

Refrigeration as a result of internal work is not possible with an ideal gas because there are no cohesive forces between the molecules. Thermodynamically the enthalpy of the ideal gas is only a function of temperature. If the throttled expansion of an ideal gas from pressure p_1 to pressure p_2 is carried out so that no kinetic energy flow is produced, that is no external work is done (Joule–Thomson expansion), the change in the internal energy with regard to the change in volume $v_1 \rightarrow v_2$ is

$$u_2 - u_1 = p_1 v_1 - p_2 v_2 \tag{66}$$

or

$$u_2 + p_2 v_2 = u_1 + p_1 v_1 = i. \tag{67}$$

This shows that with the Joule–Thomson expansion for an ideal gas the enthalpy i and also the temperature T remain constant.
In contrast as a result of the change in volume in the expansion of an actual gas internal work is performed on the cohesive forces which results in temperature changes.
The Joule–Thomson coefficient t

$$t = \left(\frac{\partial T}{\partial p} \right)_i \tag{68}$$

is decisive for the change in temperature. From the first law the following is obtained for $s = s(p, T)$ and $i = i(p, T)$

$$T \cdot \mathrm{d}s = \mathrm{d}q = \mathrm{d}i - v\mathrm{d}p = \left(\frac{\partial i}{\partial T} \right)_p \mathrm{d}T + \left[\left(\frac{\partial i}{\partial p} \right)_T - v \right] \mathrm{d}p \tag{69}$$

with

$$ds = \left(\frac{\partial s}{\partial T}\right)_p dT + \left(\frac{\partial s}{\partial p}\right)_T dp. \tag{70}$$

From (69) and (70) by "crosswise differentiation" (see (39) to (42)) and solving for the mixed derivatives the following is obtained

$$\left(\frac{\partial i}{\partial p}\right)_T = v - T\left(\frac{\partial v}{\partial T}\right)_p. \tag{71}$$

For the Joule–Thomson coefficient with $di = 0$, (69) results in

$$\left(\frac{\partial T}{\partial p}\right)_i = \frac{1}{c_p}\left[T\left(\frac{\partial v}{\partial T}\right)_p - v\right] \tag{72}$$

Depending on whether the right side of the bracket is positive or negative, a decrease in temperature at $\left(\frac{\partial T}{\partial p}\right)_i > 0$ or a rise in temperature at $\left(\frac{\partial T}{\partial p}\right)_i < 0$ is obtained.

The inversion curve

$$p_i = p_i(T) \tag{73}$$

is obtained for

$$c_p\left(\frac{\partial T}{\partial p}\right)_i = T\left(\frac{\partial v}{\partial p}\right)_p - v = 0. \tag{74}$$

Values for the change in temperature of hydrogen with Joule–Thomson expansion can be taken directly from the T–s diagram. The inversion curve for normal hydrogen is shown in Fig. 12. The inversion curve can be further approximated with the Van der Waals approximation by means of the Van der Waals equation of state

$$\left(p + \frac{a}{v^2}\right)(v - b) = RT. \tag{75}$$

From (74) the following is obtained

$$\left(\frac{\partial v}{\partial T}\right)_p = \frac{R}{\left(p + \frac{a}{v^2}\right) - \frac{2\alpha}{v^3}(v - b)}. \tag{76}$$

Replacing (76) in (74) results in the following for the inversion curve

$$\frac{2a}{v^2} - \frac{RTb}{(v - b)^2} = 0. \tag{77}$$

If the pressure, temperature and volume v are applied to the values p_k, T_k, $v_k = \frac{1}{\rho_k}$ at the critical point (Amagat variable) with

$$b = \frac{1}{3\rho_k} = \frac{vk}{3}, \quad a = 3\rho_k v_k^2, \quad R = \frac{8}{3}\frac{p_k v_k}{T_k} \tag{78}$$

and

$$\bar{T} = \frac{T}{T_k}, \quad \bar{p} = \frac{p}{p_k}, \quad \bar{v} = \frac{v}{v_k} \tag{79}$$

the reduced equation of state

$$\left(\bar{p} + \frac{3}{\bar{v}^2}\right)(3\bar{v} - 1) = 8\bar{T} \tag{80}$$

and the reduced general inversion curve with the Van der Waals approximation

$$\frac{\bar{p}}{3} = 4\bar{T}^{1/2}(2\sqrt{3} - \bar{T}^{1/2}) \tag{81}$$

are obtained. The same is shown in Fig. 12 with the values of the n-hydrogen

$$T_k = 33.18\,\text{K}, \quad \rho_k = \frac{1}{v_k} = 0.02988\,\text{g/cm}^3, \quad p_k = 12.98\ \text{bar}.$$

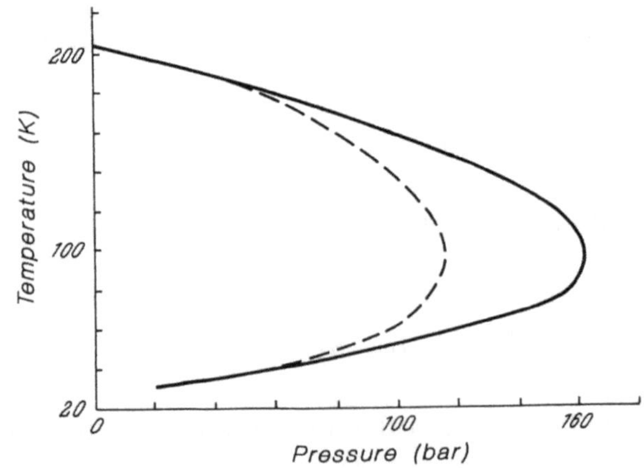

Fig. 12. Inversion curve for normal hydrogen. The dashed curve is the inversion curve obtained from the Van der Waals approximation.

Table 5. Change in temperature in K for hydrogen with Joule–Thomson expansion

	100 bar	150 bar	100 bar	50 bar	25 bar
300 K	6	5	4	2	1
200 K	2	1.6	0	0	0
100 K	−9	−9	−8	−6	−5
80 K	−16	−16	−15	−15	−6
60 K	—	—	−25	−22	−8
40 K	—	—	−20	−20	−20

Approximate results of the change in temperature of hydrogen with Joule–Thomson expansion based on different initial pressures and initial temperatures to a final pressure of 1 bar is shown in Table 5.
The drops in temperature achieved in this manner are greater than those achieved presently by isentropic demagnetization but considerably smaller than those achieved by isentropic expansion with external work.
The T–s diagram shows that starting from ambient temperature or a temperature which lies slightly beneath it, expansion with external work requires extremely high initial pressures if direct attainment of the two phase region is desired. Moreover, expansion with gas–liquid phase change in expansion engines is difficult to achieve. Besides of the Linde process therefore the Joule–Thomson expansion is used in the final liquefaction step for almost all continuously operting hydrogen liquefaction processes.

3.2 Small and Medium Liquefaction Plants

For the liquefaction of hydrogen several procedures were developed which to some extent are still in use today. They are described in detail in the technical literature, so only extraordinary characteristics or developments which are general in nature or whose initial use was revolutionary will be dealt with.
The liquefaction methods differ from one another in the thermodynamic cycle, the refrigeration process and the working medium used.

Linde–Hampson Process for Hydrogen Liquefaction

These liquefiers use a thermodynamic process which was developed from the single-stage compression Linde–Hampson air liquefaction process. Carl von Linde invented it for the first large-scale liquefaction of air.
The working medium (air) is, in the ideal case (Fig. 13), isothermally compressed to pressure p_2 at temperature T_1 and cooled along an isobar

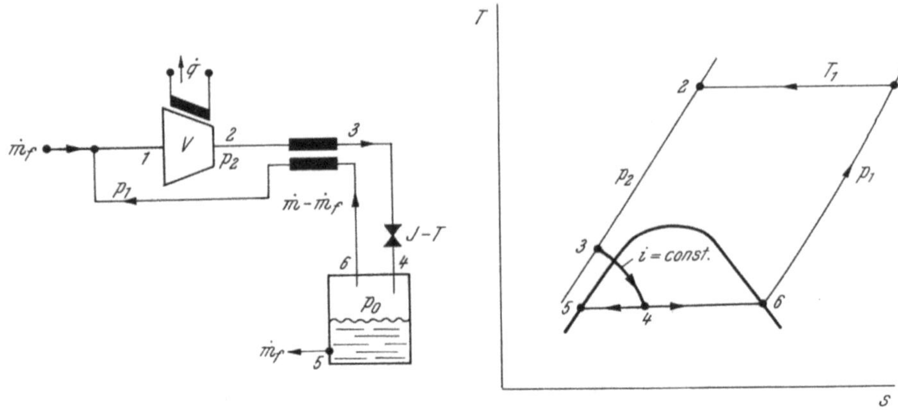

Fig. 13. The Linde–Hampson cycle, single pressure cycle

in a counterflow heat exchanger. Joule–Thomson expansion is used as the refrigeration procedure. At point 4 (Fig. 13) part of the air liquefies and is extracted, while the remaining gas is warmed up to ambient temperature along the isobar p_1 in a counterflow heat exchanger with the incoming air. With the mass flow rates and enthalpies (see (19))[3] the following is obtained from the first law,

$$(\dot{m}-\dot{m}_f)\cdot i_1+\dot{m}_f\cdot i_5=\dot{m}\cdot i_2. \tag{82}$$

For the fraction φ of the gas flow which is liquefied this results in

$$\varphi=\frac{\dot{m}_f}{\dot{m}}=\frac{i_1-i_2}{i_1-i_5}. \tag{83}$$

The ideal work of liquefaction $\alpha=\dfrac{N_r}{\dot{m}_f}$ is obtained from[4]

$$\dot{q}+N_r=-\dot{m}(i_1-i_2) \tag{84}$$

where for isothermal cooling in the compressor V at temperature T_1 (Fig. 13) the following applies

$$\dot{q}=-\dot{m}\cdot T_1\cdot(s_1-s_2). \tag{85}$$

From (84), (85) and (83) the following is obtained

$$\alpha=\frac{N_r}{\dot{m}_f}=\frac{1}{\varphi}\frac{N_r}{\dot{m}}=\frac{i_1-i_5}{i_1-i_2}[T_1(s_1-s_2)-(i_1-i_2)]. \tag{86}$$

---

[3] At point 2 an enthalpy flux $\dot{m}\cdot i_2$ enters into the counterflow heat exchanger while $(\dot{m}-\dot{m}_f)\cdot i_1$ is removed.

[4] See page 20.

Furthermore, $0 < \varphi \leqslant 1$ applies. A comparison of (83) with (23) shows that φ represents the relationship between the ideal work of liquefaction α_i and the real work of liquefaction α for this process. The reciprocal value $(\alpha_i/\alpha) \leqslant 1$ is defined as the efficiency of the process where $(\alpha_i/\alpha) = 1$ is valid for the ideal process.

Table 6 (see page 42) gives for efficiency and liquefaction energy for this process. The considerable deviation from the ideal process can be attributed to the irreversible change in entropy in the Joule–Thomson expansion since all other components such as the compressor, heat exchanger, etc. were presumed ideal, i.e., without loss and thus without irreversible changes in entropy. Taking these losses into account results in an additional reduction of efficiency and an increase in the amount of energy required for liquefaction.

Cooling by means of Joule–Thomson expansion can be accomplished only with T_1 and i_1 within the inversion curve. A maximum liquefaction rate is attained when $i_1 - i_2$ is as large as possible (i_5 is already defined by i_1, i.e., through p_1 and T_1).

Here, i_2 is given by T_1 and the maximum possible pressure p_2, where p_2 is on the inversion curve. (For air at ambient temperature p_2 is of about 400 bar, normally about 200 bar is used.)

Since the highest inversion temperature is 205 K with hydrogen, precooling is required for this process. Generally liquid nitrogen obtained from a liquefaction plant earlier in the process or from a storage tank is used for precooling.

For the fraction of the gas flow φ_{H_2} which is liquefied the following is obtained from Fig. 14 and (83)

$$\varphi_{H_2} = \frac{\dot{m}_f}{\dot{m}} = \frac{i_9 - i_4}{i_9 - i_7}. \tag{87}$$

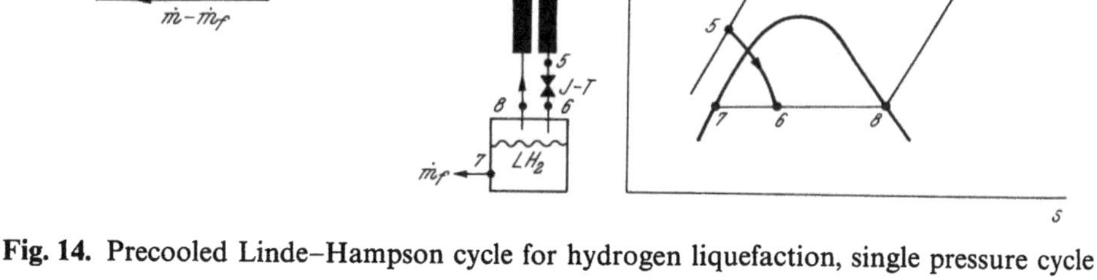

Fig. 14. Precooled Linde–Hampson cycle for hydrogen liquefaction, single pressure cycle

Analogous to (82) with evaporating nitrogen $\dot{m}_{N_2}$ and enthalpy i' the following energy balance (Fig. 14) is obtained for nitrogen

$$\dot{m}_{N_2} \cdot i'_{10} + (\dot{m} - \dot{m}_f) i_1 + \dot{m}_f \cdot i_7 = \dot{m}_{N_2} \cdot i'_{11} + \dot{m} \cdot i_2. \tag{88}$$

After substituting the ratio of the rate $\dot{m}_{N_2}$ of evaporation of nitrogen to the compressed hydrogen $\dot{m}$

$$\psi = \frac{\dot{m}_{N_2}}{\dot{m}} \tag{89}$$

in (88) the following is obtained

$$\varphi_{H_2} = \frac{i_1 - i_2}{i_1 - i_7} + \psi \frac{i'_{10} - i'_{11}}{i_1 - i_7} \tag{90}$$

and for the nitrogen consumption r, in relation to the liquefied hydrogen

$$r = \frac{\dot{m}_{N_2}}{\dot{m}_f} = \frac{\psi}{\varphi_{H_2}}. \tag{91}$$

Furthermore, (87) shows that φ_{H_2} increases with a decrease in i_4 and thus T_4. A decrease in T_4 can be attained by lowering the pressure of the liquid nitrogen bath, where the minimum value for T_4 is $T_{tr} = 63.15\,\mathrm{K}$, the triple point of nitrogen.

Similar to air liquefaction, the Linde–Hampson process, which works with precooling, can essentially be implemented with two-stage Joule–Thomson expansion (Fig. 15). The reduction in compressor work is advantageous because only the mass flow which contributes to liquefaction has to be completely compressed from ambient pressure to about 200 bar, whereas the major part of the gas flow is released at an intermediate pressure p_2 (approx. 50 bar) under cooling via Joule–Thomson expansion. (In air liquefaction plants, liquefaction is also carried out in an intermediate storage container maintained at pressure p_2.)

From the sum of the compressor outputs $N = N_1 + N_2$ and the mass flow ratio $\mu = \dot{m}_i/\dot{m}$ (see Fig. 15) the following is obtained

$$\frac{N}{\dot{m}} = [T_1(s_1 - s_3) - (i_1 - i_3)] - \mu[T_1(s_1 - s_2) - (i_1 - i_2)]. \tag{92}$$

A comparison with (86) shows a reduction in $N/\dot{m}$ caused by the right-side term in (92). For the part which is liquefied $\varphi_{H_2} = \dot{m}_f/\dot{m}$ with $\psi = m_{N_2}/\dot{m}$, $\mu = \dot{m}_i/\dot{m}$ and Fig. 15 the following is obtained

$$\varphi_{H_2} = \frac{i_1 - i_3}{i_1 - i_9} + \psi \frac{i'_{13} - i'_{12}}{i_i - i_9} - \mu \frac{i_1 - i_2}{i_1 - i_9}. \tag{93}$$

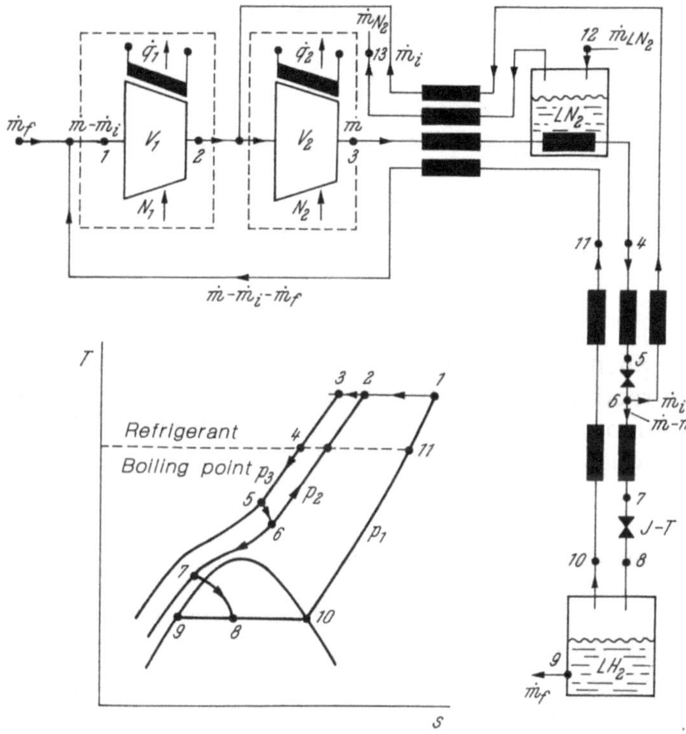

Fig. 15. Precooled Linde–Hampson dual-pressure cycle for hydrogen liquefaction

A comparison with Eq. (90) shows the slight decrease in φ_{H_2} caused by the third term in (93). This decrease, however, is offset by a reduction in required compressor output (92).

Apparently only a few hydrogen liquefaction plants were built utilizing the two-stage compression Linde–Hampson process and precooling with liquid nitrogen. This is because development in the last three decades unequivocally preferred procedures in which cooling is primarily carried out in a reversible process (expander) instead of the Joule–Thomson expansion procedure which is used exclusively in the Linde–Hampson process.

Table 6 shows the energy required per unit of mass liquefied α, as well as the liquid yield φ for hydrogen liquefaction with single-stage and two-stage compression Linde–Hampson processes, with and without nitrogen for precooling and ortho-para conversion at 80 and 20 K, without taking the losses of each component into account.

α_1 and α_2 are the work of liquefaction without and with consideration of the energy expended for nitrogen liquefaction respectively.

Table 6. Linde–Hampson process for hydrogen liquefaction

Process	α_1 (kWh/kg)	$\varphi = \frac{\dot{m}_f}{\dot{m}}$	α_2 (kWh/kg)	$\varphi = \frac{\dot{m}_f}{\dot{m}}$	$r = \frac{\dot{m}_{N_2}}{\dot{m}_f}$
Single pressure Linde	13.14	0.30	16.27	0.25	14.8
Dual pressure Linde	9.49	0.41	12.14	0.32	12.5

Despite its simplicity and reliability the Linde procedure currently appears to be less attractive due to the irreversible change in entropy in the Joule–Thomson process and the relatively large amount of energy required for hydrogen liquefaction associated with it.

Claude Process for Hydrogen Liquefaction

In the Claude process the gas is cooled in an expander by external work. The temperature drop is substantial, especially with high input pressures (see Sect. 3.1, Table 4), and far exceeds the temperature drop achieved by Joule–Thomson expansion.

In addition, the expansion process is isentropic for an ideal expander, that is, it is not characterized by an inherent irreversible increase in entropy as is the Joule–Thomson expansion.

Although, as opposed to the Linde–Hampson process, the Claude process essentially requires no precooling—in expansion with external work gas cools down in any case—the process can be improved by precooling with liquid nitrogen.

The basic Claude process for hydrogen liquefaction and precooling with liquid nitrogen is shown in Fig. 16. The compressed gas is divided into two streams after passing the heat exchanger and precooling with liquid nitrogen (point 5). Approximately 80% ($\dot{m}_E = 0.8\,\dot{m}$) is cooled in an expander with work output (point 11) and the remaining mass flow ($\dot{m} - \dot{m}_E$) is cooled by passing another heat exchanger (point 7) and is partially liquefied after Joule–Thomson expansion. Joule–Thomson expansion is basically not required in the Claude process, because liquefaction could be carried out in the expander. Even with axial-flow turbines considerable amounts of liquid are permitted to enter the gas flow. Nevertheless Joule–Thomson expansion is also used in this process in the final step in order to alleviate severe problems caused by the two-phase flow in the expander. The operating parameters can be selected, however, so that the irreversible changes in entropy are small.

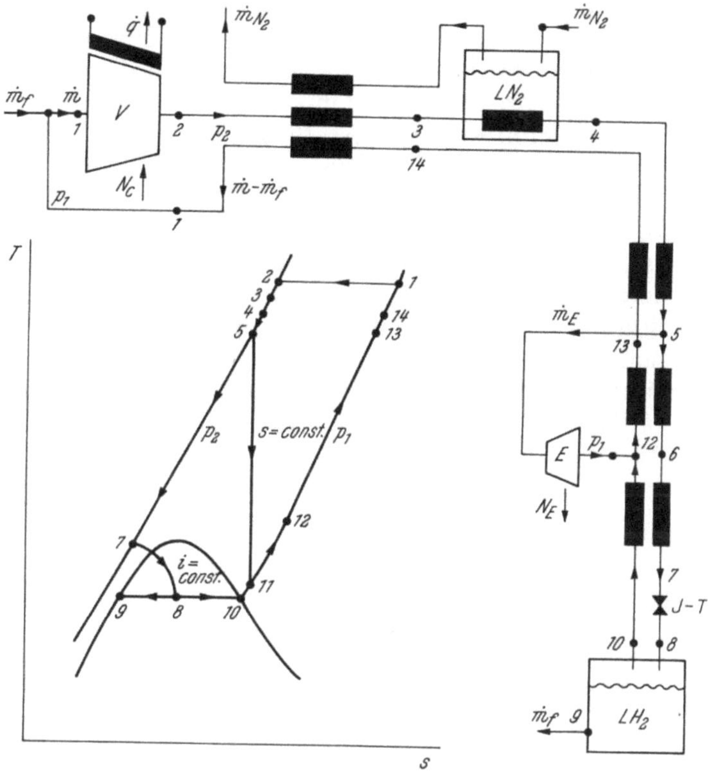

Fig. 16. Precooled single pressure Claude cycle for hydrogen liquefaction

The liquefaction rate $\dot{m}_f$ is obtained from the energy flow balance in a manner similar to the Linde–Hampson process and thus, with Fig. 16 and the enthalpies i' of nitrogen, for the liquid yield φ_{H_2}

$$\varphi_{H_2} = \frac{i_1 - i_2}{i_1 - i_9} + x \cdot \frac{i_5 - i_{11}}{i_1 - i_9} + \psi \frac{i'_2 - i'_4}{i_1 - i_9} \tag{94}$$

with $x = \dot{m}_E/\dot{m}$, $\psi = \dot{m}_{N_2}/\dot{m}$. Assuming that also the expander work is used for compression the following is obtained

$$\frac{N}{\dot{m}} = [T_1(s_1 - s_2) - (i_1 - i_2)] - x \cdot (i_5 - i_{11}). \tag{95}$$

This corresponds to the compressor work in a single-stage compression Linde–Hampson process without the expander work. Furthermore, the second term in (94) corresponds to the output as a result of the expander as opposed to the Linde–Hampson process (83).

After selecting an initial pressure p_1 and the flow fraction $x = \dot{m}_E/\dot{m}$ through the expander, a detailed examination to minimize the expenditure of energy for liquefaction results in a fix for point 5 in the T–S diagram (Fig. 16). Further improvement can be achieved in a manner similar to the two-stage compression Linde–Hampson process through partial expansion to an intermediate pressure. Figure 17 shows a two-stage compression Claude process which was examined in detail in Ref. [4]. The intermediate pressure is about 8 bar, the maximum pressure 100 bar.
Table 7 shows the energy expenditure α_1 for liquefaction including ortho-para conversion at 80 K and 20 K, the liquid yield φ_{H_2} as well as the liquid nitrogen consumption r, disregarding component losses.
The Collins helium liquefier, which was also built in a refrigerator version [1, 3], was a milestone in the development of smaller liquefaction plants. At the time of its entry to the market it provided a convenient means of

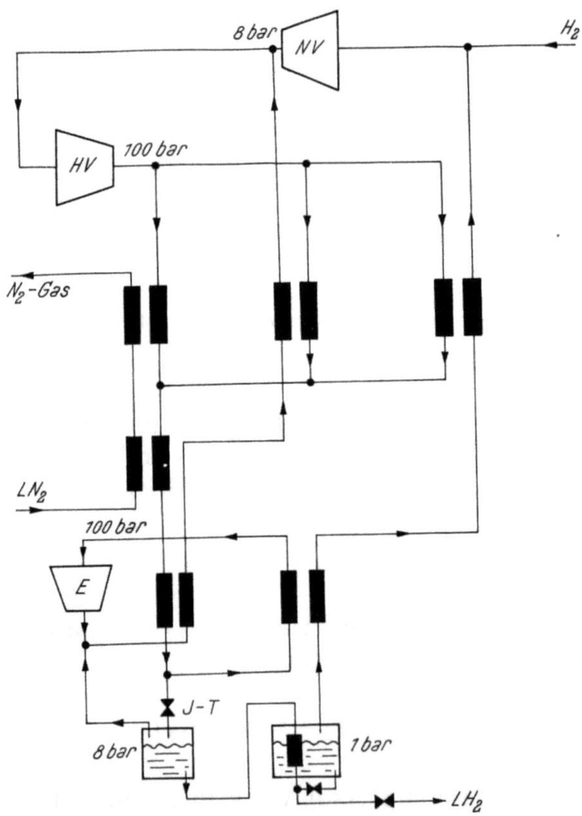

Fig. 17. Precooled dual-pressure Claude cycle with liquid nitrogen and intermediate stage for liquefaction (from [4]). The stages for the ortho-para conversion are not shown. *NV* low pressure compressor, *HV* high pressure compressor, *E* expander

Table 7. Ideal Claude process for hydrogen liquefaction

Process	α_1 (kWh/kg)	$\varphi_{H_2} = \dot{m}_f/\dot{m}$	$r = \dot{m}_{N_2}/\dot{m}_f$
Single pressure Claude	7.5	0.52	12
Dual pressure Claude	6.67	0.59	9.5

liquid helium production for many laboratories and thus contributed an integral part to the success of cryophysics and cryogenic technology. The liquefier as well as the refrigerator can be adapted for hydrogen liquefaction on a laboratory scale (Fig. 18). For the liquefier version, characterized by two independent expansion engines, a heat exchanger W_H is provided in which gaseous hydrogen at ambient pressure is cooled, liquefied and collected in a bottom container (lower part of the cryostat A which contains the heat exchanger and expander) after the ortho-para conversion. In this process, hydrogen liquefaction takes place along the low-pressure isobar (Fig. 3) so that hydrogen high-pressure compressors are unnecessary, which is desired in laboratory operation. (The helium compressors V_{He} are working at the relatively low optimal working pressure of $p_2 = 30$ bar and have a four-stage design.)

For hydrogen liquefaction, the refrigerator version can be connected in a similar manner to a condensation unit B located outside the plant. The

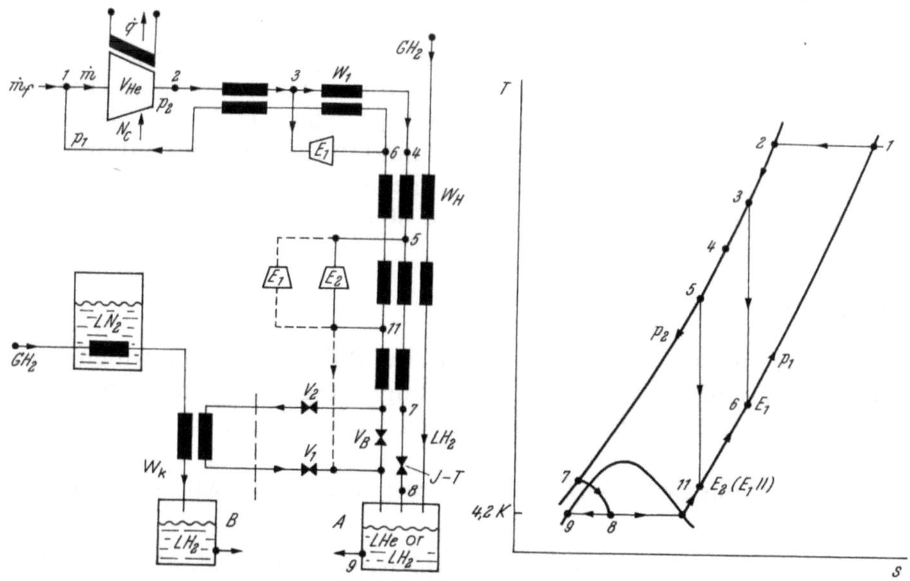

Fig. 18. Collins refrigerator and helium liquefier for hydrogen liquefaction, with helium as the working medium. Case A: liquefier; V_1, V_2, V_B not required. Case B: refrigerator; E_1 and E_2 parallel (dashed line), heat exchanger W_H and W_1 not required. The ortho-para converter is left out for clarity

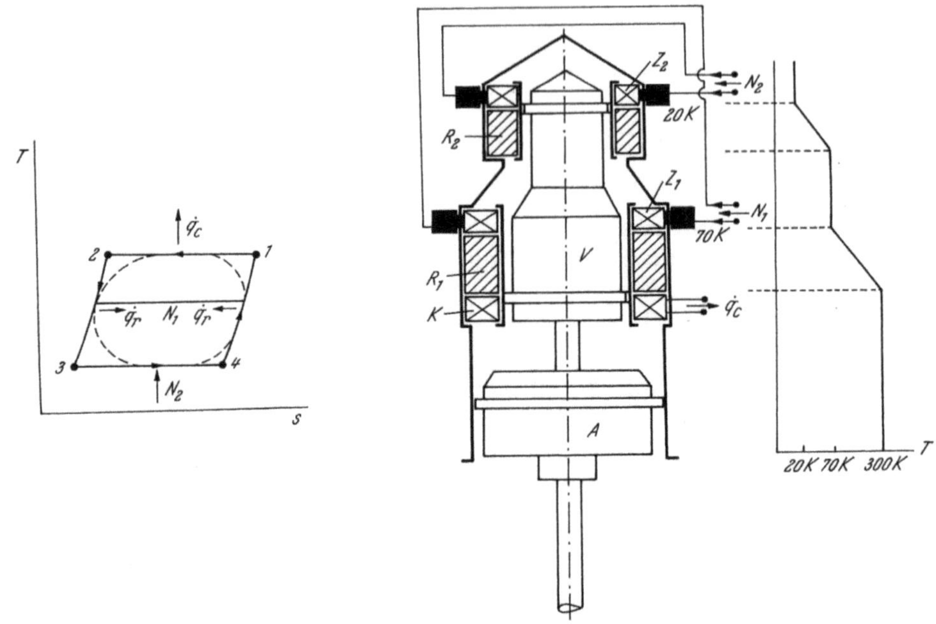

Fig. 19. Two-stage Philips–Stirling refrigerator. A, V working piston and displacer piston, K cooler, Z_1, Z_2 intermediate cooler with heat exchanger, R_1, R_2 regenerators

refrigerator supplies cold gaseous helium at low pressure via the valves V_1 and V_2 with closed bypass valve V_B and Joule–Thomson valve (Fig. 18), which flows through the heat exchanger W_K[5], where hydrogen condenses. The hydrogen, which accumulates as a liquid in the bottom container, can also be precooled with liquid nitrogen while in a gaseous state and subjected to a two-stage ortho-para conversion at 80 K and 20 K.

Stirling Process

The Stirling process plants developed by Philips [18, 19] which reach temperatures of about 70–100 K in the single-stage design and 15–30 K in the two-stage design [3] have proven successful even in cryogenic technology.
The ideal Stirling process (Fig. 19) takes place along two isotherms and two isochores:

Phase I : isothermal compression of the working medium (helium) along $1 \rightarrow 2$ and heat release $\dot{q}_c$.
Phase II : isochoric displacement and internal cooling along $2 \rightarrow 3$.

[5] Located above B in Fig. 18.

Phase III: isothermal expansion along $3 \rightarrow 4$ and heat absorption.
Phase IV: isochoric displacement and internal warming along $4 \rightarrow 1$.

The ideal two-stage Stirling process is approximated in a manner similar to the single–stage design through the movement of a working piston A and a displacer V.
In phase I, V is at top dead center while A isothermally compresses the working medium, during which the heat q_c is transferred to the water cooler K. In phase II, the working medium is forced under cooling, isochorically through the regenerators R_1 and R_2 by a downward stroke of the displacer piston V where R_1 serves as a heat sink for R_2. In phase III the working piston A and displacement piston V move downward, while the working medium in R_1 and Z_1 is cooled to about 70 K, and in R_2 and Z_2 to about 20 K. In phase IV, the working medium is displaced from the upper compression chamber through an upward motion of the displacer V and heat absorption in the regenerators R_1 and R_2, with which the initial state of phase I is again attained.
At the intermediate coolers Z_1 and Z_2, the refrigeration capacity N_1 and N_2 is available at about 70 K and 20 K.
The approximation of this actual process (dashed line in Fig. 19) to the ideal Stirling process is very good [18, 50] due to the development of high-quality regenerators which is reflected in the high thermodynamic quality of these plants.
As a result of the two-stage design further refrigeration capacity is available at about 70 K so that precooling with liquid nitrogen is unnecessary. This advantage, the simplicity of the technical concept as well as the avoidance of high-pressure circulation also make the liquefaction of hydrogen on a laboratory scale attractive although Stirling engines, as a result of their isothermal heat emission at lower temperature levels, are more suitable for refrigerator applications. With Stirling hydrogen liquefiers hydrogen is cooled and liquefied along the low-pressure isobars (Fig. 3). For this isobaric process with the enthalpies i_0 and i_f for hydrogen in a normal state and liquid state respectively, and with the refrigeration capacity N_1 and N_2 of the first and second stage of the gas refrigerator the following is obtained for the liquefaction rate

$$\dot{m}_{H_2} = \frac{N_1 + N_2}{i_0 - i_f}. \tag{96}$$

With a refrigeration capacity of $N_1 + N_2 = 240 + 900 = 1140\,\mathrm{W}$, the two-stage Stirling refrigerator type PGH 420 for example, can achieve a liquefaction rate of $\dot{m}_{H_2} \sim 1\,\mathrm{kg/h}$ (12–13 L/h) and can be best implemented where liquid hydrogen is difficult to obtain or the transportation costs are too high.

3.3 Magnetocaloric Liquefiers

These liquefiers, together with an appropriate cyclic thermodynamic process, can use isentropic demagnetization of a ferromagnetic material near its Curie point temperature as a refrigeration procedure. Magnetocaloric refrigerators were investigated and constructed for the temperature range from about 1 K to 20 K [11, 14, 17, 21–23, 59, 63, 70]. Refrigerators of this type for ambient temperature and below, for terrestrial use as well as in space projects were also proposed and tested [10, 22, 52].
Magnetocaloric liquefiers, on the other hand, have neither been built nor utilized but, because of their promising characteristics they are matter of intense investigation and will be covered here in more detail [16, 17, 54, 55].

Magnetocaloric Processes

Magnetocaloric refrigeration can basically be integrated into thermodynamic processes in a manner similar to isentropic expansion in expanders with pure gas processes. While a gas serves as the working medium in the latter, a ferromagnetic (also a paramagnetic at very low temperatures) material, i.e., a solid body, assumes this role in a magnetocaloric process [12].
Equations (33) and (35) show that pressure and specific volume of the gas process correspond to the magnetic field intensity and the magnetization respectively. Thus from the corresponding gas processes the magnetic Carnot, Brayton, Ericson and Stirling cycles can be derived [15, 23, 24].
The magnetic Carnot cycle [21] consists of two isotherms and two isentropes (Fig. 20) as in the gas cycle. Isothermal magnetization along

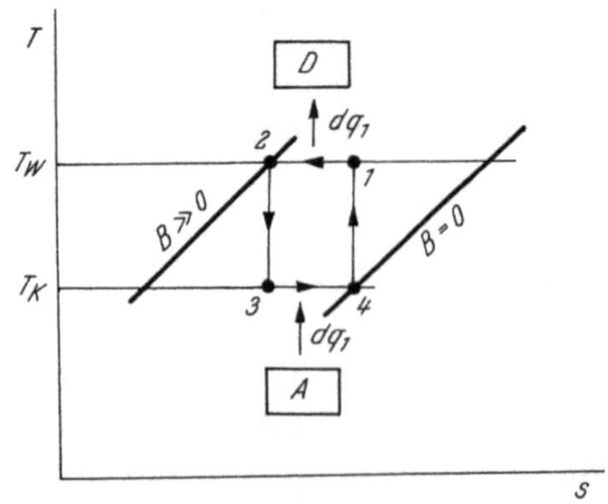

Fig. 20. Magnetic Carnot cycle in the T–s diagram

1 → 2 takes place at temperature T_w with heat release dq_1 of the magnetic working medium to the heat sink *D*. After disconnection of the thermal contact to the heat sink *D*, a partial demagnetization is carried out isentropically from 2 → 3 which cools the magnetic working medium to T_k. After closing the thermal contact with heat source *A*, the demagnetization is completed isothermally from 3 → 4 which achieves the desired cooling effect. In a final step an isentropic partial magnetization along 4 → 1 is carried out after removing the connection to the heat source *A*, which warms up the working medium to T_w thus returning to the initial state 1.

The magnetic Carnot cycle does not require internal heat exchange but in the range of the Curie point temperatures can produce only a relatively small temperature drop (5–10 K per 10 Tesla) (Fig. 11) and is very sensitive to thermal load during the isentropic phases.

The magnetic Brayton cycle [15], (Fig. 21), which takes place between two isentropes and two curves of constant field strength, requires internal heat exchange but can produce a larger temperature change than the magnetic Carnot cycle—in the ideal case about 15–20 K per Tesla. The magnetic Brayton process requires high-quality heat exchangers for internal heat exchange, which will usually be designed as regenerators. Furthermore, the Brayton cycle is also very sensitive to irreversible entropy changes during isentropic magnetization and demagnetization phases.

It is advantageous that no heat exchange is necessary during the magnetization and demagnetization phases and that heat absorption (5 → 6) and heat emission (2 → 3) do not have to be carried out isothermally so that higher temperature ranges in the external heat exchanger are possible.

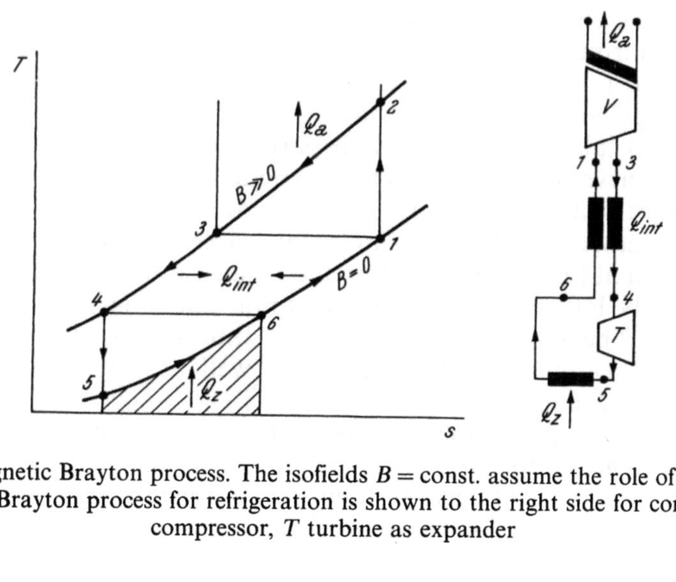

Fig. 21. Magnetic Brayton process. The isofields $B = \text{const.}$ assume the role of the isobars. The normal Brayton process for refrigeration is shown to the right side for comparison. *V* compressor, *T* turbine as expander

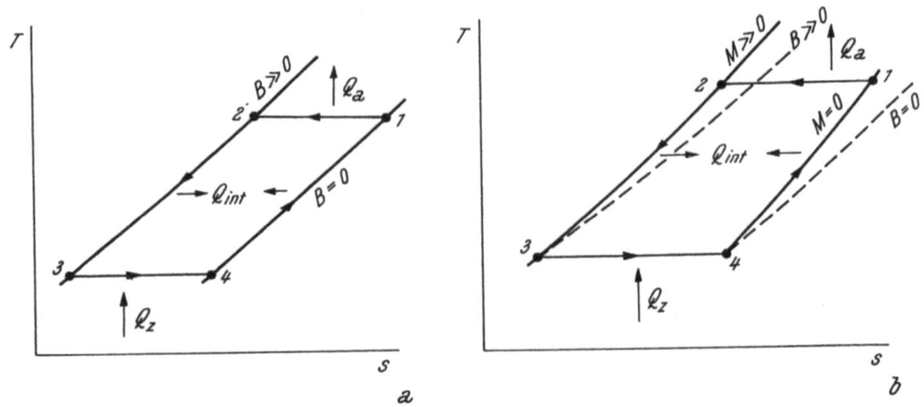

Fig. 22. Magnetic Ericson process (**a**) and magnetic Stirling process (**b**). *B* magnetic induction, *M* magnetization

During magnetization and demagnetization, the magnetic Ericson and the magnetic Stirling cycle [15, 23] (Fig. 22) have isothermal phases in common and also require internal heat exchange with high-quality heat exchangers.
As opposed to the magnetic Stirling cycle, the magnetic Ericson cycle enables the use of a larger temperature range near the Curie point temperatures. This will be demonstrated using gadolinium as an example. From Fig. 20 the small temperature range of about 5 K and the small change in reversible magnetic entropy of 0.096 R (R, gas constant) per cycle per mole for field intensities between 0 and 7 Tesla can be achieved for the Carnot cycle. This follows from the small amount of change effected in the magnetic entropy of the spin system, compared to the total entropy, which results from the magnetization and demagnetization, Eqs. (47) and (57). If Fig. 20 is transferred to a diagram with the magnetic entropy s_m (Eq. 53) (Fig. 23) it becomes evident that the magnetic Carnot cycle ($1 \rightarrow 2 \rightarrow 3 \rightarrow 4$) takes up only a small part of the area which is available between the curves 0 T and 7 T. The magnetic Stirling cyle ($1 \rightarrow 2 \rightarrow 3' \rightarrow 4'$) with constant magnetization ($s_m = \text{const.}$, $M = \text{const.}$) and a maximum temperature range of 35 K is thus more favorable. According to van Geuns [24] the thermal load is absorbed by the regenerator as a result of cooling the solid-body lattice and is returned upon reheating thus relieving the spin system. The magneticaloric effect is thus completely available for cooling and heating the spin system as well as for carrying out the refrigeration process.
Further improvement is possible if, through the execution of an appropriate process, the regenerator also takes over the cyclic cooling and heating of the spin system [15, 24] because then the complete magnetocaloric effect is exclusively available for the refrigeration process. This is particularly

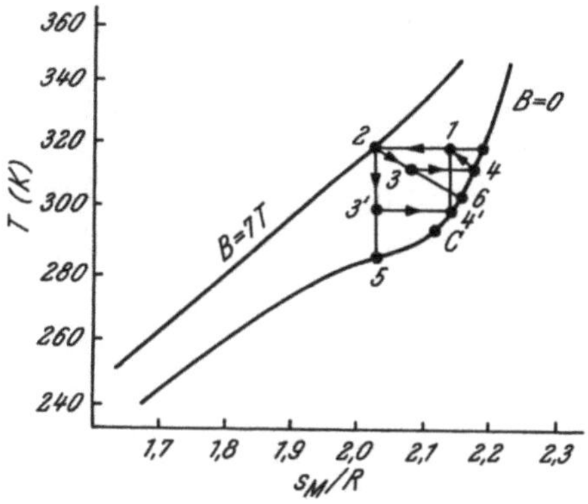

Fig. 23. Magnetic Carnot cycle ($1 \rightarrow 2 \rightarrow 3 \rightarrow 4 \rightarrow 1$) and magnetic Stirling process ($1 \rightarrow 2 \rightarrow 3' \rightarrow 4' \rightarrow 1$) in the *T–s* diagram for the magnetic entropy s_M of the spin system for gadolinium. *C* Curie point of gadolinium (from [15])

valid for the Ericson cycle, shown in Figs. 22 and 24, where regenerative heat exchange with a constant external field takes place, which for gadolinium in the ideal case enables a maximum temperature range of $\Delta T = 70$ K.

The regenerator is of fundamental importance because it must go through cyclic heat absorption and release of the solid-body lattice as well as the spin system between the temperatures of the heat source and heat sink with minimal allowable loss. A primary source of loss results from the variation of the magnetic heat capacity of the ferromagnetic working medium along the curves $4 \rightarrow 1$ and $2 \rightarrow 3$, causing different amounts of heat to be absorbed and emitted by the regenerator, which in the ideal

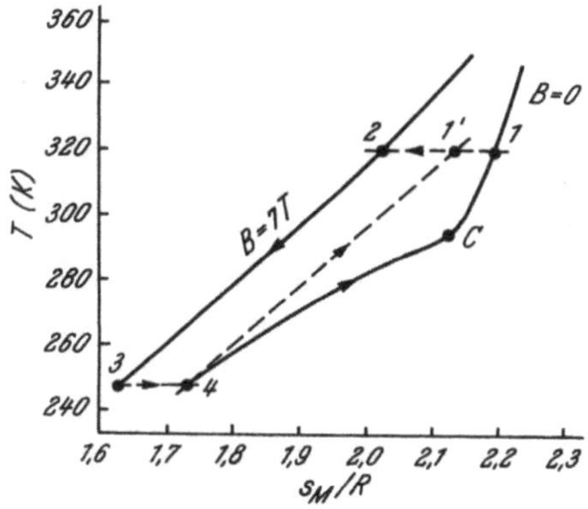

Fig. 24. Magnetic Ericson process ($1' \rightarrow 2 \rightarrow 3 \rightarrow 4 \rightarrow 1'$) *T–s* diagram for the magnetic entropy s_M of the spin system for gadolinium. *C* Curie point of gadolinium (from [15])

case must be equal. Furthermore, the changes in entropy Δs_{12} and Δs_{34} should be equivalent in the interest of maximum thermodynamic efficiency particularly in the Carnot cycle (Fig. 24). This can be attained by cyclic variation of the magnetic field during the process $4 \rightarrow 1'$ (Fig. 24), which means that after isothermal demagnetization $3 \rightarrow 4$ the magnetic field must cyclically increase during heating by the regenerator allong $4 \rightarrow 1'$.

Active Magnetic Regenerator

For the high-quality regenerators which are necessary for the Brayton, Ericson and Stirling cycles, the magnetic working medium itself can be used according to Ref. [15, 24].

The exchange of heat for the lattice as well as for the spin system is thus carried out directly at the magnetic material whereas the heat input and emission necessary for the refrigeration process is taken over by the gaseous or liquid working medium which flows cyclically through the regenerator [54, 55].

If in this process larger temperature changes and variations in entropy are required than can be attained by means of a single ferromagnetic near its Curie point temperature, it is possible to extend the regenerator into a series of steps with ferromagnetic materials of successively decreasing Curie point temperatures. This can be achieved advantageously with alloys of "rare earth metals" in the range from ambient temperature to that of liquid hydrogen (see Table 8).

Therefore, a magnetic regenerator is constructed of several porous ferromagnetic materials with gradually differing Curie point temperatures T_{c1} to T_{cn} (see Fig. 25) [16]. Its operation corresponds to that of a normal regenerator—the one in the Philips-Stirling refrigerator for example. The difference is that the temperature of the regenerator materials is increased and decreased by applying and removing a magnetic field respectively, so that a thermal wavefront travels cyclically in the regenerator by acting in combination with the cyclic flow of a gaseous or liquid medium through the regenerator. Since the switching of a strong magnetic field of the required superconducting magnet cannot be carried out without large losses because of its intermittent character, moving the regenerator into and out of the stationary magnetic field of a superconductive magnet could be an alternative procedure.

After a few cycles a temperature distribution in the regenerator as shown in Fig. 25(1) is attained. In this procedure T_k and T_w are the temperatures in the regenerator at the cold and warm side respectively. After initializing the magnetic field the temperature inside the regenerator rises constantly by $\Delta T \approx 15$–$20\,K$ per 10 Tesla (Fig. 25(2)) because ΔT is independent of the temperature T near Curie point temperature (see Ref. [17]).

Table 8. Curie point temperatures of gadolinium alloys and compounds for which the magnetic moment of the gadolinium atom (7.5 μ_B) is almost completely maintained (from [17])

Material	T_c(K)	Material	T_c(K)
$Gd_{2.71}S_4$	21	$Gd_{0.5}Ag_{0.5}$[a]	100
$GdAg_{0.8}In_{0.2}$	24	$GdAg_{0.5}In_{0.5}$	111
$Gd_{2.73}S_4$	28	$Gd_{0.68}N_{0.32}$	125
$Gd_{0.65}Th_{0.36}Al_2$	30	$Gd_{0.8}Au_{0.2}$	150
$GdNi_5$[a]	32	$GdAl_2$	153
$GdNi_2$	38	$Gd_{0.7}La_{0.3}$	185
$Gd_{2.76}S_4$	42	GdGa	200
$Gd_{0.7}Th_{0.3}Al_2$	50	$Gd_{0.67}Y_{0.33}$	211
$GdAg_{0.7}In_{0.3}$	57	$GdZn_{0.85}In_{0.15}$	224
$Gd_{2.8}S_4$	58	$Gd_{0.8}Y_{0.2}$	254
GdN	65	GdZn	268
GdNi[b]	81	$Gd_{0.9}Y_{0.1}$	281
$Gd_{0.8}Th_{0.2}Al_2$	90	Gd	293
		$GdMn_2$	300

[a] Amorphous.
[b] Crystalline.

Afterwards, a gaseous medium (helium or hydrogen for example) coming from the cold side flows through the regenerator, whose temperature at the cold side is now $T_k + \Delta T$. In this procedure the gas warms up while the regenerator cools down and then a thermal wave front with a temperature difference T is produced (Fig. 25(3)). A thermal wave front with a temperature difference of $T_w + \Delta T - T_k$ then travels in the direction of the flow toward the warm end of the regenerator where the gas exists at temperature $T_w + \Delta T$ as long as the gas temperature sinks to T_w as a result of the thermal wave front arrival. After stopping the gas flow, isentropic demagnetization takes place during which the temperature in the regenerator decreases constantly by T (Fig. 25(4)). After being cooled by ΔT in the refrigerator by heat emission the gas flows into the regenerator at temperature T_w from the warm side (Fig. 25(5)) where a thermal wave front with a temperature difference of $T_w - (T_k - \Delta T)$ travels toward the cold side of the regenerator. The gas exits the cold side at $T_k - \Delta T$ and can now be warmed up by ΔT to T_k thus producing the cooling effect at the cold side. After stopping the gas flow with the arrival of the wave front at the cold side, the cycle is complete and can start over again.
For the magnetic Brayton cycle, which is covered in the following because of its practicality, the refrigeration capacity $\dot{Q}_k$ of the system is

$$\dot{Q}_k = \dot{m}_k \cdot c_p \cdot \Delta T_k. \tag{97}$$

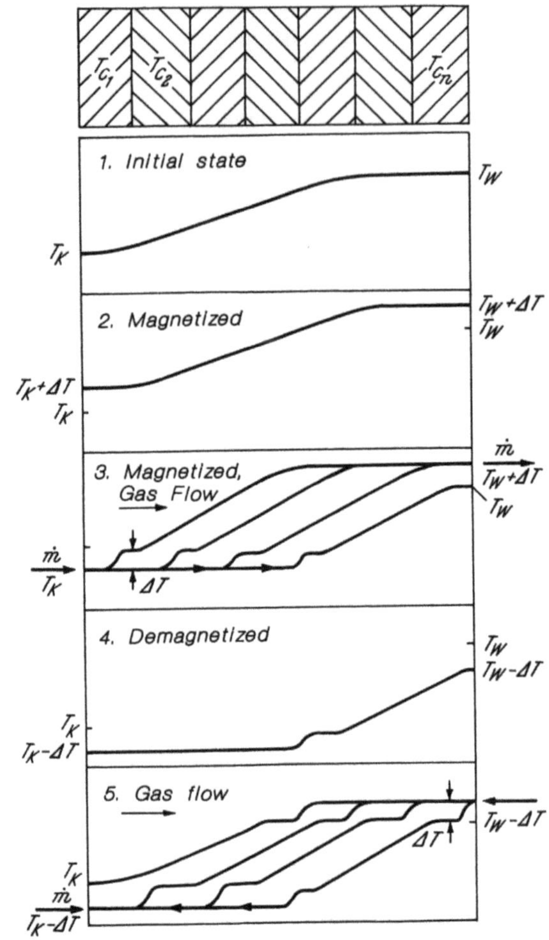

Fig. 25. Schematic of the active magnetic regenerator (AMR)

Here, c_p is the specific heat of the gas at constant pressure ΔT_k and $\dot{m}_k$ are the magnetic temperature change and the mass flow rate of the gas at the cold side of the regenerator respectively.
At the warm side the following applies

$$\dot{Q}_w = \dot{m}_w \cdot c_p \cdot \Delta T_w. \tag{98}$$

And for the ideal Carnot efficiency

$$\frac{\dot{Q}_w}{T_w} = \frac{\dot{Q}_k}{T_k}. \tag{99}$$

Assuming constant c_p from (97), (98) and (99) the following is obtained

$$\frac{T_w}{T_k} = \frac{\dot{m}_w}{\dot{m}_k} \cdot \frac{\Delta T_w}{\Delta T_k} \tag{100}$$

or in the case of $\Delta T_w \approx \Delta T_k = \Delta T$

$$\frac{T_w}{T_k} = \frac{\dot{m}_w}{\dot{m}_k}. \tag{101}$$

Equation (100) can be satisfied through an appropriate distribution of the magnetic field intensity along the regenerator and thus with different values of ΔT_w and ΔT_k. According to (99), (98) and (97) though, the refrigeration capacity is already given by $\dot{Q}_k$ and T_w at the temperature T_k so that in general ΔT_k and thus $\dot{Q}_k$ can no longer assume the maximum possible value. It is more advantageous to satisfy (100) by a change in the mass flow rates $\dot{m}_w$ and $\dot{m}_w$ at the warm and cold side respectively so that ΔT_w as well as ΔT_k can attain the maximum possible value of about 15–20 K per 10 Tesla [17]. The change in the mass flow rate and associated temperature resulting from (101) present a serious complication for magnetocaloric refrigeration. In the case of the magnetocaloric hydrogen liquefier it must be considered that continuous cooling of the gas from ambient temperature to the boiling point is required. As a result of the ortho para conversion as well as the condensation in different temperature ranges, different amounts of heat have to be removed. This can be satisfied only by variable mass flow rate $\dot{m}$ in the individual temperature ranges. Therefore, additionally satisfying (101) implies no further complications.
For the amount of heat $\dot{Q}_R$ removed per cycle from the regenerator the following applies for the Brayton cycle

$$\dot{Q}_R = 2M \cdot \nu \cdot T_m \cdot \Delta s_R. \tag{102}$$

M is the mass of the regenerator and ν represents the number of cycles per unit of time in which two regenerative phases per cycle occur. The average temperature is $T_m = \frac{1}{2}(T_k + T_w)$, where T_k and T_w denote the temperatures at the cold and warm side respectively. For the change in entropy per mass unit, Δs_R, during the regenerative phase the following is obtained

$$\Delta s_R = \left(\frac{\partial s_R}{\partial T}\right)_H (T_w - T_k). \tag{103}$$

In a rough approximation the same value $\left(\frac{\partial s_R}{\partial T}\right)_H$ can be used for the isofield phases. This is justified as long as the curves $B = 0$ and $B \gg 0$ (Fig. 21) are approximately parallel along the curve sections between $3 \rightarrow 4$ and $6 \rightarrow 1$.

The refrigeration capacity $\dot{Q}_K$ of the Brayton cycle can be expressed on

$$\dot{Q}_K = \dot{m} \cdot c_p \cdot \Delta T \tag{104}$$

with $\dot{m}$ mass flow rate gas, c_p the specific heat of the gas at constant pressure, and ΔT the temperature change of the gas in the regenerator.
With the usual adaptation of the heat capacity of the regenerator to that of the gas flow [18] the following applies

$$M \cdot v \cdot c_H = \dot{m} c_p. \tag{105}$$

With (33) and the average gas temperature $\bar{T} = T_k + \frac{\Delta T}{2}$ the following is obtained

$$c_H = \bar{T}\left(\frac{\partial s_R}{\partial T}\right)_H = \left(T_k + \frac{\Delta T}{2}\right)\left(\frac{\partial s_R}{\partial T}\right)_H. \tag{106}$$

For the relationship of the regenerator output $\dot{Q}_R$ to the refrigeration capacity $\dot{Q}_K$ from (102) to (106) and from [17] the following is obtained

$$\frac{\dot{Q}_R}{\dot{Q}_K} = \frac{T_w^2 - T_k^2}{\Delta T\left(T_k + \frac{\Delta T}{2}\right)}. \tag{107}$$

Table 9 shows the relation $\dot{Q}_R/\dot{Q}_K$ at different temperatures T_w and T_k of the warm and cold sides of the regenerator for a temperature range which results from the magnetic cooling of $T = 14$ K. It illustrates that compared to the amount of heat input and emitted, the magnetic regenerator has to store extremely large amounts of heat at higher temperatures so that even while maintaining smaller temperature differences (70 K for example) the development of appropriate low-loss magnetic regenerators is of critical importance.
From (104), (105) and (106) the following is obtained for the refrigeration capacity $\dot{Q}_k$

$$\dot{Q}_k = M \cdot v \cdot \left(T + \frac{\Delta T}{2}\right)\left(\frac{\partial s_R}{\partial T}\right)_H \cdot \Delta T. \tag{108}$$

Table 9. Ratio of regenerator output to refrigeration capacity

T_k \ T_w	300 K	230 K	160 K	90 K	20 K
230 K	10.4	0	—	—	—
160 K	27.5	11.7	0	—	—
90 K	60.3	33.0	12.9	0	—
20 K	221	139	66.7	20.4	—

For $\nu = 1$ (one cycle per second) for instance for gadolinium near its Curie point temperature with an average value of

$$\left(\frac{\partial s_R}{\partial T}\right)_H = 0.1555\ \mathrm{J/Mol\,K^2},$$

$T_k = 290\,\mathrm{K}$ and $\Delta T = 14\,\mathrm{K}$, the refrigeration capacity is $\dot{Q}_k = 650\ \mathrm{W/Mol}$, or with $\rho_{\mathrm{Gd}} = 7.9\ \mathrm{g\,cm^{-3}}$ depending upon the regenerator volume

$$\dot{Q}_k = 32.5\ \mathrm{kW/L}.$$

Therefore, from this view magnetocaloric refrigerators and liquefiers would basically be very compact machines, which in comparison to the pure gas processes can amount to a very high capacity per unit volume [17].

Principle Design of Magnetocaloric Liquefiers

Magnetocaloric liquefiers, as previously mentioned, have not been built yet. Because of the research on constructed magnetic refrigerators using liquid helium [14, 16, 21, 22] as well as those at ambient temperature (see [15, 23]) the principle design can be outlined in accordance with [17] as follows:

A magnetocaloric liquefier essentially consists of a multiple-stage system of magnetic regenerators which are coupled with heat exchangers for cooling the hydrogen which is to be liquefied.

In order to accomplish the continuous cooling process an appropriate ferromagnetic material, whose Curie point temperature is in the range of the cooling temperature, has to be cyclically magnetized and demagnetized. The heating and cooling of the solid body that arise from this magnetization and demagnetization can be transmitted to a transfer medium (fluid, gas) which flows through the porous or pelletized magnetic material.

Magnetization and demagnetization are accomplished by an oscillating movement or steady rotation through the magnetic field. Thus, the mechanical work is input to the system, whereas in the gas cooling process this role is played by the high-pressure compressor. With a linear oscillating design regenerators are required for the cyclical internal heat transfer whereas a rotating design permits quasi-continuous internal counterflow heat exchange [17, 20, 52, 54, 56]. In the following the simpler configuration of linear oscillating design are shortly discussed.

The gas flow, which cyclically surrounds the ferromagnetic material, is closed by appropriate heat exchangers and regenerators in the refrigeration cycle [17]. Locating the regenerators outside the magnetic field is known as external regeneration. The design of the magnetic material as an active magnetic regenerator (AMR) by stringing together different materials with graded Curie point temperatures permits the coverage of larger temperature ranges while avoiding external regeneration [56–58, 67, 68]. The advantage of the ease of switching the transfer gas flow is accompanied by the disadvantage of an increased need for magnetic material and magnetic field volume.

Figure 26 shows a possible schematic design of a small four-stage hydrogen liquefier from [17]. Each of the four stages, whose temperature range is about 70 K with regard to a relatively small ratio of $\dot{Q}_R/\dot{Q}_K$ (see Eq. (107) and Table 9), represents an active magnetic regenerator according to

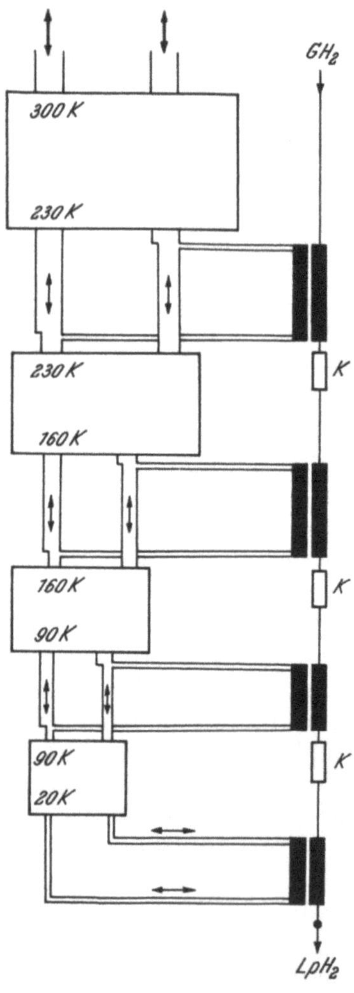

Fig. 26. Schematic of a four-stage magnetocaloric liquefier using the magnetic Brayton process and three-stage ortho-para conversion (*K*). Magnets, valves, etc., are left out for clarity (from [17])

Fig. 25. After each of the first three stages are heat exchangers in the bypass, which serve to cool the hydrogen which is to be liquefied and perform the ortho-para conversion as well as allowing the different mass flow rates of the gas flow in each stage of the regenerator for more efficient use of the regenerator materials (see Eqs. (100) and (101)).

It is intended to use a common magnetic field produced by a superconducting magnet for the four regenerators. Refrigeration of the superconducting magnet (about 8–10 Tesla) requires about 120 mW at 4.2 K and can basically be achieved by a fifth, very small regenerator stage, (see [20]). According to Ref. [20] this additional thermal load has only a very small impact on the overall efficiency.

The control of the intermittent helium gas flow in the regenerator is accomplished by means of appropriate valve arrangements and intermediate gas storage with the help of a compressor which only has to supply the energy for maintaining the gas circulation and not the energy for the liquefaction work.

A detailed examination of the irreversible increase in entropy as a result of heat conduction and gas mixing in the regenerators, flow losses, internal heat transfer in the regenerator, heat exchanger losses, etc., resulted in the following data contained in Table 10 ([17]).

Since there are no data and experience from experiments with magnetocaloric liquefiers, a comparison with existing liquefiers and refrigerators in use can be made only with reservation.

The examination of magnetocaloric refrigeration processes provides not only the cyclic path of the magnetic forces but also the deviation of the efficiency from that of the ideal Carnot cycle. The corresponding loss in refrigeration capacity has to be considered as "inherent" because it is not influenced by other parameters with a given magnetic material and temperature range. When using homogenous materials (like Gd, GdNi, etc.) these losses are typically in the range of 20–40% of the ideal Carnot refrigeration capacity.

Table 10. Small magnetocaloric refrigerator data

Cycle duration	approx. 10 s
Magnetic field	approx. 10 T
Refrigeration capacity	20 W at 20.4 K
Liquefaction rate	0.7 L/h
Liquefaction work	7.26 kWh/kg (54% Carnot)
Power demand	557 W
Heat release at 300 K	621 W
Volume	approx. 16 L
Mass	approx. 166 kg

The loss is a result of the deviation of the entropy temperature function of the magnetic working material with regard to the following idealizations:

- System with External Regeneration: The isofields[6] in the entropy-temperature diagram are parallel (see [57, 60, 64, 66, 69]).
- System with AMR: The isofields in the entropy-temperature diagram must converge in direction to the entropy axis (see [67, 68]).

Technical irreversibilities like heat transfer losses with finite temperature differences as well as flow losses which can occur in the circulation of the transfer medium during heat and cold transfer to the magnetic material as well as during regeneration phases, can be reduced to a minimum through variation of characteristic design parameters.

These parameters are: particle size and porosity of the magnetic material, pressure of the heat transfer medium, number of temperature levels with an external regenerator, time period of the regeneration phase.

With regard to efficiency[7] of the magnetocaloric refrigeration process as well as to the losses resulting from irreversibilities the following was obtained [55–60]:

- With external regeneration as well as with active magnetic regenerators, the efficiency of the magnetocaloric refrigeration process achieves results which, considering the irreversibilities and loss mechanisms mentioned, can amount to almost the maximum, $\eta = 0.8$.
- With a relatively low working pressure of the helium gas circulation (15 bar) and relatively short cycle times the attainable efficiency is just above $\eta = 0.6$.

The influence of a minimum particle size with respect to irreversible losses is of specific importance. In the case of external regeneration, according to [56, 60] the optimal efficiency for the diameter ranges from $d = 1$ to 2 mm, whereas in the case of an active magnetic regenerator a particle diameter of more than 1 mm would have a negative impact.

Comparing magnetocaloric processes with conventional gas processes, magnetocaloric processes have better efficiencies because there are no

[6] Curves with constant magnetic field or constant magnetization in the entropy-temperature diagram.

[7] The efficiency of a refrigeration process (coefficient of performance) is defined as:

$$\eta = \frac{\text{(released cold)}}{\text{(expenditure of energy)}}$$

The thermodynamic efficiency (quality factor, percent Carnot efficiency) is: $\eta_G = \eta/\eta_c$

η_c: efficiency of the ideal Carnot comparison process.

considerable irreversible entropy portions caused by losses in gas compression and expansion. A comparison of the output to volume ratio, however, does not turn out so well. Although the refrigeration capacity per volume in the magnetic material can be considerable, the heat transfer capacity has to remain relatively small in order to avoid considerable losses which arise due to flow irreversibilities as well as in the interest of small temperature drop for heat transfer.

An adaptation of the entropy-temperature function of the working material in order to improve the efficiency is possible in principle through variation in the production conditions of the material.

The influence of the axial heat conduction in the magnetic regenerators as well as the influence of the heat transfer in the porous material on the irreversible increase in entropy requires further intense investigation. Similarly, with regard to flow losses and heat produced by eddy currents (see [65, 66]) cycle times of less than one second are difficult to achieve. Thus the size of the regenerators cannot be further reduced while maintaining liquefaction capacity.

Nevertheless it is not surprising that the liquefaction work of magnetocaloric liquifiers compared to other procedures can be considerably smaller because critical components like a compressor, intermediate cooler or expander are not used, which in existing plants, despite being highly developed, contribute considerably to the irreversible change in entropy (see Table 13, page 67).

Prospect for Future Applications

Due to expected improvements in efficiency as well as no requirement for compressors and expanders, magnetocaloric refrigeration is of essential importance for the future, though reductions in energy and maintenance costs have to be measured against the necessary investment. Another positive aspect is that the large output to volume ratio achievable in the magnetic working medium leads conceptually to very compact cooling systems. Low problem occurrence and the extremely long expected lifetimes of such plants ensure essentially maintenance-free operation for long periods of time. Applications which go beyond hydrogen in the field of terrestrial energy technology include air conditioning, cooling below ambient temperature as well as the reversal of the magnetocaloric process conduction to the thermal engine in case of availability of economical low-temperature heat. Another aspect which should not be underestimated is the possibility of very effective refrigeration processes without using refrigerants based on halogen hydrocarbons or combustible low boiling hydrocarbons.

Next to current refrigeration of sensors below 2.2 K (λ-point of helium) by means of cyclic adiabatic demagnetization of paramagnetic salts (one-shot device) applications in space technology are underway for several future space missions in earth orbit or in interplanetary space. For these applications, reliable cooling systems with low vibration and operational lifetimes of about 10 years are required for the refrigeration of cryogenic fuel tanks or superconducting magnets for energy storage and shielding of charged particle radiation for example. A desirable decrease in the regular operating temperature of unmanned satellites and space flight systems to about 100–200 K in the interest of reduced electronic failure, requires reliable heat pumps in order to increase the radiator temperature. Magnetocaloric systems would be quite attractive here. In the applications mentioned configurations with a rotating magnetic working medium are preferred to those with linear oscillating working mediums.

Further detailed statements are not yet possible due to insufficient research and the lack of standards in this area. This applies primarily to the temperature range, the question of heat transfer between the magnetic working medium and the transfer gas, the regeneration and cycle time as well as the generation of the magnetic field.

When using state of the art superconducting magnets, i.e., cooling with liquid helium and using superconductors like niobium–titanium or niobium–tin, a range below about 20 K is economical for smaller plants whereas for large plants it seems more likely below 80 K. This is primarily a result of high operating costs for the cooling of helium whereas questions regarding heat transfer, regeneration and cycle time have not yet been taken into consideration. This situation can change though if developments in ceramic super-conductors with sufficient current density, high critical field and transition temperatures around 120 K to ambient temperature should be successful. In the case of a potential cooling with liquid nitrogen a decrease in operating costs for the generation of the magnetic field to about one tenth of that with helium is possible [61, 62]. Thus the temperature range above 20 K to ambient temperature is definitely attractive for magnetocaloric refrigeration without restrictions with respect to refrigerator power. In the interest of low demand on ferromagnetic material, it must be also be taken into consideration that the magnetic field intensity has to be increased with increasing Curie point temperatures in order to compensate for the associated decrease in refrigeration power output to volume ratio.

In any case the counteracting effects on efficiency of parameters like magnetic field strength, heat tranfer and regeneration losses, and magnetocaloric characteristics of the working medium, have to be adjusted carefully in order to guarantee the benefits of magnetocaloric refrigeration.

A prerequisite is further more intense support of continuous research and development in the increasingly important field of cryo-technology, which is currently only in an initial stage.

3.4 Large-scale Industrial Liquefaction Plants

As a result of the space programs, there was a considerable requirement for liquid hydrogen in the U.S. since about 1956, especially for the Apollo program. Although previous experience essentially consisted only of small to medium-size plants (maximum about 200 kg/day), within a short time large-scale plants were successfully designed and built [25–31]. At the peak of this development there was a total plant capacity of more than 200,000 kg LH_2 per day available. After the completion of the rocket engines for the Apollo program the requirement for liquid hydrogen decreased drastically so that several plants were shut down or dismantled. This trend continued after the conclusion of the Apollo program. With the start of the Space Shuttle program a reverse in this trend took place.
Table 11 shows typical large-scale plants currently in operation in the U.S. [40]. For comparison, it also shows the largest plant which had been in operation in the U.S. and the largest European plant which are currently in operation.[8]
The requirement for economy is characteristic of large-scale plants since customers who require liquid hydrogen in large quantities also want a correspondingly inexpensive product. Apart from operating costs, investment costs play an important role. Furthermore, reliable liquefaction

Table 11. Large-scale H_2 liquefaction plants

Place	Producer, operator	Capacity[a]	In operation
Mississippi, Test Fac.	Air Products	>36 t/day	1960
Long Beach, Calif.	Air Products	30 t/day	1958
Ontario, Calif.	Union Carbide, Linde Div.	30 t/day	1962
Sacramento, Calif.	Union Carbide, Linde Div.	60 t/day	1966[b]
Lille France	L'Air Liquide	10 t/day	1985
Rozenburg Netherlands	Air Products	5 t/day	1986

[a] 1 t/day = 1,000 kg/day.
[b] Not in operation anymore.

[8] In 1984 there were only two liquefaction plants in Germany with daily capacities of approximately 100 kg LH_2 and 50 kg LH_2 respectively that the author is aware of. [51] The first of which was utilized for commercial purposes. In France there was a plant with a daily capacity of 1.3 t/day in operation [32].

processes, which are easily carried out and controlled are preferred. Therefore, it is difficult to establish general criterion for the design of such plants, which, with regard to investments and costs can best be compared to chemical processing plants and petrochemical plants.

A compromise should be made between minimum investment costs and maximum thermodynamic efficiency (see Fig. 27) of the liquefaction process. Furthermore, the question of location with regard to the availability of cheap hydrogen must be taken into consideration. This suggests the location of large-scale liquefaction plants near refineries or large-scale chemical plants where hydrogen is produced in large quantities or is needed as a raw material.

Variations of the dual pressure Claude process (see Fig. 17) as a large scale liquefaction procedure has proven to be the optimal solution. In this process turbines are used as expanders—generally one or two—which, compared to reciprocating expanders, can handle larger mass flows but at a smaller pressure ratio (see [33–37]).

According to research by Baker and Shaner [138] based on experience with currently operating large-scale plants, it is possible to construct economical plants with a liquefaction capacity up to 250,000 kg LH_2/day [36]. Larger capacities can be achieved by modular design. Figure 28 shows a simplified flow chart of a large-scale plant; more detaild figures and data can be found in [38]. The process, like that outlined in Fig. 17, uses hydrogen as a working medium. Two expansion turbines for cooling the hydrogen are provided, one each for the recirculation path and the liquefaction path.

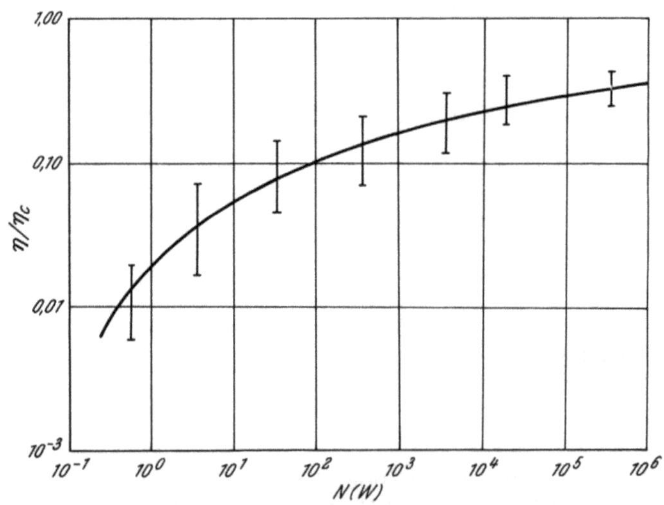

Fig. 27. Efficiency η of refrigerators and liquefiers, based on the efficiency η_c of the Carnot process, as a function of the refrigeration capacity N (from [49])

A third path serves to precool with liquid nitrogen which is supplied by a separate plant. In this process the gaseous nitrogen flowing back from the hydrogen liquefaction plant is reliquified in a refrigerator and returned. In order to cover nitrogen losses the nitrogen refrigerator is usually combined with an air separation plant.

The stage for the purification of the raw hydrogen [41–43] entering the plant will not be dealt with here. The hydrogen can generally contain nitrogen, carbon monoxide, carbon dioxide, oxygen, methane and water vapor as impurities.

In general the raw material has to be purified to a maximum impurity content of about 1 ppm in order to avoid freezing out the impurities and the problems in the liquefier associated with them. The applied cryogenic purification processes either use absorption in subcooled liquid hydrocarbons or physical adsorption with appropriate adsorbing materials, like silica gel (see [41, 43]).

After cooling the hydrogen to about 26 K the liquefaction is carried out in a manner similar to Fig. 17, by two-stage compression and Joule–Thomson expansion to 9 and 1 bar respectively. The gaseous return flow from the first stage (9 bar) is input to the low-pressure feed compressor. Thus, as in the two-stage compression Linde process (see Fig. 15) a reduction in compressor output is achieved.

The pressures and temperatures (see Table 12) corresponds to the minimum work of liquefaction taking the irreversible losses of the components into consideration. Here the input pressure at about 41 bar has a definite minimum for the liquefaction work which is due to a more favorable overlap of the operating temperature ranges of the turbines working as expanders.

Table 12. Typical data for a large-scale plant (250 t/day) for the liquefaction of hydrogen (see Fig. 29) [38]

	H_2-input	H_2-liquefaction
Pressure (bar)	1	9.2
Temperature (K)	308	20.57
Mass flow rate (kg/s)	2.625	2.625
Percentage p-H_2 (%)	25	99
Lowest pressure (bar)	2.87	
Highest pressure (bar)	41	
Expander exit temperature (K)	26	
Temperature difference at the heat exchanger (K)	8	
Efficiency of the H_2-turbine (%)	79	
Efficiency of the N_2-turbine (%)	84	
Medium compressor efficiency (%)	79	
Work of liquefaction (kWh/kg)	—	10.8[a]

[a] Corresponds to 36% Carnot.

The lowest pressure in the recirculation path is limited to about 2.8 bar in order to avoid liquefaction at the turbine exit. This turbine exit temperature of 26 K is chosen as low as possible in the interest of liquefaction work in order to achieve a minimum irreversible increase in entropy through a connected Joule–Thomson expansion. Furthermore, the four-stage ortho-para conversion used to reduce the work of liquefaction is typical for large-scale plants. This conversion can be carried out quasi-continuously using heat exchangers (Fig. 29) [41, 44–48]. For clarity this is not shown in Fig. 28.

The work of liquefaction of 10.8 kWh/kg reflects the more conservative values in [49] for liquefiers of appropriate size.

Furthermore, the percentage of loss for each component due to irreversible increases in entropy given in Table 13 [38–40] are very informative.

A comparison shows that the hydrogen compressors contribute the greatest losses, followed by the nitrogen refrigerator, turbines and then heat exchangers. Therefore, significant improvements are necessary in these areas.

The examples mentioned here provide for an improvement in the compressor and turbine efficiencies of 1%, a reduction of the liquefaction energy of each by 1.35% and 0.69% respectively.

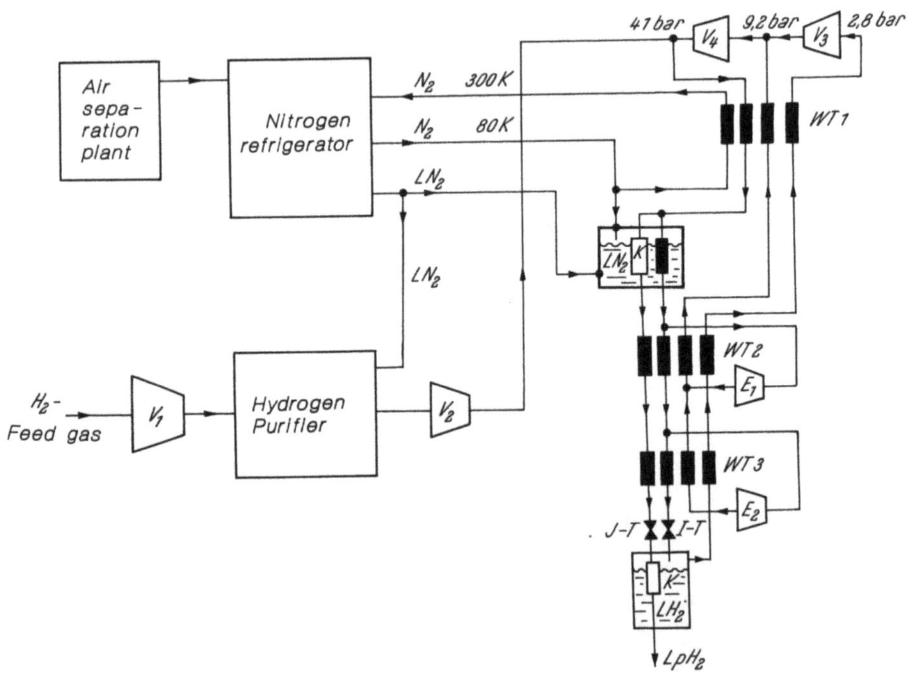

Fig. 28. Large-scale hydrogen liquefaction plant (from [38]). *V* compressor, *WT* heat exchanger, *K* ortho-para converter

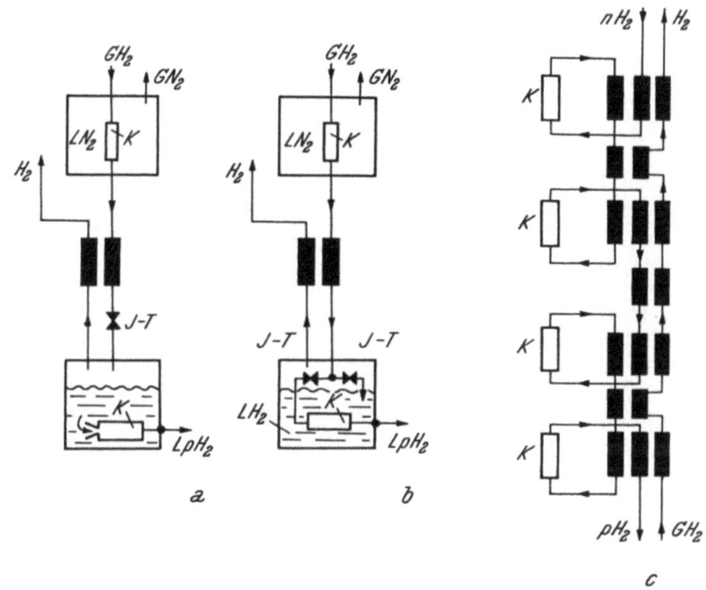

Fig. 29. Location of the catalysts K for ortho-para conversion. (**a**) single pressure cycle, Joule–Thomson expansion, two-stage conversion. (**b**) dual pressure cycle, Joule–Thomson expansion, two-stage conversion. (**c**) Quasi-continuous conversion below 70 K and GH_2 return flow for cooling of the high-pressure circulation

Table 13. Irreversible component losses of large-scale hydrogen liquefiers [38]

$W_i = 3.92$ kWh/kg $W = 10.8$ kWh/kg	$W - W_i = 6.88$ kWh/kg
Cycle compressor	29.35%
Feed compressor	8.61%
Turbines	12.96%
Heat exchanger	12.65%
O–P conversion	4.08%
LN_2-refrigerator	25.02%
Other losses	7.33%
Total	100.00%

Further reduction of the work of liquefaction to about 9.8 kWh/kg corresponding to about 40% Carnot or more is possible through the development of improved components which according to [49] corresponds to the best results attained for some plants of smaller size. Because of high investment costs such developments can be expected only with adequate demand.

References

[1] R. Barron: Cryogenic Systems. New York, McGraw Hill (1966).
[2] A. J. Westbrock: Liquefaction of Oxygen, Nitrogen and Hydrogen. In: Applied Cryogenic Engineering (R. D. Vance, W. M. Duke, eds.), pp. 170–191. New York, Wiley (1962).
[3] G. G. Haselden: Refrigeration and Liquefaction Cycles. In: Cryogenic Fundamentals (G. G. Haselden, ed.), pp. 17–89. London, Academic Press (1971).
[4] R. B. Scott, W. H. Denton, C. M. Nicholls: Technology and Use of Liquid Hydrogen. Oxford, Pergamon Press (1964).
[5] R. B. Scott: Cryogenic Engineering. Princeton, N.J., van Nostrand (1959).
[6] J. Macinko, D. B. Chelton, J. Dean: Hydrogen Liquefaction Cycles. In: Adv. Cryog. Eng., Vol. 3. New York, Plenum Press (1960).
[7] R. Plank (ed.): Handbuch der Kältetechnik. Berlin Göttingen Heidelberg, Springer (1959).
[8] Ch. Kittel: Einführung in die Festkörperphysik. München Wien, Oldenbourg (1969).
[9] F. E. Simon, N. Kurti, J. F. Allen, K. Mendelssohn: Low Temperature Physics. Four Lectures. London, Pergamon Press (1952).
[10] J. A. Barclay, W. A. Steyert: Magnetic Refrigeration for Space Applications, Report on an Design Study. Los Alamos Sci. Lab. Rep. LA-8134 (1980).
[11] T. Hashimoto, T. Numasawa, M. Shino, T. Okada: Magnetic Refrigeration in the Temperature Range from 10 K to Room Temperature: The Ferromagnetic Refrigerants. Cryogenics, November (1981).
[12] N. Kurti: The Temperature Range below 1° Absolute. In: Low Temperature Physics, Four Lectures (F. E. Simon, N. Kurti, eds.). London, Pergamon Press (1952).
[13] D. S. Betts: Refrigeration and Thermometry below 1 K. Sussex University Press (1976).
[14] J. A. Barclay, O. Moze, L. Paterson: A Reciprocating Magnetic Refrigerator for 2–4 K Operation: Initial Results. J. Appl. Phys. *50*, 5870–5877 (1979).
[15] G. V. Brown: Magnetic Heat Pumping Near Room Temperature. J. Appl. Phys. *47*, 3673–3680 (1976).
[16] J. A. Barclay, W. A. Steyert: Materials for Magnetic Refrigeration between 2 K. Cryogenics *22*, 73–79 (1982).
[17] J. A. Barclay: Can Magnetic Refrigerators Liquefy Hydrogen at High Efficiency. ASME-paper 81-HT-82, 20th Joint ASME/AICHE National Heat Transfer Conf.. Milwaukee, Wisc., August 2–5 (1981).
[18] J. W. L. Köhler, C. O. Jonkers: Grundlagen der Gaskältemaschine—Konstruktion einer Gaskältemaschine. Philips Tech. Rdsch. *11*, *12* (1954).
[19] W. L. Köhler, P. F. Stevens, A. K. de Jonge, D. C. Beuzekom: Computation of Regenerators Used in Regenerative Refrigerators. Cryogenics *15*, 521–530 (1975).
[20] J. A. Barclay, W. F. Stewart: The Effect of Parasitic Refrigeration on the Efficiency of Magnetic Liquefiers, 1166–1170. In: Proc., 17th IECEC, CH 1789-7/82/0000-1166, August (1982).
[21] W. A. Steyert: Rotating Carnot-Cycle Magnetic Refrigerators for Use Near 2 K. J. Appl. Phys. *49*, 1227–1230 (1978).
[22] J. A. Barclay: A 4 K to 20 K Rotational-Cooling Magnetic Refrigerator Capable of 1 mW at 1 W Operation. Los Alamos Sci. Lab. Rep. LA-81111 (1980).
[23] W. A. Steyert: Stirling Cycle Rotating Magnetic Refrigerators and Heat Engines for Use Near Room Temperature. J. Appl. Phys. *49*, 1227 (1978).
[24] J. R. van Geuns: Philips Res. Rep., Suppl. *6* (1966).
[25] N. P. Chopey: Industry Joins Liquid Hydrogen Scene. Chem. Eng. *61*, 164–167 (1960).
[26] P. C. van der Arend: Large-Scale Liquid Hydrogen Production. Chem. Eng. Progr. *57*, 62–67 (1961).

[27] H. L. Coplen: Large-Scale Production and Handling of Liquid Hydrogen. J. Amer. Rocket Sci. *22*, 309–322 (1952).
[28] P. C. van der Arend: Liquid Hydrogen—Ultimate Fuel. Chem. Eng. Progr. Symp. Series 57, No. *34*, 1–7 (1961).
[29] T. M. Flynn, C. N. Smith: Trends in Cryogenic Fluid Production in the United States. In: Proc., Int. Inst. of Refrigeration, pp. 241–247, Tokyo, Japan (1970).
[30] P. C. van der Arend: Large-Scale Production, Handling and Storage of Liquid Hydrogen. In: Adv. Cryog. Eng., Vol. 5, pp. 49–54. New York, Plenum Press (1959).
[31] C. R. Baker, L. C. Matsch: Production and Distribution of Liquid Hydrogen. Adv. Petrol. Chem. Refining *10*, 36–81 (1965).
[32] H. Fujita: Liquefied Hydrogen in France. Production, Storage, Transportation and Utilization. Koatsu Gasu *12*, 18–27 (1975).
[33] J. Blackford, P. Halford, D. H. Tantam: Expanders and Pumps, Chapt, 8. In: Haselden, G. G. (ed.): Cryogenic Fundamentals. London, Academic Press (1971).
[34] J. L. Smith, Jr.: A Metal Bellows Expansion Engine. Adv. in Cryog. Eng., Vol. 12, pp. 595–601 New York, Plenum Press (1967).
[35] W. A. Morain: Design of a Cryogenic Expansion Engine for Tonnage Hydrogen Liquefaction. In: Adv. Cryog. Eng., Vol. 12, pp. 585–594. New York, Plenum Press (1967).
[36] M. L. Land: Expansion Engines and Turbines for Low Temperature Processing. In: Adv. Cryogenic Eng., Vol. 2. New York, Plenum Press (1960).
[37] N. Eber, H. Quack, C. Schmidt: Gas Bearing Turbines with Dynamics Gas Bearings and their Application. Cryogenics *18*, 585–588 (1978).
[38] C. R. Baker, R. L. Shaner: A Study of the Efficiency of Hydrogen Liquefaction. Int. J. Hydrogen Energy *3*, 321–334 (1978).
[39] C. R. Baker: Economics of Hydrogen Production and Liquefaction updated to 1980. NASA Contractor Rep. 159163 (1979), also see NASA-CR-132631, NASA-CR-145077, NASA-CR-145282.
[40] R. O. Voth, D. E. Daney: H_2-Liquefaction: Effects of Component Efficiencies. In: Proc. 14th IECEC, Newark, Delaware, pp. 1356–1362, August (1975).
[41] W. Foerg: Purification of Hydrogen by Means of Low Temperatures. Sci. Technol. *15*, 18–26 (1970).
[42] C. R. Baker, R. S. Paul: Purification of Liquefaction-Grade Hydrogen. Chem. Eng. Progr. *59*, 61–64 (1963).
[43] K. Wilson: Adsorption. In: Cryogenic Fundamentals (G. G. Haselden, ed.), pp. 375–403. London, Academic Press (1971).
[44] D. H. Weitzel, C. C. van Valin, J. W. Draper: Design Data for Ortho-Parahydrogen Converters. In: Adv. Cryog. Eng., Vol. 3. New York, Plenum Press (1980).
[45] A. H. Singleton, J. F. Kucirka, A. Lapin: Investigation of the Para-Ortho Shift of Hydrogen Air Force Prop. Lab. (APEL), Tech. Rep. AFAPL-TR-66-111, 80 p, Wright Patterson Air Force Bose, Ohio (1966).
[46] P. L. Barrick, L. F. Brown, H. L. Hutchinson, R. L. Cruse: Improved Ferric Oxide Gel for Ortho-Parahydrogen Conversion. In: Adv. Cryog. Eng., Vol. 10. New York, Plenum Press (1965).
[47] A. H. Singleton, A. Lapin: Design of Para-Orthohydrogen Catalytic Reactors. In: Adv. Cryog. Eng., Vol. 10A. New York, Plenum Press (1965).
[48] A. H. Singleton, A. Lapin, L. A. Wenzel: Rate Model for Ortho-Parahydrogen Reaction on a Highly Active Catalyst. In: Adv. Cryog. Eng., Vol. 13. New York, Plenum Press (1968).
[49] T. R. Strobridge: Cryogenic Refrigerators—An Updated Survey, NBS, Tech. Note 655 (1974).

[50] A. A. Dros: Large Capacity Industrial Stirling Machine. In: Adv. Cryog. Eng., Vol. 10A. New York, Plenum Press (1965).
[51] W. Kanoldt: Selected Examples for European Cryogenic Design Practice. In: Adv. Cryog. Eng., Vol. 10A. New York, Plenum Press (1965).
[52] W. E. Keller: Worldwide Cryogenics—U.S., Cryogenics at the Lost Alamos Scientific Laboratory. Cryogenics *20*, 547–556 (1980).
[53] J. E. McCormick, J. B. Brauer: The Feasibility of Solid State Cryogenic Refrigeration to 70 K. In: Adv. Cryog. Eng., Vol. 10. New York, Plenum Press (1965).
[54] J. A. Barclay: Magnetic Refrigeration: A Review of a Developing Technology. In: Adv. Cryog. Eng., Vol. 33, pp. 719–732. New York, Plenum Press (1988).
[55] T. Hashimoto, T. Yazawa, R. Li, et al.: Recent Progress in Magnetic Refrigeration Studies. In: Adv. Cryog. Eng., Vol. 33, pp. 733–742. New York, Plenum Press (1988).
[56] S. R. Jaeger, J. A. Barclay, W. C. Overton, Jr.: Analysis of Magnetic Refrigeration with External Regeration. In: Adv. Cryog. Eng. Vol. 33, pp. 751–756. New York, Plenum Press (1988).
[57] C. R. Cross, J. A. Barclay, A. J. De Gregoria, et al.: Optimal Temperature-Entropy Curves for Magnetic Refrigeration. In: Adv. Cryog. Eng., Vol. 33, pp. 767–776. New York, Plenum Press (1988).
[58] G. Green, W. Patton, J. Stevens: The Magnetocaloric Effect of Some Rare Earth Metals. In: Adv. Cryog. Eng., Vol. 33, pp. 777–784. New York, Plenum Press (1988).
[59] G. M. Claude: Magnetic Refrigeration Study at CEN Grenoble. In: Adv. Cryog. Eng., Vol. 31, pp. 733–742. New York, Plenum Press (1985).
[60] W. Peschka, C. Carpetis: Möglichkeiten der Magnetischen Kühlung, pp. 8–13. DLR-Nachrichten 57, Juni 1989.
[61] G. Bogner: Technische Anwendungen der Supraleitung-Anforderungen an die neuen Hochtemperatur-Supraleiter, Siemens, Techn. Mitteilungen 80 (1987) 8.
[62] E. M. W. Leung, D. D. Madura, R. E. Bailey: Yttrium Barium Copper Oxide Superconducting to Normal Transition Characterization for a Solenoid Configuration. In: Ad. Cryog. Eng., Vol. 35A, pp. 633–639. New York, Plenum Press (1990).
[63] P. Seyfert: Research on Magnetic Refrigeration at CEN Grenoble. In: Adv. Cryog. Eng., Vol. 35B, pp. 1087–1096. New York, Plenum Press (1990).
[64] J. A. Barclay, S. R. Jaeger, F. G. Jr. Pregner: Operational Envelope for Magnetic Refrigerators. In: Adv. Cryog. Eng. Vol. 35B, pp. 1097–1104. New York, Plenum Press (1990).
[65] P. Kittel: Eddy Current Heating in Magnetic Refrigerators. In: Adv. Cryog. Eng. Vol. 35B, pp. 1141–1148. New York, Plenum Press (1990).
[66] E. Schroeder, G. Green, J. Chafe: Performance Predictions of a Magnetocaloric Refrigerator Using a Finite Element Model. In: Adv. Cryog. Eng. Vol. 35B, pp. 1149–1155. New York, Plenum Press (1990).
[67] C. P. Taussing, G. R. Gollagher, Jr., J. L. Smith, Y. Iwasa: Magnetic Refrigeration Based on Magnetically Active Regeneration. In: Proc., Fourth Int. Cryocoolers Conference, pp. 78–88 (1986).
[68] C. P. Taussig: Magnetically Active Regeneration, MIT, Ph.D. Thesis (1986).
[69] J. L. Smith, Jr., Y. Iwasa, F. J. Cogswell: Material and Cycle Considerations for Regenerative Magnetic Refrigeration. In: Adv. Cryog. Eng., Vol. 35B, pp. 1157–1164. New York, Plenum Press (1990).
[70] G. Green, J. Chafe: A Gadolinium-Terbium Active Regennerator. In: Adv. Cryog. Eng. Vol. 35B, pp. 1165–1174. New York, Plenum Press (1990).

Thermal Insulation, Storage and Transportation of Liquid Hydrogen 4

Hydrogen as a cryogenic liquid must be stored and transported in thermally insulated containers in order to avoid uneconomical high boil off losses. The same requirement applies for the insulation of liquid hydrogen transfer pipelines.

Insulation techniques are a part of the general problem of heat transfer, including heat transfer via flowing media in heat exchangers for example. The question of heat transfer in flowing media is fundamentally important in cryogenics in connection with the design and construction of highly efficient heat exchangers for hydrogen liquefaction procedures. Because this important subject cannot be properly dealt with in the scope of this book reference is made to relevant literature. The quotations made here refer to publications which typically serve as a lead-in to this very important subject [1–4]. The following is an overview of common low-temperature insulation procedures with regard to hydrogen requirements.

Until about 1955 liquid hydrogen was essentially needed, and used, in laboratories where primarily laboratory storage containers were shielded with liquid nitrogen. Metal cryostats and glass Dewar containers were used.

As a result of the demand created by the space programs, as in other fields of cryo-technology, developments were made with hydrogen that provided for its use on a large scale. Especially in the U.S. there are currently stationary storage containers for liquid hydrogen with capacities of about 4,000 m^3, railway transport tanks up to about 100 m^3 as well as highway transport tanks of up to about 70 m^3.

The temperature range of hydrogen lies between that of liquid helium and liquid oxygen or liquid nitrogen. Dependent upon the intended use of hydrogen, either highly efficient insulation techniques (vacuum multilayer insulation) are used, as in long-term storage of small amounts of liquid hydrogen, or—primarily on a large scale—insulation techniques are used that are appropriate for nitrogen or oxygen, like vacuum powder insulation for example. Because, as previously mentioned, the insulation technique primarily depends on the intended use as well as the quantity of hydrogen being dealt with, typical problems of thermal insulation will be briefly covered in the following section with references to technical literature for further details.

4.1 Mechanisms of Heat Transfer and Insulation Techniques

Heat transfer in general takes place via convection, heat conduction and thermal radiation. Convection predominates in uninsulated surfaces which are exposed to the air. For safety reasons this is a problem with liquid hydrogen because, due to the low boiling point, air liquefies on the outer walls of uninsulated containers and pipelines, drips and can explode if it comes in contact with flammable materials.
Nevertheless, for special applications under corresponding safety conditions uninsulated pipelines and containers can be used if it is economical due to a high mass flow rate, as in the rapid filling of special containers like rocket fuel tanks, or a very short storage time. According to [5] wind velocity dependent coefficients of heat transfer in the range of about 38–65 $W/m^2 K$ with wind velocities up to about 20 km/h were measured on uninsulated pipeline surfaces. For general use this procedure is out of the question due to safety reasons.
Conventional non-vacuum insulation in the form of mineral wool, cork, rubbery foam based on polystyrol, perlite, rigid polyurethane foam as well as foam glass and Balsa wood are used for light-weight construction—mostly in rocket technology and, if necessary, in aeronautics. In this case, the air convection is largely prevented by the porous structure of the material [6–17].
According to Fig. 30 these materials have the following characteristics:

- heat conductivity in the range of gases, especially air
- low density between 20–250 kg/m^3
- change in the heat conduction coefficient at low temperatures by a factor of 2–4
- rigid foams, wood, cork and foam glass can serve as filling material in honeycomb structures and at moderate surface pressure of about 10^6 N/m^2 they have a constant heat conduction coefficient.

Also, with hydrogen, the layout and installation of the insulation has to be accomplished in such a manner that the warm side is expected to be above the condensation temperature of the air. Furthermore, the penetration of moisture and air and the accompanying condensation in the insulation has to be prevented either though seals, fitted joints or through an appropriate foil sealant [18]. In the latter case the pores and cavities of the insulation at the warm side can be exposed to an inert gas like nitrogen or argon instead of air.

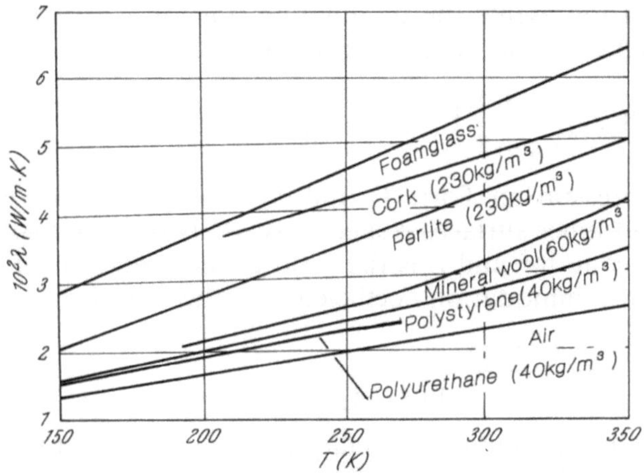

Fig. 30. Heat conductivity λ for non-vacuum insulation as a function of the absolute temperature T

Vacuum Insulation: This is the preferred method for liquid hydrogen on a large scale [19–23]. In this process heat transfer from convection and heat conduction in the gas is greatly reduced by lowering the gas pressure, i.e., through the use of a moderate vacuum. The amount of convection decreases roughly proportional to the density of the gas whereas the heat conduction of the gas is independent of the pressure as long as the mean free path L of the molecules is relatively small as compared to the gap D between inner and outer wall. In this case collisions of particles of mass m and Temperature T with the container walls are less frequent than between the particles themselves [24–27].

Thus the following applies for the pressure independent heat conduction coefficient λ with $L \ll D$

$$\lambda \sim \sqrt{\frac{T}{m}}. \tag{109}$$

With a further decrease in the pressure the mean free path eventually exceeds the gap of the vacuum jacket and the number of collisions with the container walls begins to predominate [24, 25]. In this case the following applies for the pressure dependent heat flux $\dot{q}$ with $L \gg D$

$$\dot{q} \sim \alpha_{12} \cdot \frac{p}{\sqrt{T \cdot m}} (T_1 - T_2). \tag{110}$$

Here p is the pressure, T gas temperature, T_1, T_2 the wall temperatures, m molecular mass and α_{12} the accommodation coefficient which characterizes the energy exchange with the walls.

Since hydrogen represents the dominating portion of the residual gas in the vacuum jacket in this case according to [25] the following applies

$$\dot{q} = 9.18 \cdot \alpha_{12} \left(\frac{\kappa + 1}{\kappa - 1} \right) \cdot p \cdot (T_1 - T_2) \cdot A_2 \quad (\mathrm{W/m^2}). \tag{111}$$

Here p is the pressure in Torr, T_1 and T_2 are the absolute temperatures of the outer and inner surfaces respectively, A_2 is the inner surface and A_1 the outer surface in m^2 and κ is the ratio of the specific heats.
For the total accommodation coefficient α_{12} the following applies

$$\alpha_{12} = \frac{\alpha_1 \cdot \alpha_2}{\alpha_1 + \alpha_2 \cdot (1 - \alpha_1) \cdot (A_1/A_2)}. \tag{112}$$

The individual accommodation coefficients α_1 and α_2 for hydrogen (which are between 0 and 1) are given in Table 14.
According Eq. (110) to be true, the heat conduction in the residual gas can be reduced at pressures $< 10^{-5}$ Torr to an insignificant amount as compared to the heat flux through the suspension of the inner container as well as the considerable heat flux into the inner container caused by thermal radiation.
Through the use of powdered materials of small grain size like perlite, expanded SiO_2, etc., the radiative heat transfer is considerably reduced [28, 30]. At higher interstitial pressure the heat conduction in the gas is still present in addition to the effective heat conductivity of the powder itself. The powder heat conductivity is considerably reduced due to the loose contact of the powder grains with each other as compared to pure solid-state heat conduction. At interstitial gas pressures in the powder below about 10^{-2} to 10^{-3} Torr the heat conductivity of the gas is insignificant and the heat transfer consists primarily of thermal radiation and the effective heat conduction of the powder. Therefore, a decrease in heat flux through the use of powder in the vacuum gaps can be achieved only if the reduction in the radiative heat transfer outweighs the effective solid-state heat conductivity of the powder (Fig. 31).
By adding highly reflective metal particles like aluminum powder for example, the radiative heat transfer can be further reduced whereas the

Table 14. Individual accommodation coefficients

Surface temperature (K)	α_1, α_2
300	0.3
80	0.5
20	1

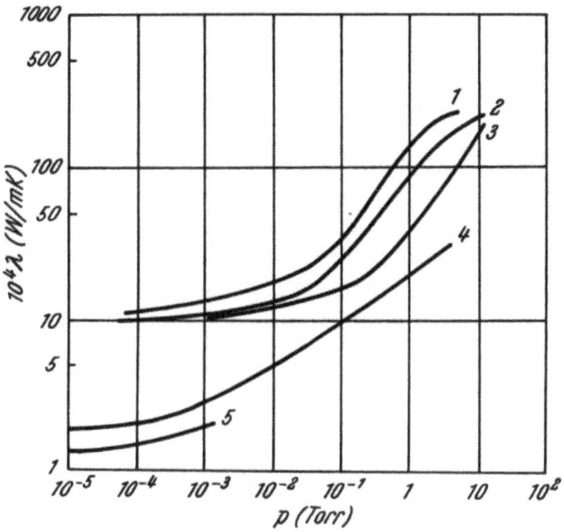

Fig. 31. Effective heat conductivity λ of powder-vacuum insulation as a function of the gas pressure p (nitrogen or helium). *1* Phenolic resin spheres, *2* perlite (300/78 K, nitrogen), *3* diatomaceous earth, *4* perlite (78/20 K helium)

heat conduction increases as a result of the greater heat conductivity of the metal powder. With certain combinations of grain size and mixes of metal powder minimum effective heat conduction coefficients can be attained which correspond to an improvement by a factor of 2–3 [28, 29].

A reduction in the effective heat conductivity of about a factor of 5–10 in contrast to vacuum powder insulation can be achieved in a vacuum (10^{-2} to 10^{-3} Torr) through the use of hollow glass microspheres, which consist of boron silicate with diameters in the range of 15–150 μm [31] in which, depending on the temperature range, nonmetallized, metallized, or a mixture of both are more advantageous [32, 33].

In addition to a reduction in heat flux and very low weight this insulation procedure has about the same advantages as conventional powder, that is ease of handling including the area around support members, suspensions and feed-throughs as well as low dependency of the thermal insulation on compressive load up to about 10^4 N/m^2 [31, 34].

Therefore, microspheres will also become increasingly important in hydrogen insulation technology as a result of these characteristics and because they are not flammable.

High-vacuum Multilayer Insulation: The heat transfer of the residual gas can practically be eliminated through the use of a high vacuum whereas the radiative heat transfer can be further reduced by means of a large number of reflective sheets surrounding the inner container [26, 35, 37].

For the energy flux density of a surface which radiates in a direction perpendicular to the surface with the emission coefficient ϵ and the temperature T the Stefan–Boltzmann law applies

$$\dot{q} = \sigma \cdot \epsilon \cdot T^4 \tag{113}$$

with the Stefan–Boltzmann constant $\sigma = 5.67 \cdot 10^{-8}\ \mathrm{W/m^2\,K^4}$.

For the heat flux density transferred between two parallel surfaces at temperatures T_1 and T_2 the following applies[1]

$$\dot{q}_{12} = \sigma \cdot \epsilon_t \cdot (T_1^4 - T_2^4) \tag{114}$$

in which the total emission coefficient ϵ_t is obtained from the individual emission coefficients ϵ_1 and ϵ_2 for the emission into the half space [36]:

$$\epsilon_t = \frac{1}{1/\epsilon_1 + 1/\epsilon_2 - 1}. \tag{115}$$

Appropriately modified results [36, 37] for Eq. (115) are obtained for concentric spheres and long coaxial cylinders.

For n reflective surfaces, each with two reflecting surfaces of the same quality ($\epsilon_1 = \epsilon_2 = \epsilon_m \ll 1$) the following is obtained from (114) for the heat flux

$$\dot{q}_{12}(n) = \frac{\dot{q}_{12}}{n+1} = \frac{\sigma \cdot \epsilon_m}{n+1}(T_1^4 - T_2^4). \tag{116}$$

Therefore, the heat flux can be reduced considerably with the help of such reflective shields as well as through small emission coefficients for the appropriate long-wave infrared radiation. Since Kirchoff's law of small emission also corresponds to small absorption and thus proper reflection, metals or rather metallized foils are appropriate for the reflective surfaces. In order to obtain proper reflection the metal has to be thicker than about 0.8 μm.

In this process a sufficiently large number of layers is applied around the inner container as insulation. These layers primarily consist of either plastic foils (Mylar), metalized by vapor deposition of the metal, or very thin aluminum foils with intermediate layers (spacers) of fiberglass or another similar material.

This type of multilayer insulation is currently the best insulation procedure available [38–40].

The reflectivities of the metal foils are of considerable importance. Typical

---

[1] The total radiation emitted into the half space with the solid angle 2π is obtained from (113) by multiplying by π.

values for the emission coefficients of different metals (from [36, 41–43]) are given in Table 15. The values for ϵ given here and in the literature are generally understood as average values. They are rarely attained with insulation because the conditions under which they are measured cannot be reproduced in a standard production environment [44, 45]. In particular it is impossible to obtain absolutely clean, i.e., grease and oxide free as well as not chemically corroded, surfaces—due to sulfide formation with copper or silver for example. Therefore open-air storage of these insulation foils may still be a problem today.

Another problem is the optical quality of the surfaces required in the use of long-wave infrared and that a visual inspection of surfaces to determine high quality cannot guarantee the proper behavior with infrared.

Furthermore, the condensation of impurities on each layer, like water vapor, carbon dioxide, hydrocarbons, etc., can increase the emission coefficient and thus deteriorate the thermal insulation properties.

Experience shows that compared to Table 15 a deviation of 100% from the values given is possible so that it does not appear practical to differentiate between the total and normal emissivity (emission into the half space and emission in a direction perpendicular to the surface respectively) or to overestimate the influence of temperature (see Table 15).

Table 15. Emission coefficient of clean metal surfaces

Material		Temperature (K)	ϵ
Aluminum	mechanically polished	4	0.011
		77	0.018
		300	0.03
	rough	300	0.07
	strongly oxidized	300	0.2–0.3
Copper	electrolytically polished	4	0.005
		77	0.008
		300	0.02
	mechanically polished	300	0.04
	unpolished	300	0.15
Silver	mechanically polished	4	0.0044
		77	0.008
		300	0.02
	oxidized	300	0.02–0.04
Gold	mechanically polished	4	0.01
		300	0.02
Stainless steel	mechanically polished	77	0.048
		300	0.08
Metallized foil (Ag, Au, Cu, Al)		300	0.02–0.04

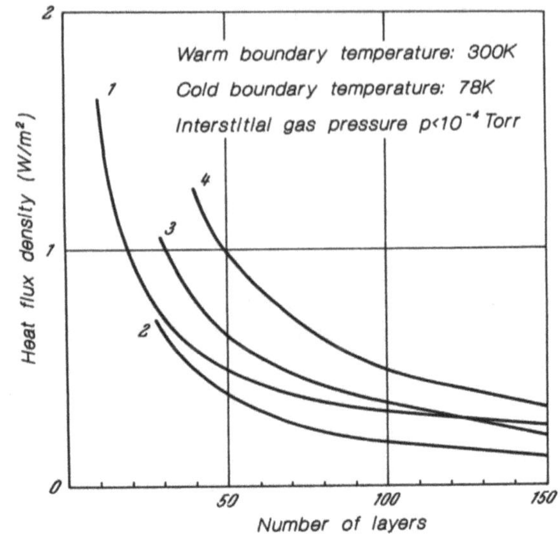

Fig. 32. Heat flux density of multilayer insulation as a function of the number of layers per cm (layer density). *1* Linde-Dimplar. 10 layers per cm. *2* NRC 2-mylar foil, one side aluminized, 15 layers per cm. *3* like *2*, 30 layers per cm. *4* like *3* and *2*, 60 layers per cm (as provided by the manufacturers)

Furthermore, it must be taken into consideration that the addition of thermal radiation shields without accompanying mutual physical contacts is not possible. Each layer is subject to physical contact which results in additional heat transfer through heat conduction which is strongly dependent on surface pressure. This is apparent because the heat flux does not steadily decrease with an increasing number of layers but, as a result of increasing compressive load which is difficult to reproduce during production, tends toward an optimum between about 60–100 layers (see Figs. 32, 33) [46].

Therefore, the temperature profile in the insulation does not correspond to that of radiation alone but lies between radiation and pure heat conduction in the range of mixed heat transfer (Fig. 34), which demonstrates the similarity of the heat transport mechanism to that of powder insulation and vacuum-microsphere insulation.

In general it can be stated that at higher temperatures, especially around ambient temperature, heat transport by radiation predominates in superinsulation whereas at low temperatures—below about 60–80 K—heat conduction, compared to radiation, starts to become predominant as a result of its strong decrease with T^4.

Several calculation methods were developed for determining the heat flux through multilayer insulation. One-dimensional calculation methods,

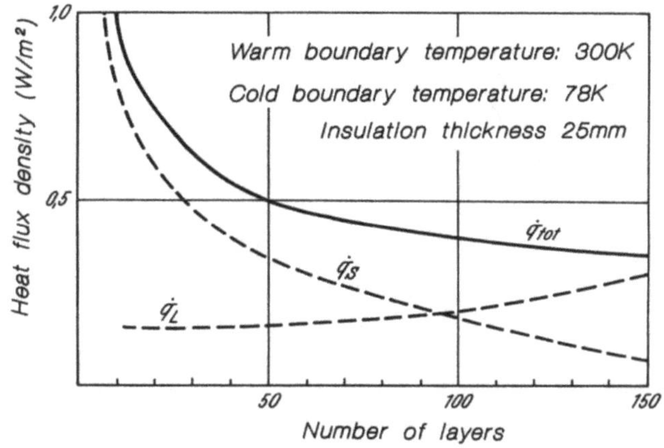

Fig. 33. Total heat flux density $\dot{q}_{tot}$ of NRC 2 multilayer insulation as a function of the number of layers with constant insulation strength (as provided by the manufacturer). Vacuum between the layers. $p < 10^{-4}$ Torr. $\dot{q}_L$ and $\dot{q}_s$ correspond to the portions of the heat conduction and heat radiation respectively as calculated by the author

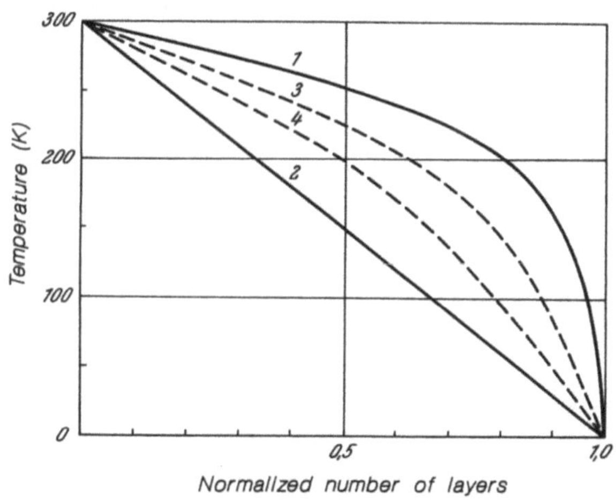

Fig. 34. Internal temperature distribution of multilayer insulation. *1* pure radiative heat transfer with $\varepsilon = \text{const.}$; *2* pure heat conduction with $\lambda = \text{const.}$; *3* and *4* multilayer insulation with small and large number of layers per cm respectively

which include only the heat flux perpendicular to the surface, are generally sufficient whereas for more stringent requirements, two-dimensional methods were developed which also include the transverse conductivity along the single layers and thus the strong anisotropic behavior of the multilayer insulation [47].

The anisotropic behavior of the multilayer insulation, which is strong compared to that of the powder or vacuum-microsphere insulation, causes considerable problems. Since the multiple layer insulation is very sensitive to errors made during installation [45], problems exist with insulation in pipelines and container suspension because contact between warm and cold insulation layers as well as radiative heat exchange between them has to be prevented [48, 108, 109].

Furthermore, it is very difficult to achieve the required vacuum of $\leqslant 10^{-4}$ Torr between the layers. On one hand pumping out the residual gas is difficult as a result of the small gaps between the layers and, on the other hand, the interstitial vacuum between the layers suffers from outgassing where, particularly at low pressures, a high rate of outgassing with regard to the gap's volume exists [49].

The residual gas which remains in the insulation and whose pressure can be 10–100 times higher than in the free vacuum [50] is evidently responsible for the residual heat conductivity of the multiple layer insulation especially for layers at low temperature (see Fig. 35). According to [51] a considerable reduction in the residual heat conductivity can be achieved by coating the layers which are made of fiberglass, paper or a similar material and which act as spacers between the metal foils, with absorbent material like activated carbon for example, in order to improve the interstitial vacuum between each layer (see Ref. [51], Scurlock insulation). In this manner heat flux can be reduced by a factor of 3 or more. It seems possible to achieve effective heat conduction coefficients in the range of $5\text{–}6 \cdot 10^{-6}\ \mathrm{W m^{-1} K^{-1}}$. The deviation of the heat flux from pure radiative heat transfer, therefore, is evidently caused primarily by the heat conduction of the residual gas in

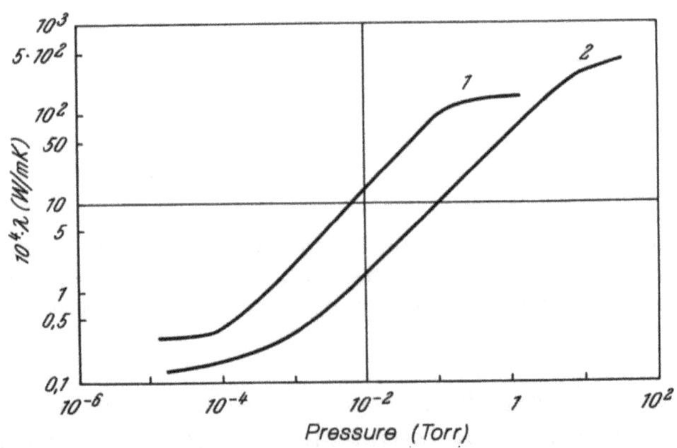

Fig. 35. Effective heat conductivity λ of multilayer insulation as a function of the vacuum between the layers (according to data from [40, 46]). *1* aluminized polyester foil, air; *2* aluminum foil and glass silk, helium

the insulation and not so much by the solid-state body contact between each layer.

The insulation vacuum between the inner and outer container is maintained for the desired service lifetime of the container (in the order of a few years) by getter materials and adsorbent materials which adsorb accumulated amounts of gas resulting from outgassing and microleakage. Getter materials are used for higher temperatures (warm wall) whereas non-flammable adsorbent materials, in particular molecular sieves, are used at the cold side.

Insulation Cooling Through Vapor Cooled Shields: A considerable reduction in the boil off rate $\dot{m} = \dot{q}/i_v$ of the inner container (i_v = enthalpy of evaporation) as a result of the heat leakage $\dot{q}$ can be achieved by cooling the insulation with the cold venting gas. This can be accomplished with metallic shields which are installed with the insulation. These shields are cooled through soldered and bonded coils of pipes by the exhaust gas. This procedure is primarily used with large containers [52].

Furthermore, multilayer insulation with foils of high transverse conductivity can cool each layer directly through direct thermal contact at appropriate places on the vent line and the neck (Fig. 36). This procedure approaches the best possible heat flux reduction especially in small containers (around 100 liter capacity and more) for helium and hydrogen [52–56].

The effectiveness of this type of vapor cooled shield is dependent on the ratio of the enthalpy difference of the gas between ambient temperature and boiling temperature to the enthalpy of evaporation, i_v, of the liquid (see Table 16).

In the case of full yield of the available vapor enthalpy, i.e., complete warming of the exhaust gas to ambient temperature, the resultant boil off rate can be easily estimated. This case applies strictly to an infinite number of vapor cooled shields but can be closely approximated through continuous cooling of a finite number of layers which, for this purpose, have to have good transverse conductivity. They must also have thermal contact at an

Table 16. Enthalpy of evaporation i_v and enthalpy difference Δi of the gas between boiling point and 300 K

	i_v (J/kg)	Δi (J/kg)	$(i_v + \Delta i)i_v$	$(\dot{m}/\dot{m}_0)_\infty$	$(\dot{m}/m_0)_1$
N_2	$199 \cdot 10^5$	$2.35 \cdot 10^5$	2.18	0.67	0.85
Ne	$8.72 \cdot 10^4$	$2.8 \cdot 10^5$	4.2	0.45	0.75
H_2	$4.48 \cdot 10^5$	$4 \cdot 10^6$	9.9	0.26	0.5
He	$2.09 \cdot 10^4$	$1.57 \cdot 10^6$	76	0.06	0.2

H_2 in ortho-para equilibrium, $(\dot{m}/\dot{m}_0)$ ratio of the boil off rate with and without vapor cooled shields (infinite number (∞) and one shield (1).

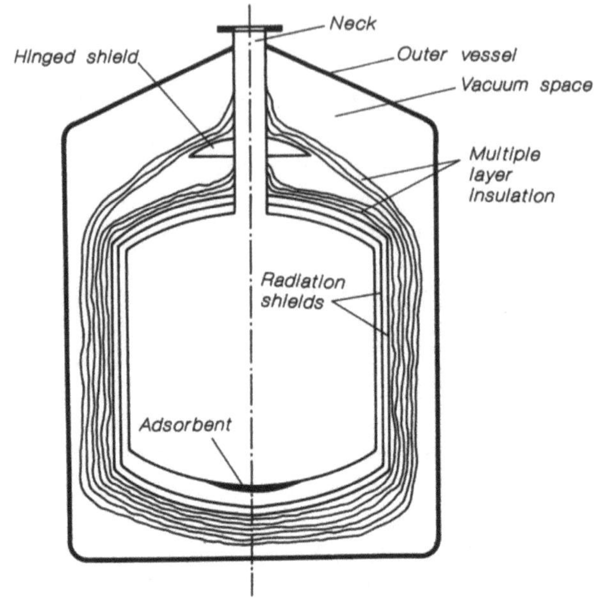

Fig. 36. Schematic of a cryocontainer with vapor cooled multiple layer insulation (from [52] as well as statements from Messer Griesheim GmbH)

appropriate place on the vent line (at the neck of the inner container for example) (Fig. 36).
For the ratio of the boil off rate with and without vapor cooled shields m and m_0 respectively (Table 16) from [52, 56] the following is obtained

$$\left(\frac{\dot{m}}{\dot{m}_0}\right)_\infty = \frac{\ln s}{s-1}$$

$$s = \frac{i_v + \Delta i}{i_v}. \tag{117}$$

The corresponding values for $\left(\frac{\dot{m}}{\dot{m}_0}\right)_\infty$ are also shown in Table 16.
From this, the considerable reduction in the boil off rate, especially for helium and hydrogen, through continuous vapor cooling of the shields of the multiple layer insulation becomes evident. For comparison the corresponding values $(\dot{m}/\dot{m}_0)_1$ for a single vapor cooled shield, which is located about in the middle of the insulation [19] is presented. The smaller values, as compared to continuous cooling, are obtained from the naturally lower yield of the enthalpy of the exhaust gas by means of only one discrete temperature level.

Such single vapor cooled shields can also be installed within vacuum powder insulation and provide a considerable reduction in the boil off rate especially when using vacuum-microspheres.
The arrangement of multilayer insulation to a small number of individual vapor cooled shields [54, 55] can also offer advantages, because in this manner a distribution of total insulation into groups of lower number of layers on individual shields is possible and thus an improved reduction in heat flux as a result of reduced layer compression. The innermost shield can here be formed as a pure thermal radiation shield surrounding the inner container. At a sufficiently low shield temperature the heat flux onto the inner container is almost reduced to the heat leakage from the suspension and tubing. Cooling the supports of the inner container by thermal contact with individual vapor cooled shields produces a further advantageous reduction in the heat leakage. The best possible conditions for smaller containers are achieved when the container neck simultaneously serves as the suspension of the inner shell. For example, continuously gas-cooled containers for liquid helium with capacities of 100 liters have a boil off rate of 0.8% per day (0.8 L helium or 9.1 kg helium per day) which corresponds to an effective heat leakage into the inner container of 25 mW [54, 56].
With the values from Table 16 an effective heat leakage for liquid hydrogen of $25 \cdot \frac{0.26}{0.06} = 108\,\text{mW}$ is obtained, which would correspond to a boil off rate of 0.0216 kg/day or 0.3 L/day, that is 0.3% per day.
By using "Scurlock insulation" (see page 80) these results can be further improved for containers of this size, resulting in a technically feasible limit for the typical maximum storage time of one year. With increasing container volume these results initially become more favorable. For example, the 680 liter container for helium II ($T < 2.2\,\text{K}$) examined in [54] with a boil off rate of about 0.3% per day (a year's supply of liquid helium for an experimental device aboard a satellite), filled with hydrogen corresponds to a boil off rate of about 0.12% per day. This means that a heat flux which corresponds to this boil off rate could be proced from the heat conduction through a copper wire of 1 m length and approximately 0.8 mm diameter situated between the outer and inner container!

4.2 Storage and Transportation of Liquid Hydrogen

The development of liquid hydrogen storage and transportation technology started at a time when procedures for liquid oxygen and liquid nitrogen had already been in use for a long time. Because the low heat of evaporation of liquid hydrogen compared to that of liquid oxygen and liquid nitrogen,

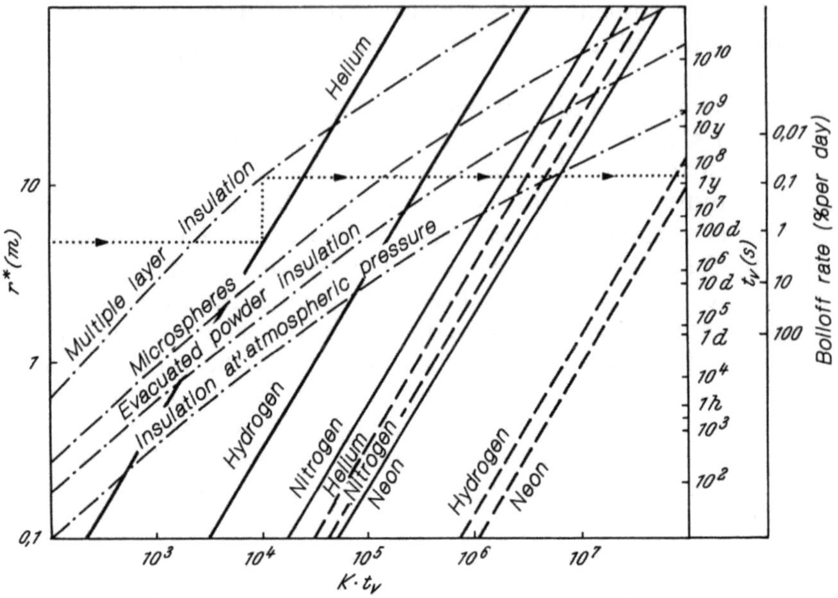

Fig. 37. Shelf life t_v of cryogenic liquid in containers in relation to size (equivalent spherical radius r^*) and method of insulation. The dashed curves correspond to ideal vapor cooled insulation. K effective heat transfer coefficient of the insulation (W/m^2K)

specialized procedures are necessary for liquid hydrogen similar to those needed for liquid helium. The development of these required techniques with their more stringent demands for liquid hydrogen and liquid helium also led to improved procedures for the handling of other cryogenic liquids.

Techniques for the storage of liquid hydrogen are currently very highly developed for all sizes.

Figure 37 provides an overview of the dependency of the attainable boil off rate of different liquefied gases on the insulation system and the size of the container. It can be seen that for sizes from 100 liters up to several thousand cubic meters and more, boil off rates of far below 1% per day can always be obtained.

Large Containers: The storage of large amounts of hydrogen is currently state of the art particularly due to intensive application in space technology as well as in nuclear research. It is primarily stored in storage tanks with perlite vacuum insulation. Many such tanks can be found in the U.S. The largest is located at Cape Kennedy (Fig. 38). It has a storage capacity of about 3800 m^3 (approximately 270,000 kg LH_2) and an exterior spherical diameter of about 20 m and a boil off rate of less than 0.03% per day.

Fig. 38. View of the largest tank in use for liquid hydrogen. This tank has a 3,800 m^3 LH_2 capacity and is located at the NASA Kennedy Space Center (Source: NASA)

Figure 39 gives an impression of the internal design and instrumentation of such containers[2].
While the outer container of such tanks can be constructed from ordinary steel, the inner container requires austenitic chrome nickel steel or aluminum alloys with good weldability. The latter are preferred because, as a result of weight reduction, it is possible to achieve a cost saving as well as a reduction in heat leakage as a consequence of smaller diameters of the support rods in the suspension of the inner container. As a result of the high thermal conductivity of aluminum, temperature differences in the wall of the inner container are kept small which reduce the tendency to thermal stratification in the liquid which, in turn, together with retardation of boiling, can considerably effect the function of the container [57–59]. Figure 40 shows the distribution of the temperature [57] in a filled hydrogen container of approximately 200 m^3 capacity, and Fig. 41 shows its influence

[2] The author would like to express his gratitude to the NASA Kennedy Space Flight Center, especially A. L. Bain, for the kind permission to publish Figs. 38 and 39, and to the Los Alamos National Laboratory, in particular Dr. F. J. Edeskuty.

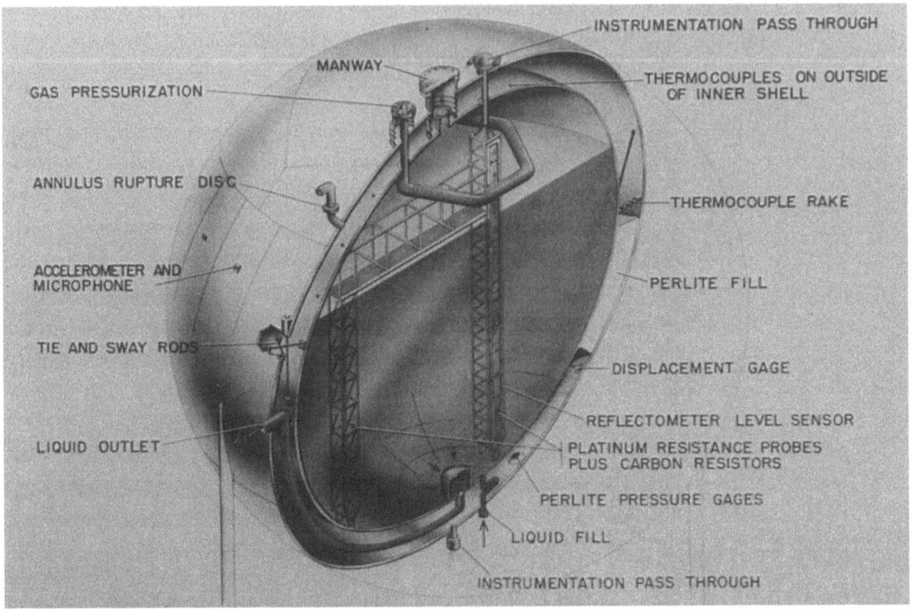

Fig. 39. Interior view of a typical large container with a capacity of 1,800 m^3 for liquid hydrogen (Source: Los Alamos National Laboratory)

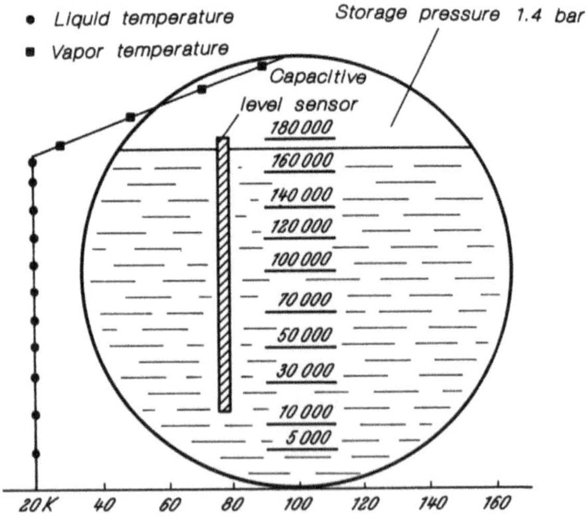

Fig. 40. Temperature distribution in a 180,000 L (50,000 Gal.) liquid hydrogen container (from [57])

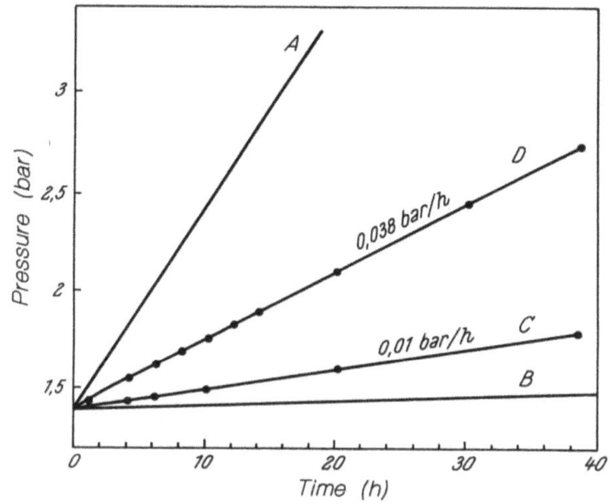

Fig. 41. Pressure buildup rate in a container in accordance with Fig. 40 as a function of time (from [57]). *A* maximum calculated pressure buildup rate as a result of the absorption of the incoming heat in the vapor space, *B* minimum calculated pressure buildup rate as a result of the heat absorption in the liquid alone, *D* and *C* measured values

on the pressure build up rate. Maximum and minimum rise in pressure (curve A and B respectively) occur when the total incoming heat is absorbed by either the vapor, or by the total amount of liquid whereas the normal operating states lie somewhere in between (curves B and C). In this case the smaller pressure buildup rate which corresponds to curve C is obtained as a result of the small amounts of gaseous helium flowing into and intermixing with the liquid hydrogen causing a reduction in tendency to thermal stratification.

Because of aluminum's high heat conductivity it cannot be used for tubing between the inner and outer container. Tubing is generally made from stainless steel and has to be connected to the inner container with vacuum sealed transition joints on the cold side. The stringent specifications for sealing these joints, especially during required cooling and warming, require particularly careful design and construction [57, 60]. The relatively high thermal expansion of aluminum can cause problems at the inner container. Due to contraction of the inner container when cooling, the perlite insulation partially fills the resulting spaces. This cannot be reversed completely when warming the inner container and thus the powder is cumulatively compacted. As a result of different positions of the inner container as well as by the forces resulting from the compaction of the insulation the danger of a failure in the suspension of the inner container exists, but this can be avoided through appropriate construction and elastic support [57].

In order to determine the operating conditions, these containers are designed with appropriate, redundant instrumentation similar to containers for other cryogenic liquids. The fluid level indicator can depend upon capacitive, optical or discrete level sensors.

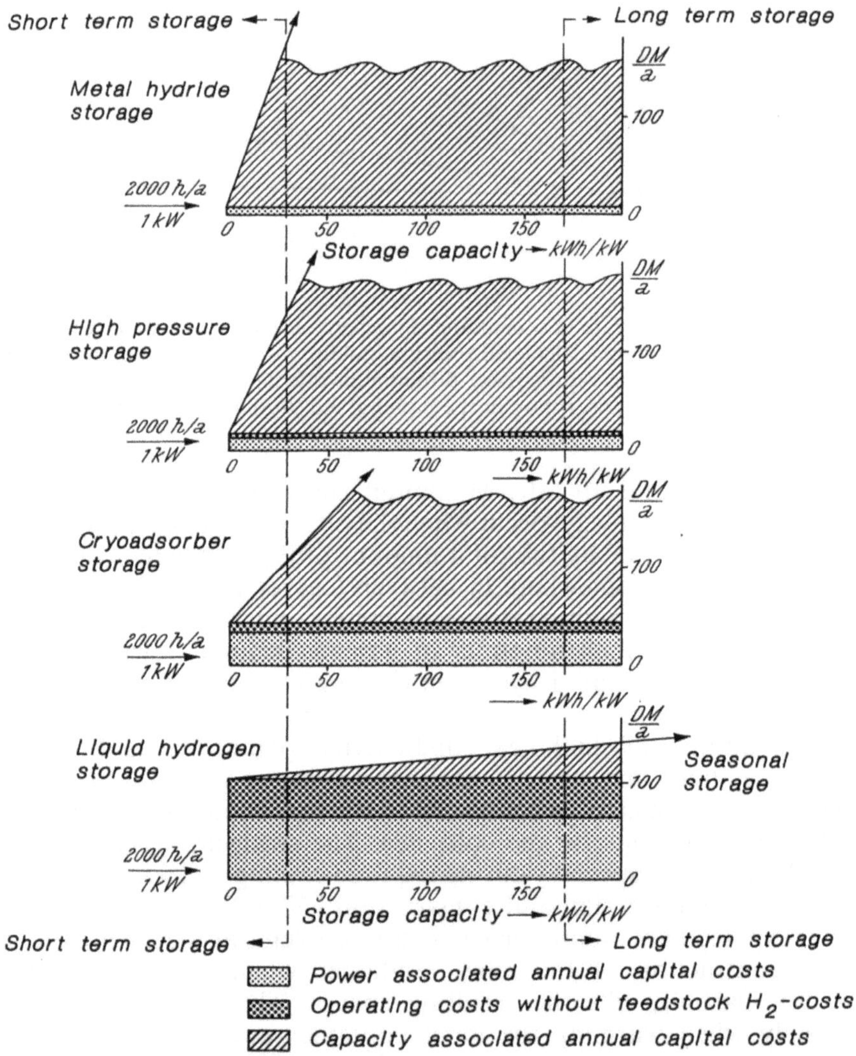

Fig. 42. Comparison of annual costs for various methods of stationary large-scale hydrogen storage. Annual utilization of 2,000 hours. The annual costs (DM/a) as a function of the storage capacity are given, defined as the stored energy, based on the charge and discharge output (kWh/kW), corresponding to the mass flow rate of the hydrogen (from Chap. 8, [48, 49])

These types of liquid hydrogen containers are also appropriate for economical long-term storage in future applications of hydrogen as an energy carrier. Based on the same amount of stored hydrogen these containers, as opposed to hydride storage containers and pressurized storage containers, have lower investment costs by a factor of about 30 which can be further reduced by the transition from production of individual containers to production in mass quantity. Thus for future application in energy technology, compared to pressurized storage containers and the similarly priced hydride storage containers, the storage of liquid hydrogen represents the most economical alternative above an output related storage capacity of about 50 kWh/kW. The very low, capacity dependent costs (see Fig. 42) of the storage containers more than compensate for the higher investment and operating costs for the liquefaction plant.

Small Storage Containers: The storage of liquid hydrogen in smaller stationary and transportable containers (<5,000 L) down to about 100 L as in liquid helium containers is state of the art. Some larger containers are constructed with perlite-vacuum insulation whereas smaller containers are almost always fitted with multilayer insulation and continuously vapor cooled shields. Thus boil off rates of below 1% per day are attained but further improvements are possible. Furthermore, there are still containers in use which are cooled with liquid nitrogen. They also have very low boil off rates (<1% per day) but are not as easy to handle as vapor cooled containers with multilayer insulation because they require periodic refilling with liquid nitrogen and are heavier. In concept liquid helium containers can also be used for liquid hydrogen (concerning the change in boil off rate see page 81), however appropriate local safety regulations must be followed.

Transportation of Liquid Hydrogen: The transportation of hydrogen [59, 61] in its liquid state is more economical, depending upon demand, and the container is not as heavy as its alternatives. This becomes immediately apparent when it is considered that with transportation via pressurized gas bottles or the similarly heavy hydride storage containers the portion of the weight of the hydrogen to the storage container is only about 1.5%. The transportation of 70 kg of liquid hydrogen for instance, requires about a 1,000 L container which weighs about 200 kg. About 5 tons of "packing" is required with the use of gas bottles or iron–titanium hydride[3].

In the U.S. tank trailers and tank trucks are in use with a capacity from about 30,000 L to a maximum of 70,000 L of liquid hydrogen and boil off

[3] Under these circumstances it is more sensible to transport hydrogen in its liquid state which is transferred into the needed low-pressure gas at the customer site with "cold vaporizers" in a manner similar to liquid nitrogen or liquid oxygen (see footnote on page 123).

rates in the range of 0.3 to 0.5% per day. Also in use are tank cars for rail transport with capacities of 105,000 L and a boil off rate of less than 0.2% per day. Both perlite-vacuum insulation and multilayer insulation are in use. With these tank trucks liquid hydrogen from liquefaction plants in California and Florida, etc., is delivered to the entire U.S.

Under normal operating conditions the time it takes to build up enough pressure to actuate the safety valves permits transportation and service intervals of up to two weeks without any discharge of gaseous hydrogen into the air. When the safety valves are actuated in some transport containers the hydrogen is vented to an expansion engine. It drives a fan and hydrogen is mixed with air below the lower ignition point (4% volume).

Furthermore, during the Apollo program liquid hydrogen was transported on barges in containers of approximately 900 m^3 capacity (Fig. 43). The experience gathered, together with the state of the art in liquid natural gas (LNG) tankers, constitute the basis for a possible development of future tankers for the transportation of liquid hydrogen overseas.

Liquid nitrogen cooled containers and containers with vapor cooled multilayer insulation with a capacity of up to about 20,000 L for liquid hydrogen are currently used for overseas transportation. Liquid hydrogen

Fig. 43. Barges with 900 m^3 storage containers for the transportation of liquid hydrogen for the Apollo Project (Source: NASA)

is shipped in this manner for instance via ship from the New Orleans area to Kourou in French Guiana for the supply of the Ariane project.

4.3 Liquid Hydrogen Pipelines

For laboratory application and on a small scale in experimental technology, vacuum-insulated pipelines, connections and valves are used for the transfer of liquid hydrogen as well as for liquid helium [75]. Vacuum-insulated pipelines are constructed so that the inner pipeline which comes in contact with the hydrogen is supported within the outer pipeline by low heat conducting spacers—fiberglass epoxy for example [62–66]. Stainless steel, which can be welded in inert gas down to very a thin wall thickness, is generally used for the inner line. The axial contraction of the inner pipeline as a result of cooling and the associated mechanical stresses can be avoided by an appropriate configuration of the pipeline. Sufficient flexibility can be provided by 90° bends for example [67–74]. Straight pieces of pipeline must have stress relief in the form of compensators (corrugated tubes) in the inner pipeline. This can cause problems because of susceptibility to leakage. If the inner pipeline is not too long it can be produced from a material with a very small coefficient of thermal expansion (invar steel for example).

Elbows and branches can either be sized so that the inner pipeline is inserted and connected piece by piece, or elbows and tees of the outer pipelines can be divided vertically in two halves and welded after installation. Connections between line sections can be made either under continuous vacuum or individual vacuum. In the case of continuous vacuum the inner pipelines are first connected and then the outer pipelines are connected by means of movable sleeves [62]. In the case of individual vacuum the inner pipeline is welded to the outer pipeline by so called labyrinths which are highly heat-resistant and then connected with the counterpart in a plug-in fashion (Fig. 44).

Flexible vacuum-insulated pipelines are fabricated in the same manner via coaxially arranged flexible hoses made of stainless steel. To improve the insulation the inner pipeline can be covered with multilayer insulation if necessary. In order to maintain the vacuum in the vacuum jacket absorbing materials, such as molecular sieves, are installed on the operating cold inner pipeline.

Manual, pneumatic, or motor operated valves are generally inserted into tees and are fitted with an extended shaft for valve actuation, which leads to the outside via a stuffing box or an airtight rotary seal. They can also be fabricated with a membrane seal without a stuffing box with an interrupted shaft which has mechanical and thermal contact with the cold side only during actuation of the valve.

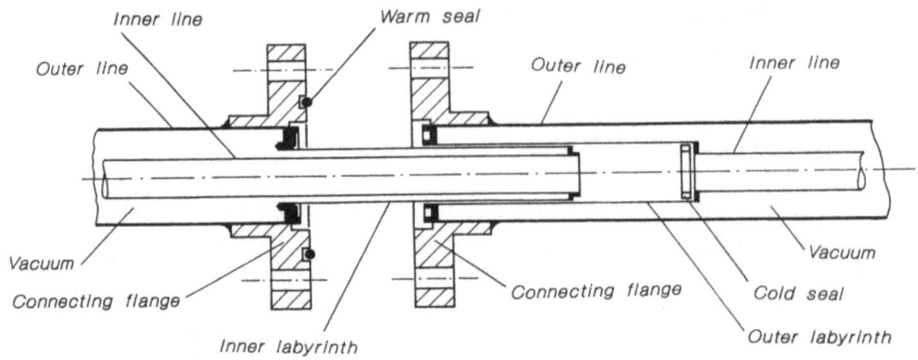

Fig. 44. Schematic of Johnston-Cox coupling for the connection of vacuum-insulated pipelines with separate vacuum (see [19])

The valve seat gasket is either made from organic material such as polyethylene or PTFE (poly-tetra-fluoro-ethylene) or it is a pure metal seal. Despite potentially advantageous applications solenoid valves are troublesome. Like mechanically operated valves they can either be made with an extended shaft and magnetic circuit at ambient temperature, with the solenoid inside or outside the vacuum or as a complete "cold valve". Problems can occur at the airtight connection of the valve shaft with the body of the valve and, with larger valve orifices at the valve seat itself. Both organic sealing materials which maintain a certain amount of flexibility down to about 200 K, as well as metal seals, require a relatively high contact pressure. Together with the pipeline pressure acting on the valves this contact pressure has to be overcome by the magnetic circuit when the valve opens. In the case of a single stage design this requires a strong magnetic circuit. With a two-stage design, in which the magnetic circuit only triggers the puppet against the pressure acting on the valve, the magnetic circuit has to be located on the warm side of the valve in order to avoid cold sealant problems.

Valves with smaller cross sections and orifice diameters up to about 8 mm can be produced from metal as a "cold valve". They are sealed by an autocentering ball with lateral clearance embedded in the valve piston against a tapered seat. Additional cooling of the solenoid in the vacuum is not necessary when operating the valve with direct current (maximum 1 W power loss). The electrical "starting voltage" required to open the valve normally is about three times higher than the "hold on". A completely welded stainless steel design with low mass and therefore low heat capacity is possible which is particularly desired with intermittent operation and small flow rates (see Chap. 6, p. 185).

Table 17. Examples of LH_2 and LNG lines

Place		Length m	ca. ID mm
Kennedy Space Center, Florida	LH_2, LO_2	500	150
Jackass Flats, Nevada, Rover Project Test Site	LH_2, 150 bar	50	400
Aerojet California	LH_2, 40 bar	160	500
Borneo	LNG	10,000	450

Flow rates of up to 250 m^3/min LH_2, 100 m^3/min LO_2, 250 m^3/min LNG and above have been attained.

Large Pipelines: A considerable number of liquid hydrogen and liquid oxygen pipelines have been developed and constructed within the U.S. space programs some of which are still in use today. Furthermore, liquid natural gas (LNG) pipelines may have problems similar to those for liquid hydrogen and to an extent can be compared with them (Table 17).

Large vacuum-insulated liquid hydrogen pipelines—which are generally fitted with multilayer insulation—require only periodic evacuation to maintain the vacuum because, as a result of cryopumping—freezing of the residual gas on the inner pipeline—the vacuum of cold pipelines is additionally improved. The heat leakage of these pipelines as well as their mechanical stress due to contraction of the inner line can be determined through standard procedures [72, 74]. Problems which occur on cooldown require relatively extensive methods of calculation [67, 68]. The pressure oscillations occurring upon cooldown, which generally do not exceed 25% of the static line pressure, but in some cases can be three or four times higher than the line pressure [67, 70], also require appropriate consideration including the continuous and intermittent thermal stresses. They have a considerable impact since the calculation of the pressure loss, heat transfer and thermal expansion which, together with stratified two-phase flow has to be made in order to avoid failure by pipeline deformation.

In the case of continuous thermal stress the axial contraction of the inner pipeline has to be taken into account. The resulting axial stress is reduced to acceptable amounts, as with small pipelines, by appropriate routing of the pipeline or by means of—more problematic—corrugated tubing (compensators) in the inner line.

Although the stress analysis of these pipelines can generally be performed easily with cryogenic systems, the effort required could be so excessive, due to complicated routing of pipeline, branches etc., consideration of thermal acustic oscillations, that it can be accomplished only with the help of computer programs.

4.4 Liquid Hydrogen Pumps

Different types of pumps have been developed for liquefied gases and also for liquid hydrogen.
Reciprocating pumps are primarily suitable for small flow rates at high pressure whereas for large flow rates multistage radial and axial pumps are in use [76, 77, 82, 103].
All pump types have in common that subcooled liquid must flow into the pump in order to avoid cavitation, vapor formation and a break down of the supply [83, 84, 86–88, 102]. Subcooling is attained by pressurization in which the liquid to be pumped is not in thermal equilibrium with the vapor phase. This "net positive suction head" (NPSH) can either be achieved through an appropriate height of the liquid column in the storage tank, by pressurization of the tank during pumping, by cyclic pressurization of the piston pump, by phase separators or through appropriate installation of superchargers at the intake section of the pump.
With turbopumps, inducers are used which are directly fastened to the impeller at the intake section [85]. These inducers are appropriately shaped and generally designed as helicoidal wings with variable pitch (Figs. 45, 46). This measure has also been successfully applied to pumps with superfluid helium [90–94].
Superchargers have proven worthwhile with reciprocating pumps. They are flat shaped and are mounted on the extended piston rod of the pump in front of the intake. They either move inside the free liquid or in a guide cylinder. By means of a disc valve driven by inertial force during the intake

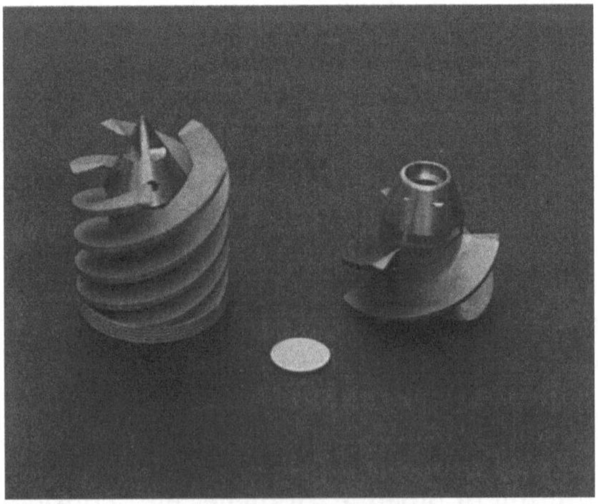

Fig. 45. Inducer of cryogenic centrifugal Pumps, diameter about 50 mm. Source: Cryomec AG

Fig. 46. Dual-stage centrifugal pump with a ring-shaped inducer (see [85]), Source: Cryomec AG

stroke of the pump, additional mechanical pressure is exerted on the liquid which flows into the pump and thus an appropriate NPSH is established [89].

Reciprocating pumps are used to pump liquid hydrogen to high pressure so that after evaporation and warming up to ambient temperature it can be stored in its gaseous state under high pressure [88].

These types of commercially available pumps which were also developed for liquid helium [86–88] generally have piston rings and piston seals made of PTFE-graphite, PTFE-bronze of PTFE-fiberglass. In the interest of small piston friction, the number of strokes is limited to about 100–250 min to promote long lifetime and low losses.

Figure 47 shows a current typical state-of-the-art pump designed with an effective NPSH = 0. This type of pump is available for pressures of up to about 800 bar and for liquefied gases like LN_2, LAr and in a vertical design with vacuum jacket also for LH_2. At the inlet this pump is fitted with a phase separator for separation of the vapor phase which occurs in the feed line of the pump. An effective NPSH = 0 requirement is attained through a supercharger.

This type of pump is recently being used for the high-pressure injection of

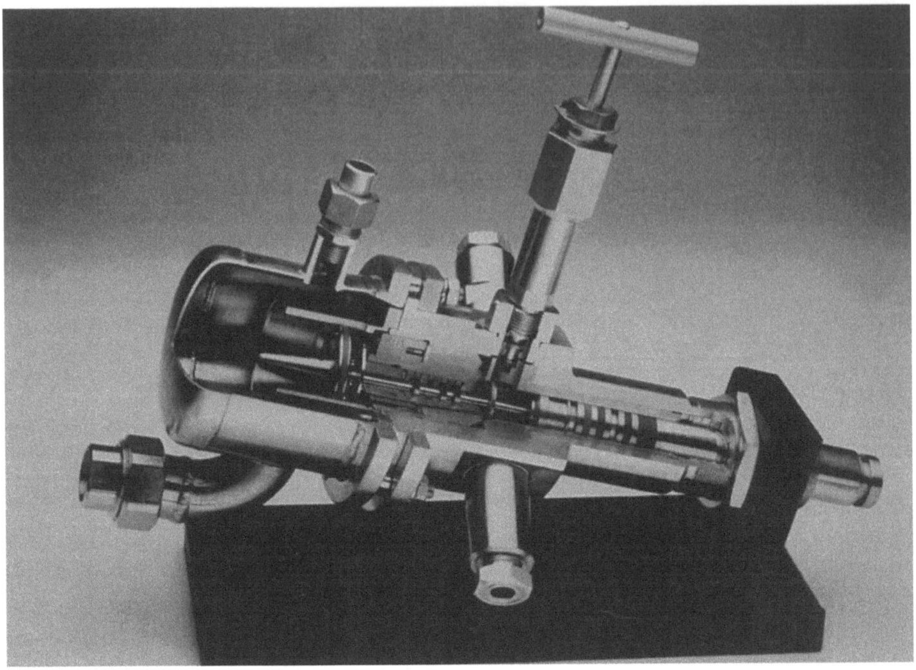

Fig. 47. High-pressure piston pump for liquefied gas with a supercharger, (See [89]), Source: Cryomec AG

hydrogen into internal combustion engines ([115], see also Chap. 6, Ref. [81]). They are also fitted with piston rings made of PTFE of PTFE-bronze. Furthermore, developments with gasfilm bearing pistons which exhibit small friction losses are being made.

Turbo pumps with flow rates of up to about 250 m^3/min and pressures of up to 500 bar were developed and successfully applied in space projects for hydrogen/oxygen rocket engines as well as for nuclear thermal propulsion [78–81].

In general, the fundamental procedures for the design of turbo pumps are also applicable to liquid hydrogen and their operational characteristics like pressure, flow rate, pump efficiency etc. can be calculated with sufficient accuracy [83–87, 95, 102].

An essential difference between pumps for liquid hydrogen and pumps for other liquids is the considerably smaller cavitation inclination of liquid hydrogen. This is demonstrated by the fact that the minimum net positive suction head which is necessary to avoid cavitation with liquid hydrogen is generally only about 60–70 mm head of liquid compared to 300 mm and more with nitrogen and water [83–86, 102].

Problems arise with the bearings of the impellers. Conventional lubricated bearings, especially ball bearings can be used at the warm side at ambient temperature if the impeller is taper bore mounted. However, the necessary rigidity of the pump shaft and the associated high heat flow from the warm to the cold side are troublesome.
With impeller bearings in the pump, ball bearings, which can be lubricated only by liquid hydrogen, are necessary at the cold side. The bearing cage proves to be a critical element of construction. Conventional bearing retainers made of steel or brass usually have a lifetime of about 1–2 hours above a linear speed of about 3 m/s [96, 99, 101, 104]. Excellent results were obtained with bearing cages made of PTFE (see Fig. 48). PTFE-fiberglass material or carbon fiber reinforced epoxy resin [98, 100].
Most bearing damage is caused by abrasion of the cage or by other powdery impurities (ice or other condensed gaseous impurities for example). They cover the bearing and can hinder the access of the cryogenic liquid which is needed for bearing lubrication and cooling. According to [100] it has been proved that ball bearings work satisfactorily in cryogenic hydrogen atmosphere with an appropriately highly forced gas flow through the bearing [96–100].
Slush Hydrogen: Homogeneous mixtures of the solid and liquid phase of hydrogen which are at thermodynamic equilibrium (slush hydrogen) are important [104–107, 110–114].
With a 50% solid phase the density and heat capacity of the mixture exceeds that of liquid hydrogen at the boiling point by 15.5% and 18.3% respectively. Therefore, slush hydrogen has better storage and transportation qualities than liquid hydrogen. Further, it can be handled like liquid hydrogen up

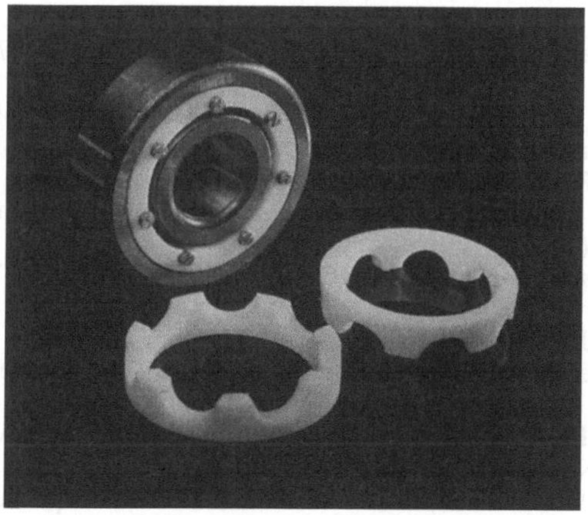

Fig. 48. Ball bearings with 40 mm external diameter and cage of PTFE for use with liquid nitrogen and liquid hydrogen (Source: DLR)

to about 60% solid phase, i.e., it can be transported through pumps, pipelines and valves. Certain storage problems still exist because grain enlargement of the solid phase is caused through sintering with materials at rest, which however does not exclude future application in air and space technology [116] and terrestrial energy technology.

References

[1] K. D. Timmerhaus: Fluid Flow and Heat Transfer. In: Applied Cryogenic Engineering (R. W. Vance, W. H. Duke, eds.), pp. 104–151. New York, Wiley (1962).

[2] J. A. Clark, R. M. Thorogood: Heat Transfer. In: Cryogenic Fundamentals (G. G. Haselden, ed.), pp. 92–197. London, Academic Press (1971).

[3] J. A. Clark: Heat Transfer. In: Cryogenic Technology, (R. W. Vance, ed. pp. 121–195), New York, Wiley (1963).

[4] H. M. Roder, R. D. McCarty, W. J. Hall: Computer Programs for Thermodynamic and Transport Properties of Hydrogen. Nat. Bureau of Standards, NBS-TN-625 (1972).

[5] D. A. van Gundy, J. R. Uglum: Heat Transfer to an Uninsulated Surface at 20 K. In: Adv. Cryog. Eng., Vol. 7, pp. 377–384. New York, Plenum Press (1962).

[6] R. L. Middleton, J. M. Stukey, J. T. Schell, et al.: Development of Lightweight External Insulation System for Liquid-Hydrogen Stages of the Saturn V Vehicle. In: Adv. Cryog. Eng., Vol. 10, pp. 216–223. New York, Plenum Press (1964).

[7] J. B. Rittenhouse: Application of an adhesively Bonded Cryogenic Insulation System. NASA-TM-X-57823, NTIS (1966).

[8] K. E. Leonhard, B. Oglin, W. F. Zimni: Determination of the Thermal Conductivity, the Specific Heat and the Weight by Volume of Insulations for Rocket Tanks Filled with Liquid Hydrogen. ELDO/ESRO Sci. Tech. Rev., Vol. 2, pp. 3–28 (1967) (In French).

[9] D. L. Dearing: Summary of the Saturn S-IV and S-IVB Liquid Hydrogen Tank Internal Insulation Development and Tehniques for Future Improvement. Bull. Int. Inst. Froid, Annexe 2, 233–246 (1965).

[10] D. L. Dearing: Development of the Saturn S-IV and S-IVB Liquid Hydrogen Tank Internal Insulation. In: Adv. Cryog. Eng., Vol. 11, pp. 89–97. New York, Plenum Press (1966).

[11] C. R. Lemons, C. R. Watts, O. K. Samassy: Development of Advanced Materials Composites for Use as Insulation for LH_2-Tanks. McDonnel-Douglas, Astronautics Co., NASA-CR-124388, NASA-CR-123928 (1973).

[12] J. J. McGraw: Cellular Insulation for Use with Low Temperature Liquids. US-Pat. No. 3.755.056 (1973).

[13] R. J. Jonke: Insulation Systems for Cryogenic Stages. Rev. Sci. Tech. CECLES/CERS *3*, 17–48 (1971).

[14] H. M. Tarid, J. C. Boissin, M. P. Segel: Thermal Insulation for Liquid Hydrogen Space Tankage. In: Adv. Cryog. Eng., Vol. 12, pp. 274–285. New York, Plenum Press (1967).

[15] G. B. Yates: PPO Foam: Liquid Hydrogen Insulation. In: Adv. Cryog. Eng., Vol. 19, pp. 327–337. New York, Plenum Press (1974).

[16] R. P. Reed, J. M. Arvidson, R. L. Durcholoz: Tensile Properties of Polyurethane and Polystyrene Foams from 76 to 300 K. In: Adv. Cryog. Eng., Vol. 18, pp. 184–193. New York, Plenum Press (1973).

[17] H. Klezath: Wärmisolierung von Speicherbehältern für tiefsiedende Flüssigkeiten. Erdöl-Erdgas *85*, 145–149 (1969).

[18] C. L. Johnson, D. J. Hollweger: Some Heat Transfer Considerations in Non-Evacuated Cryogenic Powder Insulation. Adv. Cryog. Eng., Vol. 11, pp. 77–88. New York, Plenum Press (1966).
[19] R. B. Scott: Insulation. In: Cryogenic Engineering, pp. 142–214. New York, van Nostrand (1959).
[20] R. H. Kropschot: Low-Temperature Insulation. In: Applied Cryogenic Engineering, pp. 152–169. New York, Wiley (1963).
[21] R. B. Jacobs: Thermal Insulation, Storage, Transport und Transfer of Liquid Hydrogen. In: Technology and Uses of Liquid Hydrogen (R. B. Scott, ed.) pp. 106–148. New York, Pergamon Press (1964).
[22] R. H. Kropschot: Insulation Technology. In: Cryogenic Technology (R. W. Vance, ed.), pp. 239–250. New York, Wiley (1963).
[23] W. Molnar: Insulation. In Cryogenic Fundamentals (G. G. Haselden, ed.) pp. 199–236. London, Academic Press (1971).
[24] M. Knudsen: Ann. d. Physik 31, 205 (1910); *32*, *809* (1910); *33*, *1435* (1910); *34*, 593 (1911); *6*, 149 (1930).
[25] R. J. Corrucini: Gaseous Heat Conduction at Low Pressures and Temperatures. Vacuum 7, 8 (1957).
[26] C. Gerthsen: Physik. Berlin Heidelberg, New York, Springer (1966).
[27] M. Weitz: Theorie und Praxis der Vakuumtechnik. Braunschweig, Vieweg (1965).
[28] R. H. Kropschot, W. Burgers: Perlite for Cryogenic Insulation. In: Adv. Cryog. Eng., Vol. 8, pp. 221–229. New York, Plenum Press (1963).
[29] B. J. Hunter, R. H. Kropschot, J. E. Schrodt, M. M. Fulk: Metal Powder Additives in Evacuated—Powder Insulation. In: Adv. Cryog. Eng., Vol. 5, pp. 146–156. New York, Plenum Press (1960).
[30] B. L. Knight, K. D. Timmerhaus, R. H. Kropschot: Analysis of Thermal Diffusity Evaluation under Transient Conditions for Powder Insulation. In: Adv. Cryog. Eng., Vol. 18, pp. 112–117. New York, Plenum Press (1973).
[31] G. R. Cunnington: Apparent Thermal Conductivity of Uncoated Microsphere Cryogenic Insulation. In: Adv. Cryog. Eng., Vol. 21, pp. 263–271. New York, Plenum Press (1976).
[32] G. R. Cunnington, C. L. Tien: Heat Transfer in Microsphere Cryogenic Insulation. In: Adv. Cryog. Eng., Vol. 18, pp. 103–111. New York, Plenum Press (1973).
[33] C. L. Tien, G. R. Cunnington: Recent Advances in High performance Cryogenic Thermal Insulation. Cryogenics *12*, 419–421 (1972).
[34] Al L. Nayak, C. L. Tien: Thermal Conductivity of Microsphere Cryogenic Insulation. In: Adv. Cryog. Eng., Vol. 21, pp. 251–262. New York, Plenum Press (1976).
[35] P. Petersen: The Heat-tight Vessel. Swedish Technical Research Council Rep., No. 706 (1951); see also: Satryck ur TVF *29*, 4 (1958).
[36] R. H. Kropshot: Low Temperature Insulation. In: Applied Cryogenic Engineering, pp. 152–169. New York, Wiley (1962).
[37] W. Frost: Heat Transfer at Low Temperatures. New York, Plenum Press (1975).
[38] R. P. Caren, G. R. Cunnington: Heat Transfer in Multilayer Insulation Systems. Chem. Eng. Progr., No 87, Vol. 64, pp. 67–81 (1968).
[39] P. E. Glaser: Multilayer Insulation for Large Vessels Used in Transporting and Storing Cryogenic Liquids. Mech. Eng. *87*, 23–27 (1965).
[40] K. Kutzner, F. Schmidt, I. Wietzke: Radiative and Conductive Heat Transmission Through Superinsulations—Experimental Results for Aluminium Coated Plastic Foils. Cryogenics *13*, 396–404 (1973).
[41] E. M. Sparrow, R. D. Cess: Radiation Heat Transfer. Bemond, Calif: Brooks/ Cole (1963).
[42] R. M. Coston: Handbook of Thermal Design Data for Multilayer Insulation Systems, Vol. 2. Lockheed Missiles and Space Co., Sunnyvale, Calif. Rep. NASA-CR 87485 (1967).

[43] F. Ruccia, R. Hinckley: The Surface Emittance of Vacuum-metallized Polyester Films. In: Adv. Cryog. Eng., Vol. 12, pp. 218–227. New York, Plenum Press (1967).
[44] G. Bell, et al.: Thermal Performance of Multilayer Insulation applied to Small Cryogenic Tankage. In: Adv. Cryog. Eng., Vol. 22. New York, Plenum Press (1977).
[45] F. E. Swalley, C. D. Nevins: Practical Problems in Design of High-performance Multilayer Insulation System for Cryogenic Stages. In: Adv. Cryog. Eng., Vol. 10, pp. 208–215. New York, Plenum Press (1965).
[46] R. M. Coston, T. C. Nast: Experimental Evaluation of the Equations and Parameters Governing Flow Through Multilayer Insulations During Evacuation. In: Adv. Cryog. Eng., Vol. 11, pp. 56–64. New York, Plenum Press (1966).
[47] G. C. Vliet, R. M. Costen: Thermal Energy Transport Parallel to the Laminations in Multilayer Insulation. In: Adv. Cryog. Eng., Vol. 13, pp. 671–679. New York, Plenum Press (1968).
[48] D. O. Murray: Degradation of Multilayer Insulation Systems by Penetrations. Adv. Cryog. Eng., Vol. 13, pp. 680–689.
[49] T. C. Nast: Effective Purging of High Performance Multilayer Insulation Systems. Adv. Cryog. Eng., Vol. 11, pp. 49–55. New York, Plenum Press (1966).
[50] J. W. Price: Measuring the Gas Pressure within a High-performance Insulation Blanket. In: Adv. Cryog. Eng., Vol. 13, L-1, pp. 662–670. New York, Plenum Press (1968).
[51] R. G. Scurlock, B. Saull: Development of Multilayer Insulation with Thermal Conductivities below 0.1 μW cm^{-1} K^{-1}. Cryogenics *16*, 303–311 (1976).
[52] J. A. Paivanas, O. P. Roberts, D. I. J. Wang: Multishielding—An Advanced Superinsulation Technique. In: Adv. Cryog. Eng., Vol. 10, pp. 197–207. New York, Plenum Press (1965).
[53] G. A. Bell, T. C. Nast, R. K. Wedel: Thermal Performance of Multilayer Insulation Applied to Small Cryogenic Tankage. In: Adv. Cryog. Eng., Vol. 21, pp. 272–282. New York, Plenum Press (1976).
[54] A. R. Urbach, R. N. Herring: Long-term Helium Dewar for Space Experiments. In: Proc., 6th Int. Cryog. Eng. Conf., pp. 154–156 (1974).
[55] W. F. Stewart: Operating Experience with a Liquid Hydrogen Fueled Buick and Refueling System. In: Proc., 4th Int. Hydrogen Energy Conf., Pasadena (June 1982), Vol. 3, pp. 1071–1093. New York, Pergamon Press (1982).
[56] L. R. Niendorf, S. C. Choksi: Ultra-efficient Insulation System for Solid Cryogen Coolers. In: Adv. Cryog. Eng., Vol. 12, pp. 286–299. New York, Plenum Press (1967).
[57] F. J. Edeskuty, K. D. Williamson, Jr.: Storage and Handling of Cryogens. In: Adv. Cryog. Eng., Vol. 17, pp. 56–68. New York, Plenum Press (1972).
[58] M. P. Segel: Experimental Study of Phenomena of Stratification and Pressurization of Liquid Hydrogen. In: Adv. Cryog. Eng., Vol. 10, pp. 308–313. New York, Plenum Press (1964).
[59] B. W. Birmingham, E. H. Brown, C. R. Class, A. F. Schmidt: Vessels for the Storage and Transport of Liquid Hydrogen-Research paper 2757, J. Res. Nat. Bur. Stand., A *58*, 243–253 (1957).
[60] D. H. Liebenberg, R. W. Stokes, F. J. Edeskuty: Chilldown and Storage Losses of Large Liquid Hydrogen Storage Dewars. In: Adv. Cryog. Eng., Vol. 11, pp. 554–560. New York, Plenum Press (1966).
[61] P. D. Fuller, J. N. McLagan: Storage and Transfer of Cryogenic Fluids. In: Applied Cryogenic Engineering (R. W. Vance, W. M. Duke, eds.). Section 1: Cryogenic Storage Vessels and Transport Trailers, pp. 215–237, Sect. 2: Transfer Lines, pp. 238–254. New York, Wiley (1962).
[62] F. J. Edeskuty: Nuclear Propulsion. In: Cryogenic Technology (R. W. Vance, ed.) pp. 352–374. New York, Wiley (1963).

[63] C. F. Sind: Transmission of Hydrogen. In: Selected Topics on Hydrogen Fuel (J. Hard, ed.), NBS-Sec. Publ. *419* (1975).
[64] R. B. Jacobs: Long Distance Transfer of Liquefied Gases. In: Proc., 2nd Cryog. Eng. Conf., Boulder, Co., 1956. Nat. Bureau of Stand (1957).
[65] A. J. Croft: 14 Meter Liquid Hydrogen Line. Cryogenics *10*, 167–168 (1970).
[66] J. Stuchly: Internally Insulated Cryogenic Pipelines. In: Adv. Cryog. Eng., Vol. 21, pp. 531–537. New York, Plenum Press (1976).
[67] R. S. Thurston, J. D. Rogers, V. J. Skoglund: Pressure Oscillations Induced by Forced Convection Heating of Dense Hydrogen. In: Adv. Cryog. Eng., Vol. 12, pp. 438–451. New York, Plenum Press (1967).
[68] W. G. Flieder, W. J. Smith, K. R. Wetmore: Flexibility Considerations for Design of Cryogenic Transfer Lines. In: Adv. Cryog. Eng., Vol. 5, pp. 111–119. New York, Plenum Press (1960).
[69] W. G. Steward: Transfer Line Surge. In: Adv. Cryog. Eng., Vol. 10, pp. 313–323. New York, Plenum Press (1965).
[70] R. S. Thurston: Probing Experiments on Pressure Oscillations in Two Phase and Supercritical Hydrogen with Forced Convection Heat Transfer. In: Adv. Cryog. Eng., Vol. 10, pp. 305–312. New York, Plenum Press (1965).
[71] J. C. Burke, W. R. Byrnes, A. H. Post, F. E. Ruccia: Pressurized Cooldown of Cryogenic Transfer Lines. In: Adv. Cryog. Eng., Vol. 4, pp. 378–394. New York, Plenum Press (1964).
[72] J. C. Bronson, F. J. Edeskuty, et al.: Problems in Cooldown of Cryogenic Systems. In: Adv. Cryog. Eng., Vol. 7, pp. 198–205. New York, Plenum Press (1960).
[73] O. Baker: Design of Pipe Lines for Simultaneous Flow of Oil and Gas. The Oil and Gas J. *53*, 185–195 (1954).
[74] K. Srinivasan, R. Seshagiri, M. V. Krishna Murthy: Analytical and Experimental Investigations on Cooldown of Short Cryogenic Transfer Lines. Cryogenics *14*, 489–494 (1974).
[75] C. S. Beard: Cryogenic Valves, a Survey. Cryog. Eng. News *2*, 62–68 (1967).
[76] A. E. Biermann, R. C. Kohl: Preliminary Study of a Piston Pump for Cryogenic Fluids. NASA-Memo 3/6/59E, Lewis Res. C. (1959).
[77] T. A. Carter, Jr.: Pumps for Liquid Hydrogen. Cryog. Tech. *3*, 173–175 (1967).
[78] W. H. Knuth, J. Farquhar, B. K. Lindley: Design Study of Modification of M 1 Liquid Hydrogen Turbopumps for Use in Nuclear Reactor Test Facility. NASA-CR-54422 (1965).
[79] J. Farquhar, B. K. Lindley: Hydraulic Design of M 1 Liquid Hydrogen Turbopumps. NASA-CR-54822 (1966).
[80] G. H. Ribble, Jr., G. E. Turney: Experimental Study of Low Speed Operating Characteristics of a Liquid Hydrogen Centrifugal Turbopump. NASA-TM-X-1861, August (1969).
[81] H. P. Stinson, R. J. Strickland: Experimental Findings from Zero Tank Net Positive Suction Head Operation of the J-2 Hydrogen Pump. NASA-TN-D-6824, August (1972).
[82] K. P. Martin, R. B. Jacobs, R. J. Hardy: Performance of Pumps with Liquefied Gases. In: Adv. Cryog. Eng., Vol. 2, pp. 295–302. New York, Plenum Press (1960).
[83] I. S. Pearsall: Supercavitating Pumps for Cryogenic Liquids. Cryogenics *12*, 422–426 (1972).
[84] J. F. Di Stefano, G. H. Caine: Cavitation Characteristics of Tank-mounted Cryogenic Pumps and their Predicted Performance under Reduced Gravity. In: Adv. Cryog. Eng., Vol. 7, pp. 277–290. New York, Plenum Press (1962).
[85] A. G. Cryomec, J. E. Tornare, K. Bofinger: Pumpeneinlaufkranz, Europ. Pat. EP 0 317 687 A1 (1989).

[86] G. H. Caine, L. Schäfer, D. Burgeson: Pumping of Liquid Hydrogen. Adv. Cryog. Eng., Vol. 4, pp. 241–254. New York, Plenum Press (1960).
[87] M. Morpurgo: Design and Construction of a Pump for Liquid Helium. Cryogenics *17*, 91–93 (1977).
[88] C. F. Goltzmann: High Pressure Liquid Hydrogen and Helium Pumps. Adv. Cryog. Eng., Vol. 5, pp. 289–298. New York, Plenum Press (1960).
[89] A. G. Cryomec, J. E. Tornare, K. Bofinger, Cl. Tschopp: Pumpe für kryogene Fluiden. Europ. Pat. EP 0 174 269 A3, 1984, US Pat. No. 4,639,197.
[90] P. R. Ludtke, D. E. Daney, W. G. Steward: Performance of a Small Centrifugal Pumps in the I and He II. In: Adv. in Cryog. Eng., Vol. 33, pp. 515–524. New York, Plenum Press (1988).
[91] P. M. McConnel: Liquid Helium Pumps, NBSIR 73-316, NBS, Boulder, Colorado (1973).
[92] P. R. Ludke: Performance Characteristics of a Liquid Helium Pump. NBSIR 75–816, NBS, Boulder, Colorado (1975).
[93] W. G. Steward: Centrifugal Pumps for Superfluid Helium. Cryogenics *26* (1986).
[94] P. Kittel: Liquid Helium Pumps for in Orbit Transfer Cryogenics *27* (1987).
[95] P. N. McNail, J. E. Engloud, R. H. Knoll: Design, Development, and Test of Shuttle/Centour G—Prime Cryogenic Tankage Protection Systems. In: Adv. Cryog. Eng., Vol. 33, pp. 341–348. New York, Plenum Press (1988).
[96] H. W. Scibbe: Bearings and Seals for Cryogenic Fluids. NASA-TM-X-52415 (1968).
[97] D. E. Brewe, H. H. Coe, H. W. Scibbe: Cooling Studies with High Speed Ball Bearings Operating in Cold Hydrogen Gas. ASLE-Trans., 12, No. 1, 66–76 (1969).
[98] H. H. Coe, D. E. Brewe, H. W. Scibbe: Cooling Requirements of Ball Bearings Lubricated by Glass-Fiber-Filled PTFE Retainers in Cold Hydrogen Gas. NASA-TN-D-5607, 26 pp, February 1970.
[99] W. A. Wilson, K. B. Martin, J. A. Brennan, et al.: Evaluation of Ball Bearing Separator Materials Operating Submerged in Liquid Nitrogen. Trans. ASLE *4*, 50–58 (1962).
[100] D. B. Chelton, J. A. Brennan, L. E. Scott: Dry Gas Operation of Ball Bearings at Cryogenic Temperatures. In: Adv. Cryog. Eng., Vol. 7, pp. 273–276. New York, Plenum Press (1960).
[101] D. E. Brewe, D. W. Wisander, H. W. Scibbe: Performance of 40-millimeter Bore Bearings with Lead and Lead-alloy Retainers in Liquid Hydrogen at 192 Million DN. NASA-Lewis-Res. C., Tech. Note, NASA-TN-D-6091. November (1982).
[102] R. B. Jacobs: Prediction of Symptoms of Cavitation. J. Res. NBS, *65C*, 156, July/September (1961).
[103] J. E. Blackford, P. Halford, D. H. Tantam: Expanders and Pumps. In: Cryogenic Fundamentals (G. G. Haselden, ed.), pp. 403–489. London, Academic Press (1971).
[104] C. F. Sindt: A Summary of the Characterization Study of Slush Hydrogen. Cryogenics 10/No. 5, 372–380 (1970).
[105] R. D. McCarty, J. Hord, H. M. Roder: Selected Properties of Hydrogen, NBS-Monograph 168, U.S. Government Printing Office (1981).
[106] C. F. Sindt, P. R. Ludtke, D. E. Daney: Slush Hydrogen Fluid Characterization and Instrumentation. NBS-Tech. Note No. 377, 65 pp. (1969).
[107] R. Schraewer, W. Daus: Herstellung und Förderung von Wassertoffmatsch. Forschungsbericht NT 200 des BMFT (1974).
[108] Q. S. Shu, R. W. Fast, H. L. Hart: Theory and Techniques for Reducing the Effect of Cracks in Multilayer Insulation from Room Temperature to 77 K. In: Adv. Cryog. Eng., Vol. 33, pp. 291–298. New York, Plenum Press (1988).
[109] Q. S. Shu, R. W. Fast, H. L. Hart: Crack Covering Patch Technique to Reduce the Heat Flux from 77 K to 4,2 K through Multilayer Insulation. In: Adv. Cryog. Eng., Vol. 33, pp. 299–304. New York, Plenum Press (1988).

[110] P. R. Ludtke, P. J. Storch: Survey of Instrumentation for Slush Hydrogen Systems, National Institute for Standards and Technology (NIST), NASP Report *1054*, March (1989).
[111] P. Ordin: Hydrogen Safety Standard, NASA TMX-42454, NASA CR-182200, NASA Headquarters. Washington D.C., October 1988.
[112] N. Squires et al.: Gelation, Hydrodynamics, and Heat Transfer Aspects of Gelled Cryogenic Propellants, A Progress Report, Part 1: Experiment. Grant No. NAG 3-850, Dept. of Mechanical Engineering and Applied Mechanics. The University of Michigan, Ann Arbor, MI, May 1989.
[113] D. E. Daney, V. D. Arp, R. O. Voth: Hydrogen Slush Production with a Large Auger. In: Adv. Cryog. Eng., Vol. 35B, pp. 1767–1776. New York, Plenum Press (1990).
[114] E. C. Cady, T. L. Flaska, P. K. Worell: In-Tank Thermodynamics of Slush Hydrogen for the National Aerospace Plane. In: Adv. Cryog. Eng., Vol. 35B, pp. 1755–1766. New York, Plenum Press (1990).
[115] W. Peschka: Liquid Hydrogen Reciprocating Pumps for Automotive Application. In: Adv. Cryog. Eng., Vol. 35B, pp. 1783–1790. New York, Plenum Press (1990).
[116] De Witt, R. L., Hardy, T. L., Whalen, M. V., Richter, G. P., Tomcik, T. M.: Background, Current Status, and prognosis of the Ongoing Slush Hydrogen Technology Development Program for the NASP. In: Proc. Symp. on National Aero-Space Plane and Space Applications, pp. 69–85, Hawaii: Hawaii Natural Energy Institute, Univ. of Hawaii 1990

Liquid Hydrogen as a Rocket Propellant 5

For the final velocity u of a rocket related to the launching system with a launch mass m_0, mass m at the final velocity u and exhaust velocity v relative to the vehicle the following applies

$$\frac{m_0}{m} = e^{u/v}. \tag{118}$$

In order to overcome earth's gravity a velocity of 11.2 km/s is needed, for obtaining a low earth orbit approximately 8 km/s is required. For the exhaust velocity v and reaction enthalpy per unit of mass i_0 of the rocket propellant the following applies

$$v = \sqrt{2 \cdot i_0} \quad \left(\frac{\mathrm{m}}{\mathrm{s}}\right). \tag{119}$$

With the thrust F and the mass flow rate $\dot{m}$, for the specific impulse I_{sp} the following is obtained

$$F = \dot{m} \cdot v \quad I_{sp} = \frac{F}{\dot{m}} = v \quad \left(\frac{\mathrm{Ns}}{\mathrm{kg}}\right). \tag{120}$$

The most energetic chemical rocket propellants such as hydrogen/oxygen or hydrogen/fluorine correspond to a mass ratio of 16.4 at an attainable exhaust velocity of about 4,500 m/s to overcome earth's gravity and to a mass ratio of 5.7 to achieve a low earth orbit. Furthermore, it seems to be evident that in the interest of maximum payload the rocket propellants have to be stored with the lowest possible mass. In the case of hydrogen, fluorine and oxygen storage in liquid form is required which resulted in a forced development of liquid hydrogen technology together with rocket engine technology.

5.1 Chemical Rocket Propulsion

Although the importance of liquid hydrogen as a rocket fuel was recognized very early[1], it could not initially be used in space flights due to insufficiently developed technology [1–5].

The realization of high-quality space programs—especially the Apollo-project—since 1960 required the development of hydrogen/oxygen rocket engines because the needed high specific impulse could be achieved only in this manner.

This led to the development of second- and third-stage hydrogen/oxygen rocket engines for the Saturn V-rocket in the Apollo project. With these engines a payload of up to 100,000 kg could be transported into low earth orbit[2] [6].

The high-pressure rocket engine currently used in the Space Shuttle program is of the highest technical stage of development of hydrogen/oxygen rocket engine technology. Although hydrogen/oxygen rocket engines have been developed for other space flight programs—Ariane etc.—and are in use, only the development of three rocket engines of the U.S. Space Flight program will be briefly discussed here because new developments in liquid hydrogen technology were necessary with these engines.

The first rocket that used a hydrogen/oxygen engine was a NASA-Centaur. It served as the upper stage of an Atlas booster and was launched successfully for the first time on the 27th of November 1973. The Centaur stage used two RL-10 engines developed by Pratt and Whitney with a thrust of $6.75 \cdot 10^4$ N each and a specific impulse of 4,250 Ns/kg in a vacuum.

Figure 49 shows a simple diagram of the engine's operation mode, which would later provide a six clustered $40 \cdot 10^4$ N thrust in the fourth stage of the Saturn I rocket [2, 7].

It should be noted that liquid hydrogen was used for the regenerative cooling of the rocket engine and the heat energy transferred to the hydrogen was used to drive the turbopumps for liquid hydrogen and liquid oxygen.

After expansion, gaseous hydrogen reaches the combustion chamber via a cryogenic two-phase turbine made of aluminum. At the injector head it is mixed with oxygen so that a large number of streams of liquid oxygen are produced which are concentrically surrounded by gaseous hydrogen. Thus, an intimate mixture is achieved and complete combustion is obtained.

---

1 See H. Oberth: Die Rakete zu den Planetenräumen, 92, p. München: Oldenbourg 1923 and 1925. K. E. Tsiolkowsky: Works on Rocket Technology (M. K. Tikhonrarov, ed.), NASA-TT-F243 (1965). F. H. Goddard: Papers of Robert H. Goddard, 3 Vols. (Esther Goddard, G. Edward Pendray, eds.). New York, McGraw Hill (1970).

2 See J. L. Sloop: Liquid Hydrogen for Space Flight—The Long Step from Proposal to Reality. J. Astron. Sci. *29*, 373–381 (1981).

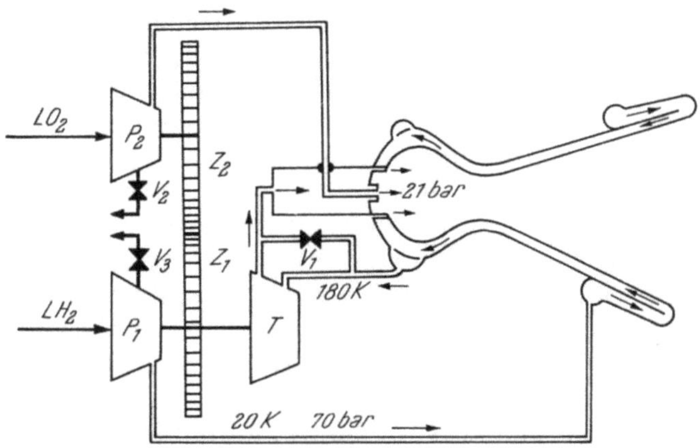

Fig. 49. Schematic of the RL-10 rocket engine (from [7]). P_1, P_2, pumps for LH_2 and LO_2 respectively; Z_1, Z_2 gear drives; V_1 turbine bypass valve; V_2, V_3 valves for precooling of the pumps

Through a turbine bypass valve V_1 (see Fig. 49) which is regulated by the pressure in the combustion chamber the thrust of the rocket engine can constantly be controlled. The RL-10 rocket engine was designed as an upper stage rocket engine for launch and restart in space in which the low normal operating pressure in the fuel tanks is sufficient for initiating the starting procedure. In this process, a small amount of hydrogen and oxygen first flows out for about 20 seconds via the valves V_2 and V_3 in order to precool the LH_2 and LO_2 pumps. Then hydrogen flows through the engine cooling system. There it is vaporized and starts to drive the turbine T. The heat capacity of the nozzle's cooling jacket and that of the combustion chamber at a temperature of approximately 55 K suffice for the supply of gaseous hydrogen which is needed to initiate the launch in which the full thrust of the engine is attained in approximately one second.

Initially for turbopumps and turbines conventional oil lubricated bearings were used which are kept at appropriately high temperature. In order to guarantee for several missions a safe restart of the engine in space even after long periods of inactivity it was necessary to replace the bearings by ball bearings which were cooled by forcing cold gaseous hydrogen through the bearing in order to eliminate frictional heat. Lifetimes of more than 7.5 hours were obtained for the ball bearings with a diameter of 35 mm and rates of revolutions of up to 30,000 rpm. This was more than sufficient for the required engine lifetime of about 1 hour.

Further development of this engine within the Apollo program led to the J2 engine of Rocketdyne which represents the most powerful hydrogen/oxygen upper-stage engine to date with a specific impulse of 4,250 Ns/kg and a thrust at $103 \cdot 10^4$ N [6].

The principle design is similar to that of the RL-10 engine with the exception of the pump bearings which are lubricated by the respective propellant component and the turbines which drive the pumps which are supplied by a hydrogen/oxygen gas generator. In this process, the combustion of oxygen and hydrogen occurs with a high excess of hydrogen. This gas is used to drive the turbine and afterwards is routed into the combustion chamber of the engine. The engine can be operated for thrust control over a wide range of the hydrogen–oxygen mixture ratio with excess hydrogen.

This engine type with a 1 hour lifetime under normal operating conditions was successfully used in the third stage of the improved version of Saturn V and as cluster of 5 rocket engines in the second stage of Saturn V within the Apollo program.

The peak of the development is the Space Shuttle Main Engine (SSME), a high-pressure engine from Rocketdyne for the Space Shuttle [8, 9]. As a cluster of three engines, each with a thrust of 2.1 MN it works together with the two solid boosters during launch phase and serves to reach earth's orbit after the solid boosters are separated.

With full use of the current state of the art technology this engine meets several additional needs. In particular it must demonstrate equivalent operating characteristics in the lower atmosphere—an area which was reserved for boosters for example the F1-engines of the Saturn rocket—as well as in the upper atmosphere and in a vacuum—areas which were reserved for the RL-10 and J2 engine. The necessary specific impulse of 3,614 Ns/kg at sea level as well as 4,529 Ns/kg in the upper atmosphere and in a vacuum required the very high combustion chamber pressure of 215 bar and an expansion ratio of the nozzle of 77.5:1 (compared to approximately 40:1 with LR-10 and J2 engines). Furthermore, regarding reuse an operational lifetime of more than 7.5 hours including 100 launches was required. The propellant supply and pump system were further improved as compared to the J2 engine. Combustion of about 20% of the propellant initially takes place in a gas generator with excess hydrogen and at a temperature of 960 K. The gas is then fed into the combustion chamber where complete combustion takes place. The turbopumps, whose bearings are lubricated by the respective propellant component, provide a pressure of 570 bar for liquid oxygen and 500 bar for liquid hydrogen. Each propellant component has two pumps, one for the lower pressure of about 20–30 bar as well as a high-pressure pump. This double arrangement provides for the operation of the high-pressure pump a considerably higher rotational speed, which reduces size and weight. The specific pump power with regard to its mass is about 150 kW/kg which corresponds to four times the amount of the J2 engine (pump power about 75 MW with SSME compared to about 6 MW with the J2).

Thus, the H_2/O_2-rocket engines and liquid hydrogen technology have reached a stage of development which on one hand permits a further development from Space Shuttle to rocket airplanes for the transport of payloads into low orbit with conventional takeoff and landing[3]. On the other hand it completes the circle of the air breathing supersonic jet engines and combination engines (air breathing engines with rocket engines) [10–16, 47].

5.2 Nuclear Rocket Propulsion

Since the reactions H_2/O_2 and H_2/F_2 provide the highest reaction enthalpy with a gaseous final product with regard to unit of mass, a further increase in the specific impulse and payload with the same final velocity u of a rocket (see Eq. (118)) is not possible with more energetic propellant combinations. The most energetic chemical reaction with regard to unit of mass is given by the recombination of atomic hydrogen

$$2H \rightarrow H_2 + 59.9\ \mathrm{kWh/kg}$$

with a specific impulse[4] of $I_{sp} = 20\ \mathrm{kNs/kg}$.

For a lunar mission, this would require only a mass ratio of $m_0/m = 2$ which corresponds roughly to the ratio of takeoff load to landing load realized in current aviation with intercontinental flights [17]. For this reason attempts to use this reaction for rocket propulsion having been going on for a long time.

The key problem of stabilizing hydrogen in its atomic state at high concentration of the atomic part has basically been demonstrated by tests that were scarcely noticed by the pertinent research groups [18–22].

The required conditions, that is temperatures of 1 K and magnetic fields of 7 Tesla exclude an application in space projects in this form but enable experiments in quantum physics of multiparticle systems, for which atomic hydrogen and helium in a triplet state are fundamental.

Since according to Eq. (118), the final velocity u is only proportional to the logarithm of the mass ratio but is directly proportional to the exhaust velocity v an increase of the latter is a more efficient way to obtain higher final velocity and higher payload.

Since nuclear reactions have energy transformations with regard to unit of mass which exceed those of chemical reactions by about six orders of magnitude it is basically possible to increase the exhaust velocity and specific impulse Eq. (118) by three orders of magnitude. In this case the energy which is set free through nuclear reaction has to be transferred at

[3] Also see: The Space Shuttle—Its Current Status and Future Impact. Proc. 1980 SAE Aerospace Congress, 92 pp, 1981.

[4] See A. Dadieu, R. Damm, E. W. Schmidt: Raketentreibstoffe, 7 pp. Wien, Springer (1968).

an appropriately high temperature to a jet mass which delivers a maximum exhaust velocity.

One may consider, to release the energy which is set free through nuclear reaction—nuclear fission or nuclear fusion—at an adequately high temperature in the gaseous or plasma state [23–36]. In the case of nuclear fission the fissionable material cannot serve as a jet mass because, apart from the problems of nuclear radiation, the loss in fissionable material would be prohibitively high.

In the case of controlled nuclear fusion, which essentially takes place in the hydrogen plasma, this would be possible. Neither processes of plasma containment in a magnetic field nor laser induced nuclear fusion are sufficiently attractive as a result of the high amount of required mass and energy for such applications. According to more recent results the use of high-energy particle jets (especially protons) at correspondingly high current intensities offer a better opportunity for possible application (see Chap. 7.5).

A possible method which was intensively and successfully tested in a large research and development program is the development of a high-temperature solid-state nuclear reactor as a high-temperature heat exchanger and use it to heat hydrogen to a high temperature bounded basically by material limits and to expand it through a nozzle [37].

From Eq. (119) and with a perfect gas with the specific heat per unit mass c_p, the specific heat ratio κ and the temperature T_0 and enthalpy $i_0 = c_p T_0$ the following applies

$$I_{sp} = v = \sqrt{\frac{2\kappa}{\kappa - 1} \cdot \frac{R}{M} \cdot T_0} \tag{121}$$

where R is universal gas constant and M the molecular weight.

In the interest of a high specific impulse therefore efforts are made to obtain maximum temperatures in the reactor and minimum molecular weight of the jet mass.

For hydrogen which has the jet mass with the minimum molecular weight ($M = 2$) an exhaust velocity and a specific impulse of about 10 kNs/kg as well as a thrust to weight ratio of the engine which is considerably larger than 1 is obtained for temperatures of about 2,500 K, which can be controlled by the currently known solid state reactors.

In 1955, a program for the development of this type of engine was started in the U.S. under the direction of the Los Alamos Scientific Laboratory and the U.S. Atomic Energy Commission. Later it was taken over by a corresponding NASA-AEC joint program and its main emphasis was placed in the field of solid-state high-temperature reactors as well as liquid hydrogen technology. For this purpose, a test center for nuclear rocket propulsion was established at Jackass Flats in Nevada. In addition, plants

for the storage, handling and operation of the rocket propulsion with liquid hydrogen were established in which the techniques of liquid hydrogen were applied on a large scale for the first time in the U.S.
The reactor and propulsion technology mainly concentrated on graphite moderated reactors with Uranium–Carbide as the nuclear fuel element and external beryllium neutron reflector (see Fig. 50). Here, liquid hydrogen supplied by means of a turbopump at a pressure of about 60–80 bar is used for regenerative cooling of the nozzle and neutron reflector. Next hydrogen enters the reactor zone in its gaseous state and leaves the propulsive unit through the nozzle after additional heating in the reactor.
Apart from the nuclear problems at this early stage considerable developments in liquid hydrogen technology were necessary in order to control the unsteady two-phase flow in pump feed lines and heat exchangers with respect to heat transfer that was necessary at start-up, during thrust, and at shutdown of the engine [38–41]. In the scope of this successful program [42], several rocket engines were tested during 1955 to 1972, which covered a range of capacity from 1,000 MW to 4,000 MW [45]. In the Rover project this consisted of the KIWI reactors [43] and in the NERVA Program its modification to the Phoebus reactor [44] and the

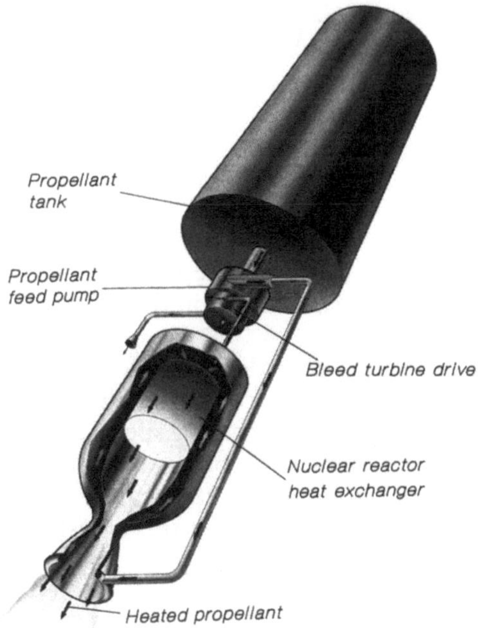

Fig. 50. Schematic of a nuclear-thermal propulsion system (Source: Los Alamos National Laboratory)

Fig. 51. KIWI-B-IA reactor at the test site in Jackass Flats, Nevada (Source: Los Alamos National Laboratory)

NRX reactor. The latter ones were further developed into the final NERVA reactor where this program, which in the next phase as a RIFT program should have lead to its use in a second and third rocket stages (Figs. 51 and 52)[5].

Because of the positive results, considerations for the use of thermonuclear rocket propulsion are not over. Especially small nuclear thermal systems with an output range of about 1 MW and a specific impulse of about 8,000 to 10,000 Ns/kg have the potential to be used for interplanetary missions particularly in the outer planetary belt, because compared to electrical propulsion systems the mass required for the transformation of heat into electricity can be avoided [46].

[5] The author expresses his gratitude to the Los Alamos National Laboratory and especially to Dr. F. J. Edeskuty for the permission to publish Figs. 50–52.

Fig. 52. NERVA engine mock up (Source: Aerojet General Corporation)

References

[1] K. B. Schönberger: Rocket Experiments with Hydrogen and Oxygen. Weltraumfahrt *4*, 80–81 (1950).

[2] J. L. Sloop: Liquid Hydrogen as a Propulsion Fuel. The NASA History Series, BASA-SP-4404, TL 785S58, Stock Number 033-000-00707-8 (1978).

[3] J. Dardare: Propulsion by Liquid Oxygen and Liquid Hydrogen, Review. Pure Appl. Cryogenics *5*, 135–157 (1966).

[4] D. Hurden: Cryogenic Liquids for Rocket Engines. Inst. Refrig. Proc. *55*, 147–165 (1958).

[5] J. L. Sloop: Cold Propellants for Hot Performance. Astronautics *3*, 28–30, 96–97 (1958).
[6] O. S. Silliams et al.: Liquid Rockets in Perspective. Part 1: Developments in the 1960s. Astronautics and Aeronautics, March 1976. Part 2: Propulsion for the 1970s and 1980s. Astronautics and Aeronautics, April 1976.
[7] R. C. Mulready: Liquid Hydrogen Engines. In: Technology and Uses of Liquid Hydrogen (R. B. Scott, ed.), pp. 149–180. New York, Pergamon Press (1964).
[8] F. W. Kirby: Space Shuttle Main Engine Program Status. AIAA Preprint 73-1177, November 1973.
[9] R. Salkeld: Mixed Mode Propulsion for the Space Shuttle. Astronautics and Aeronautics, August 1971.
[10] D. R. Sanchini, F. M. Kirby: The Future Look in Rocket Engines. Amer. Astronaut. Soc. Paper, pres. 11th Goddard Memorial Symposium, March 1973.
[11] E. Sänger, I. Bredt: Über einen Raketenantrieb für Fernbomber. Deutsche Luftfahrtforschung VM 3538 (August 1944), vgl. ferner: Sänger, E., Raketen-Flugtechnik. München, Oldenburg (1933).
[12] W. R. Dornberger: The Rocket Propelled Commercial Airliner. Univ. Minn. Inst. of Techn. Res. Rep. No. 135, November 1956.
[13] R. Salkeld: Orbital Rocket Airplanes, a Fresh Perspective. Astronautics and Aeronautics, April 1976.
[14] P. Bono: Pegasus—A Design, Concept for a V.I.P. Orbital/Global Rocket Transport. SAE National Aeronautics and Space Engineering Meeting, SAE paper No. 760687, Los Angeles, Calif., October 1964.
[15] S. Z. Pickney: Internal Performance Predictions for Langley Scramjet Engine Module. NASA-TM-X-74038 (1977).
[16] R. A. Jones, W. Huber: Toward Scamjet Aircraft. Astronautics and Aeronautics, February 1978.
[17] W. Peschka: Über die Verwendung von atomarem Wasserstoff als Treibstoff für Flüssigkeitsraketen. In: Proc., 9th Int. Astronaut. Congr., pp. 137–147. Wien, Springer (1959).
[18] R. Hess: Atomic Hydrogen. ESRO-TT-42, 1974, siehe auch: Atomarer Wasserstoff, DFVLR-Forschungsbericht, DLR-73-74 (1973).
[19] R. Hess: Atomic Hydrogen Stabilization by High Magnetic Fields and Low Temperatures. In: Adv. Cryog. Eng., Vol. 18, pp. 427–434. New York, Plenum Press (1973).
[20] R. W. H. Webeler: Behaviour of Atomic H in Solid H_2 from 0.2 to 0.8 K. NASA-TMX-71732 (1975).
[21] G. Rosen: Upper Bound on the Equilibrium Concentration of Atomic H in Solid H_2. Phys. Letts. *61A*, 1063 (1977).
[22] W. Peschka, G. Sänger, G. A. Hietkamp: Results of Experiments with Spin-Stabilized Hydrogen and Hydrogen Compounds. J. Physique *41*, 165–176 (1980).
[23] I. L. Kerrebrock, R. V. Meghreblian: An Analysis of Vortex Tubes for Combined Gas-Phase Fission Heating and Separation of the Fissionable Material. ORNL, CF-57-11-3 (1959).
[24] N. Grey: A Gaseous-Core Nuclear Rocket Utilizing Hydrodynamic Containment of Fissionable Material. ARS-Preprint 848/59.
[25] F. Winterberg: Die Erreichung von Ausströmgeschwindigkeiten bis 20,000 m/s durch isotherme Expansion in Kernraketen. In: Proc., 9th Int. Astronaut. Congr. Wien, Springer (1959).
[26] R. G. Ragsdale: NASA Research on the Hydrodynamics of the Gaseous Vortex Reactor. NASA-TN-D-288 (1960).
[27] R. V. Meghreblian: Gaseous Propulsion Reactors. Nucleonics *19*, 95–99 (1961).

[28] D. F. Spencer: The Plasma Core Reactor. NASA-Contract NaSw-6, Tech, Rep. No. 32-1-4, JPL (1961).
[29] W. Peschka: Kernenergie und Wärmeübergang durch Strahlung. Astronautica Acta *8*, 278–302 (1962).
[30] R. A. Gross, K. O. Kessey: Magnetohydrodynamics Species Separation in a Gaseous Nuclear Rocket. AIAA J. *2*, 126–134 (1964).
[31] Anon: Research on Uranium Plasmas and Their Technological Application. NASA-SP-236, 421 pp (1971).
[32] C. Roman Ward: High Temperature UF_6 RF Plasma Experiments Applicable to Uranium Plasma Core Reactors. NASA Contr. NAS1-14329, Contar. Rep. 159159, UTC (1979).
[33] G. H. McLafferty, H. E. Bauer: Studies of Specific Nuclear Light Bulb and Open-Cycle Vortex Stabilized Gaseous Nuclear Rocket Engines. NASA-CR-1030 (1968).
[34] G. M. McLafferty: Survey of Advanced Concepts in Nuclear Propulsion. J. Spacecraft *5*, 1121–1128 (1968).
[35] M. L. Thorpe: Radio Frequency Plasma Simulation of Gas-Core Reactor. J. Spacecraft *6*, 923–928 (1969).
[36] W. Peschka: Hochtemperatur-Energiesysteme unter Verwendung von Plasmareaktoren und induktiven magnetoplasmadynamischen Wandlern. DLR-FB-67-59 (1967).
[37] R. W. Bussard, R. D. De Lauer: Nuclear Rocket Propulsion. New York, McGraw Hill (1958).
[38] F. J. Edeskuty: Liquid Hydrogen in Nuclear Rocket Testing. LA-DC-7170, Los Alamos Sci. Lab., 32 pp. NTIS (1965).
[39] E. F. Hammel: Cryoengineering in the Nuclear Rocket Program. In: Adv. Cryog. Eng., Vol. 9, pp. 11–19. New York, Plenum Press (1964).
[40] F. J. Edeskuty, J. B. Henshall, J. R. Bartlit: Cryogenic Applications in the Nuclear Rocket Program.
[41] W. E. Keller: Worldwide Cryogenics—U.S., Cryogenics at the Los Alamos Scientific Laboratory. Cryogenics *20*, 547–556 (1980).
[42] L. C. Corrington: The Nuclear Rocket-Program—Its Status and Plan. J. Spacecraft *8*, 465–470 (1960).
[43] R. E. Schreiber: Kiwi Tests Pay Way to Rover. Nucleonics *19*, 77–79 (1961).
[44] S. V. Gunn, C. Dunn: Feed Systems for Phoebus Reactor Experiments. J. Spacecraft *7*, 769–777 (1969).
[45] W. E. Durkee, F. B. Damerval: Nuclear Rocket Experimental Engine Test Results. J. Spacecraft *7*, 1397–1401 (1970).
[46] W. Peschka: Neue Energiesysteme für die Raumfahrt. Munchen, Goldmann (1972).
[47] F. Suttrop: Überschallverbrennung; Zweck und eigene Versuchseinrichtungen, WGLR-Jahrbuch, pp. 366–376 (1963).

Liquid Hydrogen as Fuel 6

For more than a century fuels used for transportation have been almost exclusively based on hydrocarbons from oil. Today, transportation without hydrocarbon based fuels is almost inconceivable. Only in times of shortage for instance, in both the World Wars or as a consequence of the exertion of political influence by the—relatively small number of—non-European oil exporting countries does the general public become aware of the enormous dependency on oil. Paradoxically intensive efforts in the middle of the last century resulted in hardly any use for oil other than petroleum lights and for medicinal purposes.

The development of color chemistry, the invention of the four-stroke internal combustion engine by Nikolaus Otto in 1876 as well as the first automobiles (Karl Benz 1885, Mannheim; Gottlieb Daimler 1886, Cannstatt) fundamentally changed this situation. The high heat value as well as the obviously simple handling soon made gasoline (the German name "Benzin" originates from the English benzene, the name for benzol or low boiling hydrocarbons containing benzol) indispensable for the automobile. With the successful construction of a light and efficient gasoline fueled airplane engine in 1903 by the Wright brothers, Orville and Wilbur, hydrocarbons found their way into air transport while with the development of heavy oil engines, including the Diesel engine which was originally designed for the combustion of coal dust, hydrocarbons were utilized in stationary engines and later especially in the field of commercial vehicles. The invention of the jet engine as well as the development of the high-speed diesel engine and the kerosene-based fuels the diesel engine required complete this picture.

The impending exhaustion of the earth's oil deposits, which is increasingly more foreseeable, forced consideration of the development of alternative fuels, drawing parallels to previous times of shortage as in Germany during World War II for example where, aside from coal and hydrogen, synthetic hydrocarbon based fuels were used. The considerable amount of fossil fuels which have to be replaced in the future and the economical and ecological questions concerning their production make comparisons to these earlier situations appear unrealistic.

It must be pointed out that a considerable amount of the energy set free with the combustion of hydrocarbons or alternatives like methanol or ethanol is derived from the combustion of the hydrogen that they contain. Therefore, in a way, hydrocarbons and alcohols can be considered as chemical storage for hydrogen. However, the carbon which serves as the "storage material" is responsible for several environmental problems in the production and combustion (CO_2 and CO emission, soot emission as well as carcinogenic substances like benzo[*a*]pyrene in the exhaust gas of diesel engines for example as well as the production of nitrogen oxides). A transition to fossil fuels with gradually increasing amounts of hydrogen can reduce these environmental problems. The increased use of liquefied petroleum gas (LPG) (i.e., propane or butane) and liquid natural gas (LNG) as fuel is in line with this thinking. According to Marchetti [1–5] this shows a tendency to replace current hydrocarbons with pure hydrogen as in the past when wood was replaced by coal as a consequence of shortage of raw material.

However, it seems unlikely that hydrogen will merely be developed as an energy carrier for transportation needs because transportation consists of only a small part of the total consumption of the earth's petroleum. Apart from special applications this will change when hydrogen, commonly used in future energy systems, also begins to be utilized in other applications such as heating.

Moreover, hydrogen as a candidate among several alternate fuels, has real possibilities to be used in a gradually increasing manner when, aside from the problem of production—which hydrogen has in common with all synthetic fuels—R&D addressing the establishment of the required infrastructure is finished. Current activities regarding the application of hydrogen in the field of civil aviation, highway and rail transportation must be considered from this perspective.

6.1 Air Transport

As an alternative fuel for air transport, liquid hydrogen must be compared to contemporary fuels and other alternative fuels with regard to their characteristics properties. The energy consumption in air transport is currently of minor importance compared to total energy consumption. The introduction of alternative fuels can only be triggered by concrete, compelling arguments. The increase in oil prices in anticipation of the reduction in usable oil reserves in the next century is just such an argument.

6.1.1 Limiting Factors

With respect to fuel costs, which already average about 30% of the direct operational costs, the question of economical alternative fuels for air transport has received attention for some time. Nevertheless, it cannot be assumed that a worldwide reorientation to other aircraft fuels will occur in the near future due only to higher fuel costs, because the resulting conversion for air transport and large industrial companies requires an enormous capital investment. Considering that the development of new aircraft requires about ten years and that aircraft are used for about twenty years, the costs for fuel in the 90s must be taken into account in current plans for the next generation of aircraft. Thus, certain constraints exist for the next generation of aircraft. From this point of view hydrogen will be used as an alternative fuel on a large scale at the earliest two generations from now, i.e., around 2010. Therefore it is necessary to perform timely R&D on an international scope because of the size of the task which must be accomplished. In this respect air transport is especially well suited for the pioneer, large-scale use of hydrogen due to its international network, the logistics potential of current airports as well as the ability of the air transportation industry to accomplish difficult management tasks of large technical developments in a relatively short time [6–10].

Table 18 is a comparison of liquid methane and liquid hydrogen, which are considered as fuel alternative to synthetic aircraft fuel based on hydrocarbons (Synjet). Hydrogen's advantages are its high lower-heat value per unit of mass as well as its high specific heat. In contrast hydrogen has the smallest lower heat-value per unit of volume compared to other fuels. Liquid methane lies somewhere in the middle, between contemporary aircraft fuel, specifically Synjet, and liquid hydrogen. On one hand its calorific characteristics are closer to those of contemporary aircraft fuel while on the other hand it is comparable to liquid hydrogen as a cryogenic liquid.

Table 18. Comparison of fuels for air transport

	Synjet[a]	LCH_4	LH_2
Composition	$CH_{1.93}$	CH_4	H_2
Molecular weight	168	16.04	2.016
Density at boiling point (g/cm^3)	0.8	0.423	0.071
Boiling point (K)	440–539	111.7	20.27
Melting point (K)	223	91	13.8
Specific heat (J/gK) at boiling point	1.98	3.5	9.69
Heat of condensation (J/g)	360	510	446
Lower heat value (kWh/kg)	11.9	13.8	33.3
(kWh/l)	9.5	5.8	2.36

[a] For simplicity sake the same characteristics will be assumed for Synjet and normal aircraft fuel (Jet A).

Table 19. Comparison of LH_2 and Jet A as a function of the aircraft size (from [15, 22])

	Size: number of seats/distance (km)		
Total weight	130/2,800	400/10,200	400/19,600
LH_2 (kg)	45,500	169,000	250,000
Synjet (kg)	48,500	232,000	476,000
(1-LH_2/Synjet) (%)	6.1	27.3	47.6

The goal of several studies and tests carried out or sponsored by NASA was to determine the influence of the characteristics of these fuels on aircraft design [11–18], also see [19–21][1].

Table 19 is a comparison of LH_2 and Synjet (specifically Jet A) for aircraft in the upper subsonic range ($M > 0.85$) for various ranges and sizes. The latter is defined as the ratio of the number of seats to range (see [15, 22, 23]). Thus liquid hydrogen considerably reduces the takeoff weight for large-capacity aircraft in the subsonic range for identical distances and payloads. This is particularly evident at the extreme distance of 19,600 km. However, it must be emphasized that even for short-range aircraft liquid hydrogen provides a slight advantage compared to Synjet and Jet A [15].

As a result of the low total mass, shorter takeoff runs and steeper climbs (lower noise level) and/or reduced engine thrust requirements (less noise produced) are obtainable. Figure 53 is a comparison of the required amount of primary energy in metric tonnes of coal (t SKE) equivalent of Synjet and LH_2 according to [15, 22]. In order to make a comparison on a similar basis, a production process based on coal was also assumed for hydrogen. It is the steam-iron process which was briefly described in 2.1. Liquid methane is not included in this diagram. However it is evident that the data for methane lies between liquid hydrogen and Synjet, and closer to the Synjet values.

The estimation of the required amount of primary energy is based on assumptions for the thermal efficiency of the production of fuel from coal according to [16, 17]. The production of Synjet was based on the CSF-process (Consolidated Synthetic Fuel Process) [15, 23].

The savings on primary energy over long distances is obvious but in accordance with Table 19 under the assumptions being made here, liquid hydrogen is slightly superior even for shorter distances.

Since, under these assumptions, liquid hydrogen provides advantages primarily in the upper subsonic range particularly over long distances,

---

[1] The author would like to express his gratitude to the Lockheed California Company and especially G. D. Brewer for support and consultation.

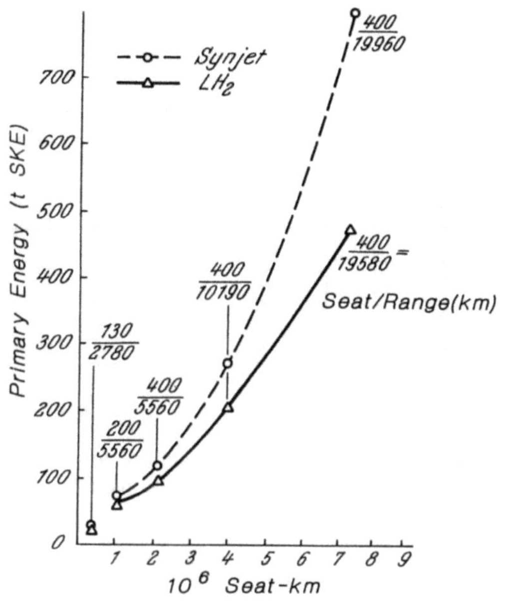

Fig. 53. Amount of primary energy in equivalent metric tonnes of mineral coal for Synjet and liquid hydrogen as a function of the transportation capacity in 10^6 seat-km (from [15, 22])

hydrogen fuel should be even more advantageous in the supersonic range, which, however has not been further examined. For comparison Fig. 54 shows the fuel portion of the total mass of a large-capacity aircraft with a payload typical of a high subsonic range aircraft and a Mach 3 supersonic aircraft over similar distances according to [19, 20]. Although there is currently no economic civilian supersonic air traffic, applications in the future for long intercontinental routes cannot be excluded due to the advantages of liquid hydrogen. It has to be taken into consideration that, as a result of the considerably lower fuel weight at takeoff, steeper climbing paths can be attained. As a result of the combustion characteristics of hydrogen flight altitudes for supersonic aircraft (above 30,000 m) considerable reduction in environmental pollution can be attained with a substantial reduction of sonic boom.

Several studies were carried out regarding the comparison of costs between the different fuels which showed that under current conditions liquid hydrogen is not yet economical as compared to Synjet (see [15, 29, 24]). A comparison of the direct operational costs in cents/seat-km as a function of the fuel costs is informative[2]. Figure 55 shows the relationship for a

[2] Within the scope of the data presented here a comparison of the different costs is of primary importance whereas the values themselves are of secondary importance.

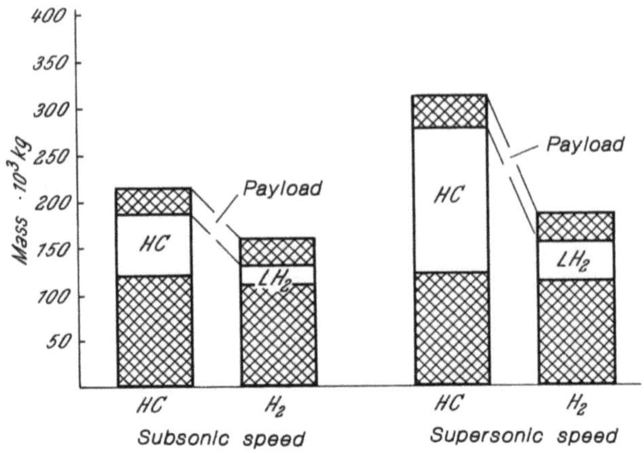

Fig. 54. Comparison of the payload and fuel fraction for aircraft at subsonic and supersonic speed. *HC* fuels based on hydrocarbons, LH_2 liquid hydrogen

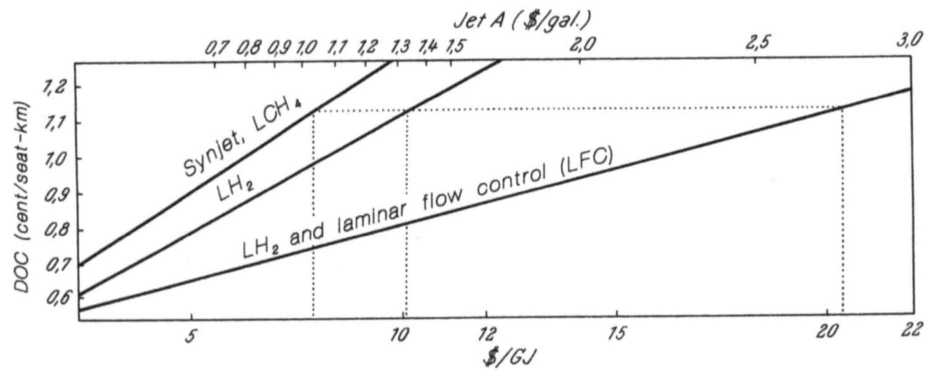

Fig. 55. Direct operating costs (DOC) as a function of the fuel costs for a large-capacity aircraft (400 passengers) with $M = 0.85$ and 10,200 km cruising range (from [22])

large-capacity aircraft according to [22, 23]. Assuming a price for Synjet of $1.08/Gal, the price paid for Jet A in March 1982 on the North Atlantic route, the equivalent energy costs are $8/GJ based on 1.2 cents/seat-km. Liquid methane could be competitive with Synjet below that price whereas airports without liquid hydrogen (middle curve) may not be more expensive than $11/GJ. According to [22, 23] the production price for Synjet, LCH_4 and LH_2 from coal delivered to the airport is $8.04/GJ, $8.39/GJ and $10.45/GJ respectively.

According to [24] under similar assumptions a price of $6.2/GJ is estimated for Synjet from oil shale. However, it must be assumed that the amount of effort to solve the accompanying environmental problems and the problems of the production process should be included in this price.

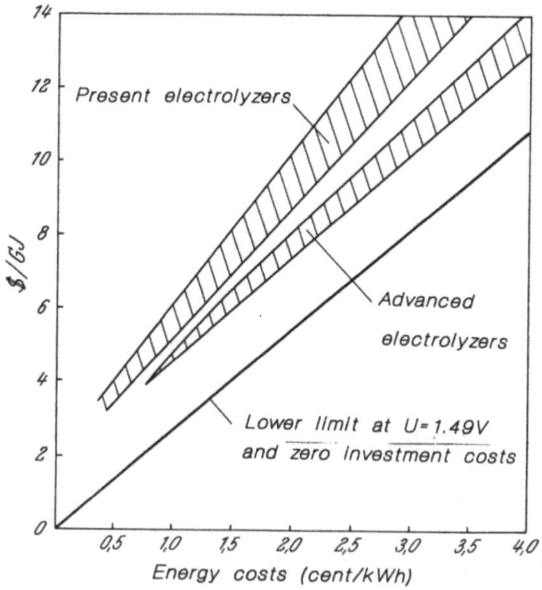

Fig. 56. Costs for gaseous electrolysis hydrogen based on the costs for electrical energy in 1982 (from [23])

The current production price of liquid hydrogen, at the rate being paid within the scope of the U.S. space projects (production of GH_2 via steam-reforming) amounts to about \$18/GJ to \$25/GJ (\$0.60 to \$1.00/Gal). The actual liquefaction costs of large-scale plants (30 metric tonnes per day or more) are about \$5/GJ. For future large-scale plants of approximately 200 metric tonnes LH_2 per day according to [16, 17] values between \$7/GJ and \$11/GJ are given for LH_2 including transportation costs.

Since hydrogen production from coal represents only a temporary solution [24, 26], the acceptance of future electrolytic hydrogen production is more feasible. According to [26] and [23] Fig. 56 shows the current and future costs for gaseous hydrogen dependent on the cost for electrical energy. With electrical costs of about 20 mills/kWh (approx 4.5 Dpf/kWh), which corresponds to the eastern provinces of Canada, costs in the range of about \$16/GJ will be obtained and after about 1990, with improved electrolysis processes, costs of about \$14/GJ will be obtained if liquefaction costs and transportation costs (approx \$2–3/GJ) are included.

If European data is taken as a basis according to [25] costs of about \$18/GJ (45 Dpf/Nm^3) are obtained for hydrogen gas produced by electrolysis[3]. At current prices in the U.S. for liquefaction, transportation and storage of liquid hydrogen—a corresponding European market does

[3] If hydrogen gas is delivered in common pressure bottles for comparison sake a price of approximately DM 5/Nm^3 is obtained, which is primarily determined by the high transportation costs of the "container" (see page 89).

not yet exist—costs of about \$25–27/GJ are obtained. These costs can be reduced to about \$23–25/GJ as a result of improved electrolytic procedures in the 1990's.

The higher price level for LH_2 in Europe partially caused by higher costs of electricity, could stimulate LH_2 imports from Canada and other counteries with cheaper hydroelectric power like Brazil, Africa or Greenland (using the run off from melting glaciers) for example within the scope of a future hydrogen economy and an accompanying energy market (see Chap. 8).

Based on those costs, LH_2 would first be more economical when Jet A and Synjet reach a price level in the range of about \$2–3/Gal, specifically two to three times higher than current prices. This type of cost increase does not seem very likely. Even if it does, in fact, occur it is unlikely that production costs for hydrogen will remain unaffected by it [26–29].

Considerable reduction of direct operational costs (cents/seat-km) is possible for all fuels through the reduction of aerodynamic resistance of the aircraft and the accompanying reduction of fuel consumption. The long known procedures for boundary layer precompression and boundary layer control [30, 31] are technically complicated and additionally require a relatively large amount of energy as a result of the large air mass which is involved. Aside from the development of improved low resistance profiles, procedures for promoting laminar flow in the boundary layer (LFC—Laminar Flow Control) provide even greater potential for future developments. Due to its low temperature and considerable heat sink capacity, liquid hydrogen enables laminar flow control without additional energy expenditure by means of cryogenic cooling of portions of the outer surface. This procedure has been examined only in theory and was confirmed in principle by a few very simple wind tunnel experiments [32, 33], but requires further intensive testing including the development of appropriate wind tunnel experimental techniques. The expected reduction in air drag and fuel consumption based on current results is shown in the lowest curve, LH_2–LFC, in Fig. 55. Thus, under these assumptions LH_2 would be competitive below a price level of about \$21/GJ based on a price of \$1.08/Gal (\$8/GJ) for Jet A and Synfuel.

The possibilities shown here with the considerable potential for development in the field or cryogenic aerodynamics should therefore offer sufficient incentive to motivate appropriate research and convince representatives of the scientific community of the necessity of establishing an international R&D program.

6.1.2 Current Stage of Development

After intensive preliminary trials [11, 12] at the NACA Lewis Flight Propulsion Laboratory (currently the NASA Lewis Research Center), an

Experimental Hydrogen - Convertible Aircraft

IN-TESTING AT NACA LEWIS, CLEVELAND 1956-57

Source: NASA-Lewis

Hydrogen System For Experimental Aircraft

TESTED IN 1956

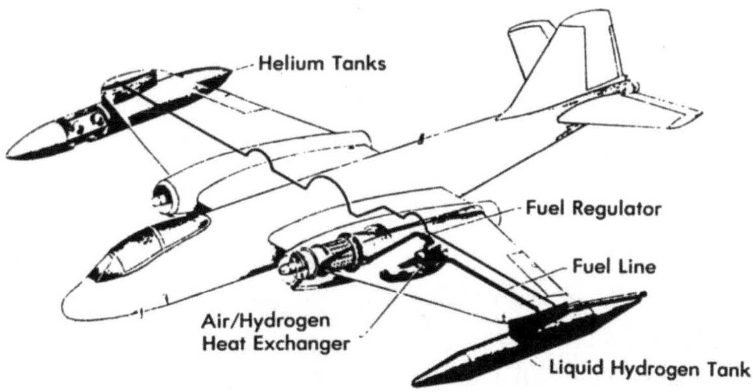

B-57 "Canberra" Engine Convertible To Hydrogen Source: NASA-Lewis

Fig. 57. B57-Canberra. NASA liquid hydrogen test aircraft. This was the first successful liquid hydrogen aircraft (NASA-Lewis Research Center)

in-flight test of a twin-jet B57-Canberra aircraft (Fig. 57) was carried out (see [18]) in 1955/56 under the auspices of a project on hydrogen propelled aircraft. Tanks were installed on both wingtips, one for liquid hydrogen and one for gaseous helium for the pressurization of the LH_2 tank. During the flight at high altitude (typical flight conditions $M = 0.75$, altitude 16,000 m) one of the two J65 engines was switched from normal operation with JP4 fuel to operation with hydrogen. In this process the JP4 fuel regulator was modified for hydrogen operation. Liquid hydrogen was warmed and vaporized in a forward facing heat exchanger and supplied to the engine. In a later modification, instead of pressurizing the liquid hydrogen, an externally driven reciprocating pump was mounted in the tank. The J65 engine performed successfully during the test flight under hydrogen operation without any significant problems (see [18]).
In 1956–58 a test program was started at Pratt and Whitney sponsored by the U.S. Air Force. Its goal was the development of a hydrogen jet engine for the Lockheed CL400 reconnaissance plane (Mach 2.5, 30,000 m altitude, 4,000 km range) [20, 34]. A J57 jet engine was adapted for hydrogen operation (Fig. 58) as a forerunner of the Pratt and Whitney 304 "Suntan"

Fig. 58. Comparison between a J57-Jet engine with conventional fuel (lower part) and the modified version for hydrogen. The considerable reduction in total length due to the shorter total length of the combustion chamber as a result of hydrogen's very good mixture behavior can be seen. (Source: Pratt & Whitney)

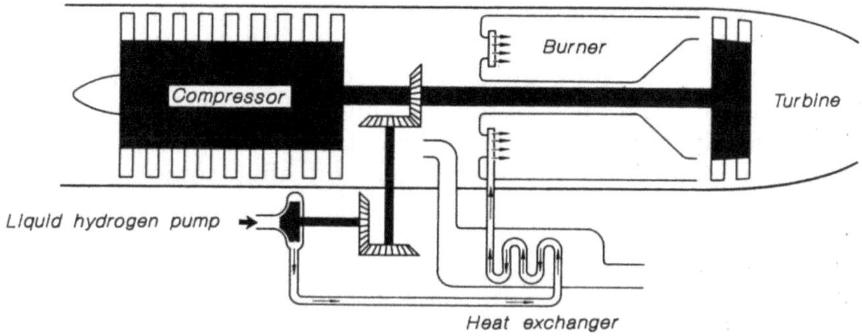

Fig. 59. Schematic of the B57-Jet engine modified for hydrogen (from [34])

jet engine. These modifications were very basic. Liquid hydrogen was led through an air-hydrogen heat exchanger serviced by a compressor via a single-stage radial pump. There it was warmed, vaporized, and forced into the combustion chamber via simple gas injectors (Fig. 59). As should continue to be the case, liquid hydrogen repeatedly proved to be a bettter fuel than initially expected. Hydrogen's wide range of flammability with air enabled stable combustion chamber operation far beyond the limits of hydrocarbons. The mixing abilities of hydrogen with the combusting air proved to be so good that a considerably shorter combustion chamber could be used compared to operation with hydrocarbons. This resulted in a marked reduction in the construction length and the weight of the jet engine (Fig. 58).

For the Pratt and Whitney 304 jet engine, which was designed for dependable operation at high altitudes (Mach 2.5, 33,000 m) with a high thrust-to-weight ratio and low specific fuel consumption, gaseous hydrogen was used to fuel the turbine (Fig. 60). Liquid hydrogen which was supplied via a centrifugal pump (1.8 kg/s, 70 bar) was heated to about

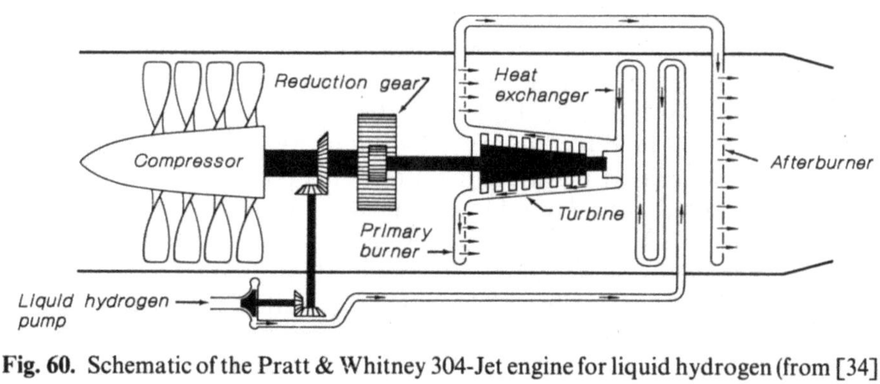

Fig. 60. Schematic of the Pratt & Whitney 304-Jet engine for liquid hydrogen (from [34])

1,000 K in a heat exchanger which was located after the turbine. After work by the turbine (12,000 HP shaft power, 25,000 rpm, 18 stages) it was partially mixed with the combustion air and ignited while the remainder of the combustion air was supplied to the afterburner [18, 34].

Seventeen months after the program had started this jet engine could be successfully evaluated on the test stand and after some modifications delivered design data within five months. This was remarkable since liquid hydrogen technology was in its initial stage and many new technical problems still had to be solved. This project was continued until about the middle of 1958 at a time when it became clear that the capabilities required for this type of high altitude jet engine were also satisfied by the J58 engine using JP4 as fuel for which there were no supply problems as opposed to the existing limited production capacity for liquid hydrogen. Therefore, the Lockheed CL400 project was stopped as it became evident as a result of technology advances made with the titanium technology, that aircraft with this speed were no longer developed based on the aluminum alloys which were planned to be used in the CL400.

As a result of the oil crisis in 1973 and the increasing inquiry for alternate fuels, hydrogen again began to receive attention. Mutual American and European activities at the end of the seventies which attempted to establish an international project unfortunately did not obtain enough public or political interest. Then in the first six months of 1988 a decisive but relatively unnoticed breakthrough took place in the Soviet Union in which a three-engine Tupolev 155 completed its first 21 minute flight partially with LH_2 after 9 years of continuous R&D. Since 1988 studies have been started in Germany with the goal of a version of the Airbus A320 converted to LH_2 for short distances.

With regard to the advancement in jet engine technology in the subsonic range in 1970–1980, the following situation was obtained for hydrogen according to [35, 36]:

For the basic process as it is currently implemented in bypass engines, the transition from Jet A and JP4 to hydrogen has definite advantages due to the fuel characteristics of hydrogen. In the case of more complex processes, including compressor intercooling which utilizes the cryogenic characteristics of hydrogen as well as regenerative heat exchange at the turbine exit in combination with turbine operation, further improvements are obtained which can be offset however by the increased complexity and the accompanying increased maintenance problems [65–74, 79].

The ideas presented thus far refer specifically to aircraft in the subsonic range and their possibilities for hydrogen application. It should be pointed out that hydrogen's characteristics make it a perfect candidate for supersonic applications. Its high heat value per unit of mass, its wide range of flammability in H_2-air mixtures, the quick ignition as well as the heat sink capacity associated with the high specific heat are some of those characteristics. Before it enters the combustion chamber, hydrogen can be warmed in the engine from 20.4 K to a maximum of about 500 °C and can absorb about 3.6 kWh/kg whereas methane can absorb about 0.75 kWh/kg and conventional aircraft fuel can only absorb about 0.2 kWh/kg at the maximum allowable warming to about 200 °C [37].

Within the hypersonic range ($M > 5$, stagnation point temperature about 1,200–1,300 °C), hydrogen can provide very effective cooling of the engine—which is not possible with air—as well as the critical parts of the outer surface, like the front section of the fuselage as well as the front edge of the wings for example. Furthermore, only ramjet engines with supersonic or external combustion are considered as air-breathing engines at this velocity, for which only hydrogen is suitable.

6.1.3 Technical Concepts and Specific Problems

Supersonic Range. Initial applications of hydrogen in the supersonic range will be in the military field, whereas in space flight the development of carrier systems with combination engines is foreseeable. In both cases the required exploitation of hydrogen's heat sink capacity and the technical developments involved will open up additional application fields in aviation in the low and medium supersonic range as well as in conventional stationary energy technology. Cooling gas turbine engine blades with hydrogen is particularly important because it increases thermodynamic efficiency, as well as the turbine shaft power as a result of the higher allowable temperature of the gas admitted into the turbine. On one hand this will lead to more compact and lighter aircraft engines and on the other hand in stationary energy technology it will result in more efficient energy systems with increased conversion efficiency (see [37–45]).

With respect to the use of liquid hydrogen at low to medium supersonic range, only a small number of project studies has been carried out for civil applications, especially for Mach numbers ranging from the limits of $M = 2.2$ to $M = 2.7$ (see [38–40, 79]). A comparison of the characteristics of LH_2, Jet A and Synfuel for a supersonic carrier at Mach $M = 2.7$ is shown in Table 20 (from [15]).

Although data in Table 20 are based on a project study from 1975 and improved concepts for supersonic carriers with Jet A, and not LH_2, are being developed in ongoing research, it can be generally said that LH_2-fueled supersonic carriers will offer shorter take-off runs with a considerable reduction in flight noise including the "sonic boom" as a result of the very low fuel weight for the same range and the lower engine thrust requirement compared to Jet A engines.

It can be inferred from the current state of affairs that start-up of R&D projects, which will lead to a second generation of supersonic aircraft,

Table 20. Comparison of supersonic aircraft LH_2—Jet A (234 passengers, 7780 km range)

		LH_2	Jet A	Jet A / LH_2
Takeoff mass	(kg)	179,000	345,000	1.93
Empty mass	(kg)	111,000	144,000	1.29
Fuel	(kg)	38,000	150,000	3.88
Wing area	(m^2)	739	1,031	1.39
Wing span	(m)	34.4	40.7	1.18
Fuselage length	(m)	103.7	90.5	0.87
Lift to drag ratio (cruise)		7.42	8.65	1.17
Specific fuel consumption (cruise) (kg/h)/N		0.059	0.153	2.59
Thrust per engine (N)		$2.35 \cdot 10^5$	$3.86 \cdot 10^5$	1.64
Takeoff distance (m)		2,377	2,893	1.22
Fly-over Noise (EPNdB)		102.2	108.0	1.057
Sonic Boom Overpressure (N/m^2)		63.2	89.1	1.41

cannot be expected before the late 90's. The current advantages of liquid hydrogen compared to hydrocarbons will therefore at the latest at this time require the use of liquid hydrogen as fuel for supersonic aircraft.

Subsonic Range. Several R&D projects were carried out with the intention of converting existing aircraft to liquid hydrogen whereas direct designs for liquid hydrogen with all the advantages of this fuel were reserved for a follow-on phase. Primary emphasis in future developments will be in the areas of LH_2 fuel tanks, fuel conditioning systems, jet engines, and cryogenic laminar flow control as well as in ground facilities and safety.

Fuel tanks: Conventional fuel is currently stored inside the wings in sack-like containers which can adjust to the available space (Fig. 61). In contrast cryogenic fuels like LH_2 and LCH_4 must be stored in rigid, thermally insulated containers. In the interest of low weight and low boil off losses, an attempt is made to attain minimum surface area and maximum volume. It is not possible to install these tanks, which have to be three times the volume of those for Jet A for the same distance (see Table 19), inside the wings with the current relatively flat wing profiles. The installation of covered external tanks above the wings or on the wing tips in a manner similar to NASA test aircraft (see Fig. 54), was repeatedly proposed. Aside from an considerable increase in air resistance, this installation places an increased static load on the wings which is not desirable, particularly when the aircraft is on the ground and no lift is being provided.

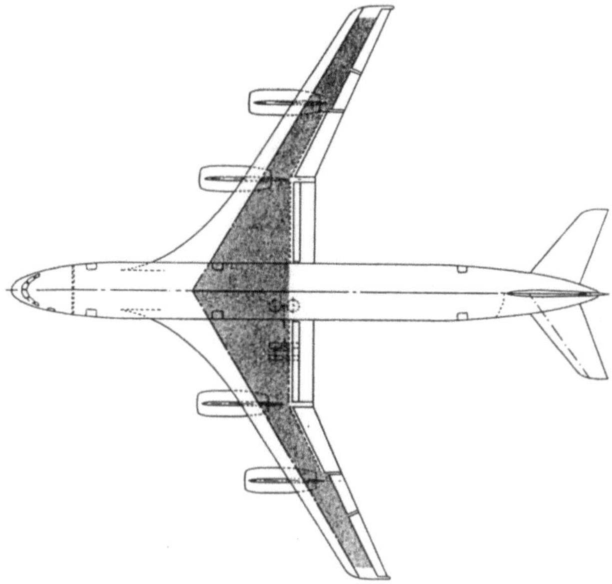

Fig. 61. Location of conventional fuel tanks (from [47])

The same can be said for the installation of the tank alongside the fuselage above or below the passenger compartment. With an installation above the passenger compartment, passengers are not safe during an accident on the ground as a result of possible liquid hydrogen leakage from the necessarily long, relatively flat liquid hydrogen tanks. With an installation below the passenger compartment the passengers are directly endangered by the rising gaseous hydrogen which in all probability will catch fire. This situation is also not alleviated by the installation of several single tanks—a disadvantage in weight—which incidentally excludes integration with the fuselage structure.

Several studies (see [46]) resulted in a preference for the installation of the tanks in a lengthened fuselage in which air resistance is only minimally increased. The installation of two tanks shown in Fig. 62, one each in the front and rear, proved to be a good alternative with regard to safety and direct operational costs. To complete the picture it must be mentioned that the ideas presented here are even more applicable to supersonic aircraft.

With regard to structure as well as thermal insulation of the fuel tanks two concepts are practical according to [47–49, 79]. These are sealed rigid foam insulation as well as insulation using hollow microspheres [50].

Rigid foam insulation is directly applied to the outer wall of the tank which is either embedded in the structure of the fuselage or is designed as an integral part of it. Outside the rigid foam insulation (Fig. 63) is an airtight

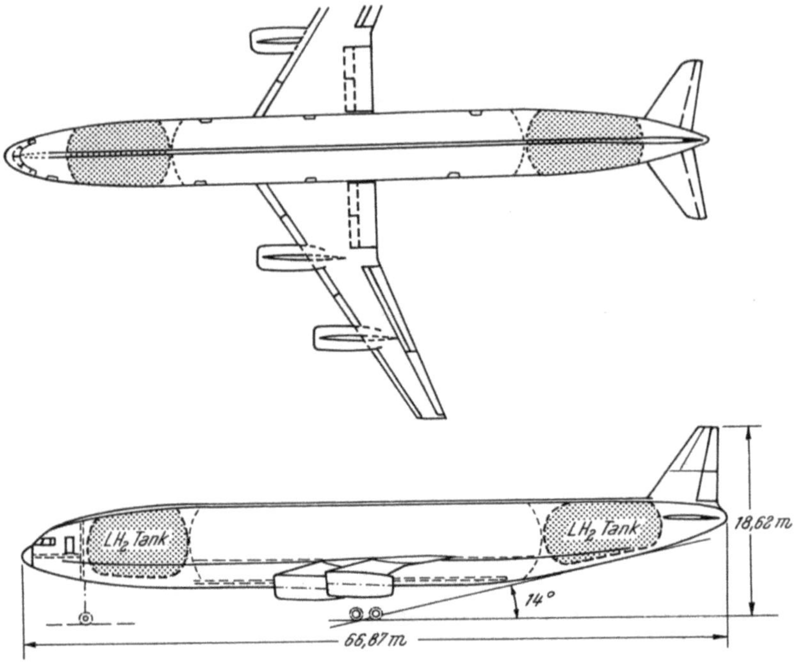

Fig. 62. Storage of liquid hydrogen in the fuselage (from [47])

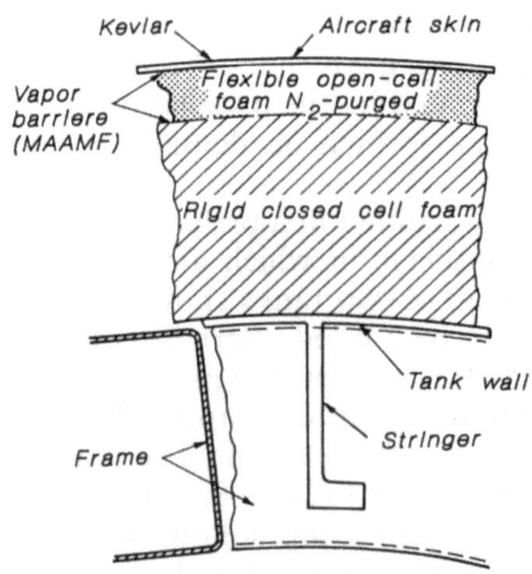

Fig. 63. Schematic of the sealed rigid foam insulation for a LH_2 fuel tank integrated with the fuselage (from [47])

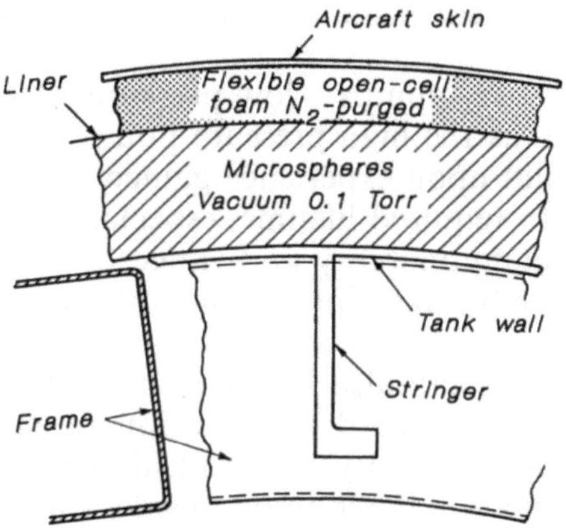

Fig. 64. Schematic of the glass microsphere insulation for a LH_2 fuel tank integrated with the fuselage (from [47])

covering which consists of a sandwich-like combination of Mylar-aluminum/aluminum-Mylar foil (MAAMF) in order to avoid the condensation of external air in the cold insulation. Over that is an unsealed flexible soft foam which functions as a cushion. This soft foam is connected to the outer skin of the aircraft through another intermediary airtight sandwich foil. The internally and externally airtight soft foam ring is filled with gaseous nitrogen to ensure that it is inert.

With glass microsphere insulation (Fig. 64), hollow spheres of boron silicate glass (diameter up to 100 μm and surface thickness of about 5–10 μm) take the place of the rigid foam insulation. They are located in an annular gap, formed on one side by the tank and on the other side by a thin stainless steel wall, which is under a constant vacuum of about 0.1 Torr. Over that is a nitrogen filled soft foam cushion similar to the rigid foam insulation, which is additionally sealed on the outside with sandwich foil (MAAMF) and supports the outer skin of the fuselage. Within the tank section the outer skin of the fuselage is constructed of Kevlar. On one hand the glass microsphere insulation is more effective than the rigid foam insulation and provides a reduction in the thickness of the insulation of about 100 to 70 mm. On the other hand it requires more effort and is more expensive. Although several other tank structures and their insulation can be used in aviation [20], the two configurations described here are primarily considered for future development in this field for the best possible integration with the structure of the fuselage.

Fuel Conditioning System: This includes pumps for the supply of the 0.15 MPa hydrogen in the tanks to the combustion chambers of the jet engines (30–50 bar) and the cryogenic feed lines and valves. Using cryogenic laminar flow control a heat exchanger for cooling the outer skin is required [22, 23, 77].

The development of LH_2 pumps—low pressure boost pumps[4], integrated into the lines as well as the high pressure pumps (40 bar) installed on the jet engines—can build upon experience gained from rocket technology in which, with regard to required lifetime, especially the bearings (about 10^4 hours), reliability, maintainability and range of operability of mass flow rate and pressure must be taken into consideration in aviation requirements [48].

The cryogenic fuel lines can also be based on experience gained in space flight. In particular with regard to vibration resistance and mechanical strength including high flexibility without leakage (accident situations) two types of thermal insulation come into consideration. They are the double wall design with vacuum jacket (internal line of stainless steel, approx. 25 mm diameter, outer pipe aluminum or stainless steel, approx. 50 mm diameter) or the double wall design and gaps filled with sealed polyurethane foam (outer diameter approx. 100 mm) [47].

Jet Engines: With regard to future developments Jet engines for subsonic speeds can be developed based on currently existing concepts, where turbine blade cooling (use of the heat sink capacity of hydrogen via appropriate heat exchangers), modified combustion chambers and fuel injectors will be the main points of emphasis [51–55] with respect to further reduction of NO_x-emissions. Furthermore, with respect to the transition phase from hydrocarbons to hydrogen there is interest in jet engines which can be operated either with hydrocarbons [62] or with hydrogen.

In general it remains to be seen if subsonic jet engines using conventional fuels have reached a technical level which will permit further improvement in lifetime, reliability, environmental pollution, operating characteristics and economy only in minimal amounts and then only within the scope of sufficiently extensive research [79].

Laminar Flow Control Through Cryogenic Cooling: It is known that the laminar boundary layer shows the lowest aerodynamic resistance with no separation along the wing and fuselage profile which is the ultimate goal of "laminar profile" shaping. However, boundary layer separation at high pressure gradients cannot be avoided without extra precautions.

One possible means of avoiding this separation is maintaining a vacuum at the boundary layer, initially proposed by L. Prandtl [30]. Since reducing aerodynamic resistance not only inhibits boundary layer separation but also shifts the laminar-turbulence transition point of the boundary layer as far as possible downstream, interest in laminar control methods is increasing [31]. Tests have shown that continuous boundary layer control

---

[4] Recent development of LH_2-Booster pumps for automotive vehicles (see p. 195) also may be of importance here.

with its associated effect on velocity profiles is particularly well suited for this application. Through the resulting increased stability of the laminar boundary layer, the transition point of the boundary layer is shifted downstream in which the critical Reynolds number for the boundary layer transition from laminar into turbulent state can be increased by one to two orders of magnitude. The power requirement for providing the vacuum is critical because here the advantage gained by the reduction in resistance can be diminished or even totally negated.

According to [32] boundary layer cooling functions in a manner similar to boundary layer control. The theoretical and experimental tests carried out in the past—mostly on the flow along flat plates—did not have much of an impact due to the lack of a practical means of cryogenic cooling of large surfaces. This situation changed however as soon as liquid hydrogen was seriously considered for use as aircraft fuel and when tests showed that the required cooling to about 155 K of approximately 75% of the upper surface of the wing and 20% of the front section of the fuselage as well as the engine pods could basically be accomplished in this manner [33].

The concept outlined in [15] entails the use of gaseous nitrogen as a coolant which flows through the cooling vanes and is cooled by liquid hydrogen via a heat exchanger.

According to [15, 33] tests demonstrated considerable savings in fuel. This results in a reduction in the takeoff weight of $169 \cdot 10^3$ kg to $156 \cdot 10^3$ kg for a large capacity aircraft such as in Table 20 for example as well as a reduction in fuel from $21.6 \cdot 10^6$ kg LH_2 to $15.3 \cdot 10^3$ kg and a reduction of the direct operational costs which become evident from Fig. 55 (curve LFC).

A direct prerequisite for the use of cryogenic laminar flow control is a cruising altitude for most of the flight above about 8,000–10,000 m because the condensation of the water vapor in the air and ice formation on the outer skin is not expected there.

Since cryogenic laminar flow control with liquid methane does not offer any advantages due to low specific heat and higher boiling temperature, this procedure will be used exclusively with liquid hydrogen in order to save energy while taking advantage of the cryogenic characteristics of this fuel to include the liquefaction energy. The establishment of appropriate R&D programs is dependent on the development of corresponding test techniques including the availability of extremely low turbulence wind tunnels with a fluctuation of less than 0.2% with respect to the flow velocity.

Ground Facilities: Beside the aeromechanical problems which have to be solved the development of an infrastructure for aviation, i.e., ground installations and the requirements of the airports are of fundamental importance when using hydrogen. About 1,000 to 1,500 metric tonnes LH_2 per day (33–50 GWh/day) is needed based on current aviation fuel requirements of large airports (like Frankfurt or San Francisco) of about 4,000 to 5,000 tonnes per day. This requires corresponding production and

liquefaction plants as well as storage tanks and cryogenic pipeline system for distribution.
If the amount of liquid hydrogen used in state of the art space projects is taken into consideration, as at the Kennedy Space Center for example, then an increase of more than one order of magnitude is currently needed to cover the requirements of large airports.
Although these problems can be applied to every alternative fuel, the problems become especially evident in aviation from an energy standpoint because here correspondingly large amounts of fuel, i.e., secondary energy, has to be made available in a relatively small space[5].
Several project studies (see [56–58, 75, 76, 26]) show that the delivery of gaseous hydrogen via a pipeline system to the airport where liquefaction takes place, is an optimal solution. Liquid hydrogen is stored in tanks with perlite vacuum insulation with an operational reserve of about two to three days.
The storage tanks and the terminals where refueling takes place are connected with each other via vacuum-insulated pipelines running through a covered canal system. The pipeline system is kept operational through a continuous circulation of liquid hydrogen.
The aircraft can be refueled at the terminals according to the plan shown in the sketch in Fig. 65 proposed in [67]. The connection is made with the supply lines, which are imbedded in the ground and fitted with tap connectors, via a refueling vehicle with two flexible vacuum-insulated hoses. The one line carries liquid hydrogen while the second serves to recover the cold gaseous hydrogen which builds up in small amounts during refueling. This returned gaseous hydrogen is routed back to the liquefaction plant for reliquefaction. Further an aircraft can basically also be refueled by tank vehicles in a manner similar to Jet A. In any case the expected refueling time with liquid hydrogen corresponds to about that for Jet A.
The 1975 NASA sponsored project studies ([57] and [26]) which assumed a hypothetical conversion of the San Francisco and Chicago airports to liquid hydrogen provide an indication of the costs associated with the conversion of the equipment on the ground. For the Chicago airport a conversion of all air routes of more than 2,780 km (1,500 nautical miles)

[5] If, however, it is considered that in the case of synthetic fuel production these enormous amounts of energy must first be produced from primary energy, it then becomes evident how advantageous and convenient it is to use the solar energy stored in fossil fuels over the course of millions of years. But if the long period of time it takes for the creation of fossil energy carriers in equilibrium with the environment, from atmospheric carbon dioxide and water, exposed to energy in the form of sunlight, is considered and if it is compared to the extremely short period required to exhaust it, then the extent of the damage to the environmental equilibrium caused by mankind and the consequences associated with it become evident.

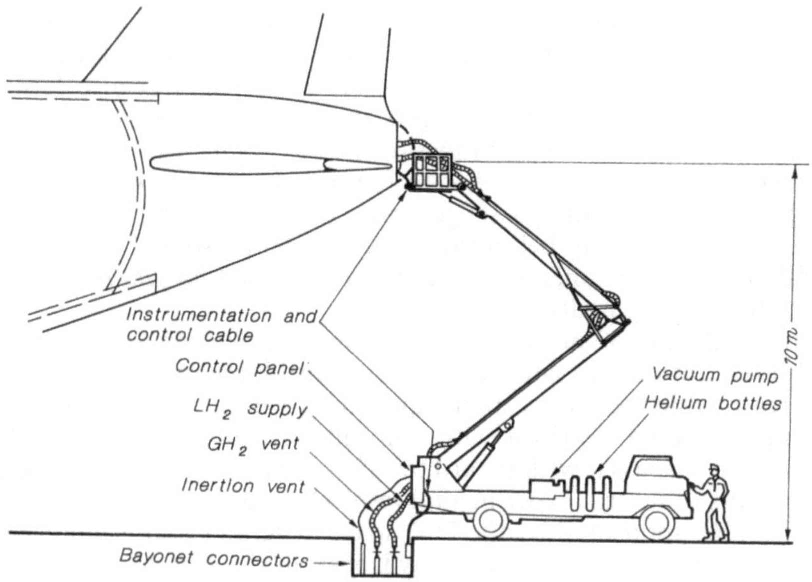

Fig. 65. Schematic of an aircraft refueling system (from [57, 79])

to liquid hydrogen was presupposed. Including a 1,000 tonnes/day liquefaction plant, five storage tanks with a diameter of 21.6 m, a vacuum-insulated pipeline system to the terminal as well as refueling vehicles, but without the hydrogen production plant, investment costs amounted to 340 and 460 million $, which would correspond to about 0.3 cents/km for passengers. A recalculation in 1980 showed that the capital investment would pay for itself in about 20 years [23]. Even if these figures are judged with skepticism it can be concluded that despite the high required investment, a conversion of the large airports basically appears to be possible with a tolerable time for return on investment at reasonable air fares.

6.1.4 Accidents

Investigations on the behavior of hydrogen in accidents are of utmost importance because liquid hydrogen is often classified by experts as a dangerous energy carrier compared to conventional aircraft fuel or liquid methane. In view of the positive experience with liquid hydrogen in space flight this is most probably the result of a dim recollection of the accident of the LZ 129-Hindenburg on the 6th of May 1937 at Lakehurst, New Jersey. In this respect it is remarkable that rigid airships with hydrogen located in gas bags acting as a buoyant gas—which would currently most likely be unacceptable with regard to safety—could successfully serve as freight and passenger transportation for over 25 years. It also has to be emphasized that many of the passengers and the crew (62 out of 97) survived the accident and the deaths of most of the 35 victims can not be directly attributed to hydrogen but rather to jumping out of the dirigible from a high altitude. Most of the passengers who remained onboard the dirigible survived. Another victim belonged to the ground personnel and was killed by a falling Diesel engine. Most burns were caused by leaking Diesel oil and not by burning hydrogen[6].

With the high rate of survival of the Lakehurst accident—a ratio which would be exceptional in current accidents with large capacity aircraft—it seems reasonable to assume that liquid hydrogen as an aircraft fuel can stand up to the comparison with conventional aircraft fuel, specifically CH_4, or even surpass it with regard to safety. This assumption is supported by the results of several experimental tests which were carried out between 1955 and 1965 sponsored by the U.S. space programs and is confirmed by the positive experience with liquid hydrogen in the Apollo program.

Further theoretical and experimental studies sponsored by NASA [47, 48] were recently carried out which led to remarkable results but made further experimental programs at their original large-scale indispensable.

In the following only events of marked interest which refer to direct effects of flight accidents on the ground where fuel tanks burst are dealt with whereas Chap. 8 deals with general aspects of the safe handling of liquid hydrogen.

The location of the LH_2 tanks in the fuselage (see Fig. 62) is advantageous with regard to safety according to [47] because here the tanks are protected from the front and from behind by the fuselage structure which can shield them from considerable amounts of stress. Investigations showed that in an accident the tank would only be affected after the destruction of the entire front section of the fuselage. The underside of the tank is supported by a sufficiently strong, approximately 45 cm thick fuselage support structure which has to withstand the stress in case the front section strikes the ground, the tail touches the ground or landing without landing gear takes place. Moreover, the tanks are separated from the passenger compartment and the cockpit by additional transverse bulkheads.

In contrast conventional fuel is stored inside the wings between the front and the back wing spar (see Fig. 61) and is therefore vulnerable if the wings

[6] See R. W. Knight: The Hindenburg Accident, Rep. 11, Safety and Planning Div., Bureau of Air Commerce, Dept. of Commerce (Aug. 1938), part 2 contains a translation of the German investigating committee.
Also see John Toland: The Great Dirigibles: Their Triumphs and Disasters. New York, Dover Press (1972) and Douglas H. Robinson, Giants in the Sky. Seattle, Univ. of Washington Press (1973).

are damaged by striking an obstacle and fuel leakage occurs. With liquid hydrogen the wings do not contain any fuel with the exception of lines which run between the wing spars and supply the jet engines. The flow of hydrogen to these lines can be shutoff by valves located in the cockpit.

Whereas conventional fuel tanks can be vented from the outside for pressure balance and a combustible mixture can form inside the tanks, cryogenic tanks are always under a small overpressure as a result of the vaporization of the fuel. Therefore, there is no air or oxygen in the tank which rules out the formation of a combustible mixture and contributes to the safe use of cryogenic fuels. The structure of the LH_2 tanks located inside the fuselage are considerably more rigid than the fuselage as a result of their design and their operating pressure. Independent of the type of fuel used, the passenger compartment is designed to an absolute pressure of about 1.25 bar (18 psi), the LH_2 tanks to about 2.1 bar (30.2 psi) [14, 22, 47]. This value is a compromise between the pressure at which the liquid hydrogen is stored, the pressure lockup time in the LH_2 tank until the actuation of the safety valves as well as the weight of the tank structure.

Thus it becomes obvious that cryogenic hydrogen tanks will not be severely damaged in an accident with a chance of survival which does not entail extensive damage to the fuselage.

Taking these facts into consideration and as a result of accident statistics, according to which severe tank damage is more unlikely to occur in accidents with a chance of survival than damage to wing surfaces, it seems to be clear that conventional tanks in the wings are more vulnerable to damage which leads to fuel leakage than those in the fuselage.

A very important question results from the possibility, which can not be ruled out, of LH_2 discharge from the tank into the passenger compartment and the cockpit in severe accidents, because even brief contact with LH_2 causes serious burns and the exposure of large parts of the body to cryogenic temperatures is deadly. It is always possible to specify accident situations in which large amounts of fuel are released and enter the passenger compartment when the tank bursts. According to what has been said so far, it is improbable that there is any chance of survival in the last case since the tank itself as well as the transverse bulk-head to the passenger compartment and the cockpit would have to be destroyed. In order to investigate this question further, larger test programs have to be carried out with aircraft that have been taken out of service for example.

The comparison between different fuels regarding spreading in soil, vaporization and dissipation into the atmosphere is very informative. In [47], several typical accident situations were investigated, some where the discharged fuel ignited and others where it did not. The analytic models are based, among other things, on results of spill experiments[7] as well as

---

[7] "Spills", see Chap. 8.

on investigations of the distribution of gaseous clouds in the atmosphere from [58–61].

For illustration purposes, the following discussion deals briefly with results obtained from [47]. Table 21 shows results of calculations using the Fay-Model [64] with the amount of hydrogen spilled $m = 12.6 \cdot 10^3$ kg LH_2 compared to equivalent amounts LCH_4, Jet A and JP4. The spill rate for an aircraft at rest at sea level, a temperature of 297 K and a wind velocity of 2 m/s was set at $\dot{m} = 900$ kg/s for liquid hydrogen whereas the amounts for the other fuels corresponds to the equivalent values for the same size of leak.

As expected this shows that compared to JP4 and Jet A the cryogenic fuels not only evaporate more quickly, but also do not spread as much in the ground. Hydrogen appears as the safest fuel in this comparison. The time required for complete evaporation is considerably reduced when ignition and flame formation take place because heat is radiated to the liquid on the ground from the flames and thus the rate at which the liquid evaporates is increased.

The very short time, compared to conventional fuels, for complete evaporation of large amounts of fuel is of utmost significance. In this case for $12.6 \cdot 10^3$ kg LH_2 it takes 32 seconds without fuel ignition or 15 seconds with fuel ignition, and for LCH_4 117 respectively 38 seconds.

The calculated evaporation time for the gasoline-like JP4 and kerosene-like Jet A without ignition seem to be relatively short but can be explained by the flat ground which prevents the formation of puddles in depressions. Distribution and spreading of burning JP4 and Jet A were not calculated within the scope of these tests. From experience with spill experiments and accidents this amount of Jet A would burn for about 1 hour. For comparison purposes the aviation accident in Teneriffe on the 27th of March 1977

Table 21. Radial spill, spreading and vaporization of fuel (from [47])

	m (kg)	$\dot{m}$ (kg/s)	V_s (m/s)	r_m (m)	t_m (s)	$\dot{m}_v$ (kg/s)	$m/\dot{m}_v$ (s)
LH_2	12,600	900	$2.5 \cdot 10^{-3}$	35	32	530	24
LH_2[a]			$6.0 \cdot 10^{-3}$	26	15	900	14
LCH_4	34,398	2,457	$1.1 \cdot 10^{-4}$	61	117	550	63
LCH_4[a]			$1.1 \cdot 10^{-3}$	35	38	1,710	42
JP4	42,210	3,015	$2 \cdot 10^{-6}$	143	785	100	
Jet A			$7 \cdot 10^{-6}$	331	4,180	18.8	

V_s = rate at which the level of the liquid drops, r_m = maximum spill radius of the liquid on the ground, $\dot{m}_v$ = maximum vaporization rate,[a] with ignition of the fuel, t_m = total fuel evaporation time, $m/\dot{m}_v$ = ratio of spilled fuel mass to fuel mass evaporation rate.

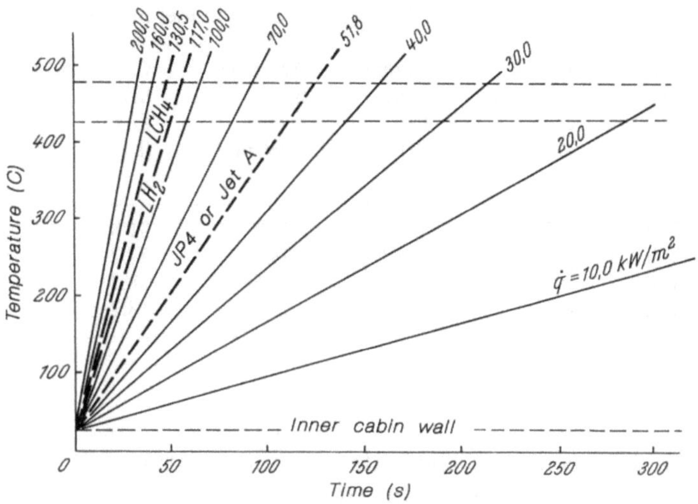

Fig. 66. Temperature of the heated fuselage as a function of time and heat flux density (from [47])

(collision of two fully tanked Boeing 747's) is cited, in which the largest amount of Jet A to date was set free and the fire which started lasted about 10 hours.

Assuming an identical accident situation, together with the short evaporation time and duration of burning of cryogenic fuels, the resultant heating of the fuselage structure is of extreme importance for the carrying out of rescue operations. In the case of Jet A or JP4 experience shows that within a few minutes the support structure of the fuselage is heated up to a temperature in the range of 430 to 480 °C. This caused a loss in tensile strength and thus the collapse of the supporting structure from its own weight. Therefore, passengers can only survive if they can be rescued from the aircraft as soon as possible after the fire breaks out.

Figure 66 is the result of tests carried out in [47]. It shows the rise in temperature of the outer supporting structure of the fuselage as a function of time. The accident situation given in Table 21 was used as a basis in which the fuselage is situated directly in the flames. Heat radiated from the flames as well as turbulent convection (23 W/m^2 K) was considered. Although the bare aluminum of the outer skin has a radiation emissivity $\epsilon < 0.1$, an increased emissivity of $z\epsilon = 0.9$ was assumed as a result of paint and corrosion caused by the flames. The radiation emissivity for the fuselage insolation was assumed to be about 0.5 (see Table 22).

With the assumptions from Table 22, Fig. 66 shows that methane flames lead to a collapse of the fuselage structure from its own weight within 40 seconds, hydrogen flames after about 50 seconds, JP4 and Jet A flames

Table 22. Heat flux into the fuselage structure in fuel fires (from [47])

Fuel	Flame temp. (K)	Flame emission coeff. ϵ_F	Flame heat flux (kW/m^2)	Absorbed heat rate (kW/m^2)	Convection heat flux (kW/m^2)	Total heat flux $\dot{q}_{tot}$ (kW/m^2)
LH_2	1700	0.4	189.4	85.2	31.8	117.0
LCH_4	1500	0.8	229.6	103.3	27.2	130.5
JP4	1100	0.9	74.7	33.6	18.2	51.8
Jet A	1100	0.9	74.7	33.6	18.2	51.8

after about 120 seconds. The internal temperature of the passenger compartment increases during this time by at most a negligible 2 °C.

If this is compared with the duration of the fire given in Table 21 it becomes obvious that for LH_2 the time until the collapse of the fuselage structure of about 50 seconds considerably exceeds the 15 second duration of the fire. For LCH_4 the times of 38 and 40 seconds are about the same, whereas under these assumptions for JP4 and Jet A the duration of the fire in any case greatly exceeds the "time till collapse" of 120 seconds. Under these conditions the fuselage structure would survive the hydrogen fire without collapsing whereas this would be uncertain with methane. With Jet A and JP4 the collapse of the fuselage structure is a certainty.

With these assumptions it can be summarized that a fire in an LH_2 fueled aircraft on the ground caused by an accident would be limited to a small area and would be very short compared to conventional fuels. In an accident with a chance of survival it is safer for the passengers, in the case of a hydrogen fire, to remain in the compartment until the fire is over whereas for fires caused by all other fuels it is safer to leave the aircraft immediately. Considering the events which take place when large amounts of fuel are released, including the spreading, evaporation, distribution and incendiary behavior, liquid hydrogen is obviously more advantageous compared to liquid methane, Jet A and JP4, the implications of which must be demonstrated by further experiments.

6.2 Ground Transport

In contrast to stationary energy users or railbound electrical vehicles, in order to move, automotive vehicles and watercraft need onboard storage of considerable amounts of energy. This is mainly solved today through the use of hydrocarbon based fuels.

Table 23. Comparison of alternative fuels and their storage (see [6, 9–11])

	Storage capacity (% wt H_2)	H_2 density (g/L)	Energy kWh/kg	Density kWh/L	Working temp. °C	α^f
Methyl-cyclohexane	6.1	47	2.03	1.56	>300	0.29[b]
Magnesium hydride	7	101	2.33	3.36	>350[c]	0.32[b]
Magnesium–nickel hydride	3.16	81	1.05	2.69	>250[c]	0.27[b]
Iron–titanium hydride	1.75	96	0.58	3.18	>−10[c]	0.14[b]
Liquid hydrogen	100	71	33.3	2.36	−250	0.04[d]
Methanol	12.5	99	5.6	4.46	>0	0.06[e]
Liquid methane	25	105	13.8	5.8	−161	0.02[e]
Ammonia	17.6	144	5.14	4.21		
Hydrazine	12.5	124	4.6	4.65	>0	
Isooctane[a]	17.3	117	12.7	8.76	>−40	0.007[e]

[a] Representative of gasoline.
[b] The heat which is required for the decomposition of the hydride and for the transformation of methylcyclohexane into toluene related to the lower heating value of hydrogen.
[c] Hydrogen pressure ⩾2 bar.
[d] The heat required to evaporate and warm up to ambient temperature related to the lower heat value of hydrogen.
[e] Heat of evaporation related to the lower heat value.
[f] See b, d, e.

6.2.1 Limiting Factors

Table 23 shows the results of using hydrogen in automotive vehicles compared with other chemical energy carriers. It shows the needed tank capacities, fuel mass as well as the lower heating values for typical unconventional fuels compared to isooctane (C_8H_{18}) which corresponds to the typical characteristics of gasoline which is a mixture of different hydrocarbons (see [1–11]). Ammonia and hydrazine are cited here only for comparison purposes, they are not considered applicable for automotive vehicles. The comparison of energetically equivalent amounts of isooctane, methane, methanol and hydrogen shows that liquid hydrogen requires the largest tank capacity but the lowest fuel mass. Methane in its liquid state corresponds most nearly to gasoline whereas methanol requires noticeably larger tank capacities and as a result of the considerably lower heating value it also requires considerably larger fuel masses.

An essential prerequisite for the use of hydrogen as a fuel is the proof of practical solutions for its storage and handling. While in aviation, hydrogen can only be used in its liquid state in an automotive vehicle it is possible to store it as a pressurized gas or in its chemical state combined as a metal-hydride or in ammonia (NH_3), hydrazine (N_2H_4), methanol (CH_3OH) or methyl-cyclo-hexane for example.
Ammonia and hydrazine are ruled out because of their toxicity and other undesirable characteristics for mobile applications (see [12]). For safety reasons storage in pressurized containers is rather inappropriate because of the required high pressures. Furthermore, the amount of stored hydrogen is only about 1% of the tank weight even at pressures of about 200 bar and thus corresponds to the behavior of iron–titanium hydride in which the tank mass is about twenty times more than that of an energetically equivalent tank of gasoline.
Moreover, the comparison in Table 24 of the mass necessary for the storage of energetically equivalent amounts of hydrogen and methanol compared to gasoline (isooctane) is informative. With the same stored energy—corresponding to 25 kg (36.2 L) of gasoline—methanol requires about double the volume and double the total mass whereas liquid hydrogen needs a little more than triple the volume compared to gasoline and a perceptibly lower mass compared to methanol. With respect to weight, hydride storage has about the same volume as pressurized storage containers but 10 to 15 times more mass is required compared to liquid hydrogen.
With the exception of the salt-like hydrides like lithium hydride, sodium hydride, or calcium hydride, with metal hydrides, hydrogen atoms are

Table 24. Mass required for the storage of hydrogen and methanol compared to energetically equivalent amounts of gasoline

	Fuel mass (kg)	Hydride mass (kg)	Hydride volume (L)	Storage vessel (kg)	Total mass (kg)
Isooctane	25	—	36.2[b]	17	42
Methanol	56.6	—	72[c]	22	78.6
Liquid hydrogen	9.5	9.5[b]	136[c]	55	64.5
Magnesium hydride	9.5[a]	135	94	81	216
Magnesium–nickel hydride	9.5[a]	301	117	180	481
Iron–titanium hydride	9.5[a]	543	99	325	868

[a] Mass of stored hydrogen.
[b] H_2 is "hydrogen-hydride".
[c] Fuel volume.

integrated into the crystal lattice of metals and metal alloys in the form of intermetallic compounds [13–16]. Each lattice atom can combine with up to two and in some cases up to three hydrogen atoms. This has the effect that when using metal alloys like iron–titanium only up to 1.7% by weight of the stored material can be hydrogen ($FeTiH_{1.95}$) because in this case the iron and titanium together already represent about 98.25% of the substance by weight. Improved percentages of stored hydrogen are achieved by using lighter atoms as lattice material, like magnesium for example (see [13–18]). A problem, however, is that the hydride formation and the liberation of hydrogen can only take place quickly enough to produce a sufficiently high hydrogen pressure at a relatively high temperature. By using catalysts like anthracene for example, it is possible to accelerate the hydride formation even at low temperatures (see [18])[8].

The temperature can be further reduced by using appropriate additives, primarily nickel. However, the heat of formation for hydrides which must be removed when filling the storage container and resupplied to the storage container when liberating hydrogen remains very high. It is about 30% of the lower heating value of hydrogen (see Table 23). This 300 °C heat cannot be provided by the exhaust gas under all engine operating conditions so—apart from the materials—there are considerable problems with this concept. A combination of a MgH_2 storage vessel and a $FeTiH_2$ storage vessel has recently been proposed. Thus, hydrogen is set free at low temperatures from the iron–titanium hydride and ignited to heat the MgH_2 high temperature storage vessel. Apart from technological problems, the additional combustion of fuel outside the engine, which must be transported and is therefore costly, can hardly result in reasonable fuel economy and the required low storage mass.

Similar conditions regarding the stored hydrogen portion, heat of decomposition and decomposition temperature also apply for the decomposition of methylcyclo-hexane into toluene and hydrogen (see Table 23) which has been proposed for the storage of hydrogen [19, 20, 137]. Furthermore, compared to hydride storage the development of this concept represents a far more difficult task which can be solved only with a considerably higher technical effort and without much expectation of far-reaching advantages or improvements over hydride storage.

6.2.2 Propulsion Techniques

Electric Propulsion with Fuel Cell Batteries. The development of fuel cells for automotive application to date was carried out under the premise that

[8] See B. Bogdanovich: Katalytische Synthese von Magnesiumhydrid unter milden Bedingungen. Angew. Chem. *92*, 845–846 (1980). A further publication was to be released in 1984. See further [136, 159].

in competing with the internal combustion engine it should use the same fuel, i.e., hydrocarbons. This prerequisite was not applied in the use of hydrogen as an alternate fuel. Thus the more easily accomplished task of developing a hydrogen fuel cell for automotive application attains increased importance because beside being exceptionally environmentally safe it is basically an improved use of the fuel. Of the conventional fuel cells which were primarily developed for the space programs, the hydrogen–oxygen cells are the most highly developed and exhibit efficiencies in the range of 50–60% with lifetimes of several thousand hours (see [21–25]).
Methanol cells, some of which operate with pure oxygen, some with air, are among the tested fuel cells. A disadvantage of the methanol cells is that the electrode reactions inevitably lead to the formation of carbonates in which the electrolyte is consumed. Furthermore, only relatively low current densities at the electrodes are obtained at ambient temperature which results in a high specific mass[9] [24]. For this reason these cells, as well as ammonia (NH_3) and hydrazine (N_2H_4) cells are of secondary importance for automotive vehicles. However, it is possible to split catalytically methanol and suitable hydrocarbons into hydrogen and carbon monoxide and use this mixture in hydrogen–air cells, which have to exhibit low sensitivity to CO_2. In this case, methanol and hydrocarbons serve as the hydrogen storage.
The use of fuel cells in a vehicle is still problematic despite the recognized advantages because they have several undesirable characteristics which are of utmost significance when the vehicle is operating. Before they are put into operation fuel cells have to be warmed up until the heat produced from the cell is at equilibrium with the heat loss to the environment, which causes additional problems especially in winter. Catalysts which must be partially noble metals or metals of the platinum group must be added to the electrodes yield an increase in output current density which is directly proportional to the amount of catalyst. A reduction in the power density is synonymous with a decrease in the specific power[10], which for the best hydrogen–oxygen cells is about 66 W/kg (15 kg/kW) and for some methanol cells 3.3 W/kg (300 kg/kW) (see [10, 25, 26]).
Since the use of hydrogen–oxygen fuel cells is only considered for special applications due to the complexity of the necessary refueling equipment with two fuel components as well as for safety reasons, an effort should be made to use hydrogen–air fuel cells. The problems are mostly the relatively large amounts of nitrogen given off as well as the removal of the carbon dioxide from the air which is supplied to the cell in order to prevent carbonate formation in alkaline cells. The cell could also be adversely affected by pollutants which are in the air such as sulfur dioxide and nitrogen

[9] Defined as mass per unit of power output (kg/kW).
[10] Power output per unit of cell mass (kW/kg).

oxides for example, which can contaminate the catalysts located on the electrode. Furthermore, with the same gas flow rate the power density and thus the specific mass will be lower than with pure oxygen operation.
In addition to the specific power of the fuel cell, the electrical propulsion system including the power conditioning system has to be taken into consideration, which currently provide to about 100 W/kg (10 kg/kW) [10, 26, 27] and can be increased to about 150 W/kg (6 kg/kW). This implies that at best a specific power of about 50 W/kg and a specific mass of 20 kg/kW can be obtained for electrical propulsion with fuel cells (Fig. 67). This approximately corresponds to the highest specific power attainable for electrical propulsion with lead batteries. In the case of electrical propulsion with fuel cells and liquid hydrogen storage the stored energy per unit of mass of the system (fuel cells, fuel storage, electrical motor) is from about 3.3 kWh/kg to about 6.0 kWh/kg. With the use of hydride storage 0.33 kWh/kg to about 0.55 kWh/kg can be obtained. If this data are compared with the 0.03 kWh/kg for lead batteries, it become apparent that electrical propulsion by fuel cells could be used instead of, or in combination with battery electrical propulsion.

Internal Combustion Engines. The I.C. engine is basically very well suited for hydrogen operation and thus currently represents the best possible compromise between production costs, fuel economy, power to mass ratio, road performance and longevity.

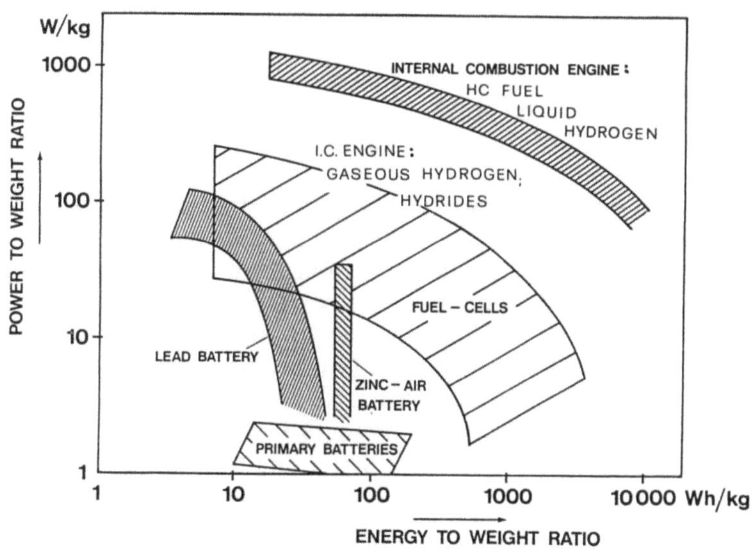

Fig. 67. Energy to mass ratio and power to mass ratio for different vehicular propulsion systems, source: Bosch

Moreover, it has great development potential especially coupled with digital control systems.
Initial applications were attempted in the Zeppelin dirigible LZ 127—Graf Zeppelin, because here hydrogen vented during the trip was additionally intended to be used as fuel [29, 30]. In this process, the five 12-cylinder Maybach engines, which are usually fueled by a 80/20 mixture of benzene and gasoline (570 HP, 1400 rpm, compression ratio 6:1) were converted to a compression ratio of 7:1 and driven by coal gas mixed with 5% to 30% hydrogen. In an 82 hour nonstop flight over the Mediterranean in 1928 a 14% improvement in fuel economy was achieved.
In 1935, five Daimler-Benz 16-cylinder Diesel engines were installed in the LZ 129 and operated in a manner similar to the LZ 127. Similar measures were intended for the English dirigible R 101 prior to its tragic accident [30].
H. R. Ricardo [32] performed the first tests in which hydrogen was used in I.C. engines [31]. They were carried out with external mixture formation and quality governing (mixture control). They were followed by further tests on "Ricardo E35" engines by A. F. Burstall [33].
Rudolf Erren, who first worked in Germany and then in Great Britain carried out intensive tests with hydrogen–air, hydrogen–oxygen as well as multi-fuel engines. He designed and demonstrated engines with internal mixture formation, i.e., hydrogen injection at the beginning of the compression stroke [30, 34–36].
In 1935, Gerrish and Foster worked on dual-fuel Diesel engines which were supplied by a hydrogen–air mixture during the intake stroke [37] and used Diesel injection during the expansion stroke. In 1942, Oehmichen [38] published results on one-cylinder engines with internal mixture formation in which the hydrogen injection took place early in the compression stroke. Further works followed by R. O. King et al. [39, 40] which in turn were followed by new activity since about 1970 (see [41, 42]). A detailed description can be found in [42, 43].
Otto-cycle-engines operating with hydrogen under partial load can achieve about the same efficiency as conventional Diesel engines [28]. Compared to conventional fuels the hydrogen operation of internal-combustion engines provides a drastic reduction in toxic engine emissions. The exhaust gas does not contain any carbon-dioxide, sulfur-dioxide or soot, nor does it contain hydrocarbons or lead oxide. Nitric oxides are the only pollutants which could be emitted, the formation of which can be suppressed to a large extent through appropriate measures [163].
If, for example, hydrogen is ignited in the engine with an excess of air[11],

---

[11] See R. F. Stebar, F. B. Parks: Emission Control with Lean Operation Using Hydrogen Supplemented Fuel, Gen. Motors Res. Publ., GMR-1537 (1974).
Also see K. S. Varde, G. H. Frame: A Study of Combustion and Engine Performance Using Electronic Hydrogen Fuel Injection. Int. J. Hydrogen Energy 9, 327–332 (1984).

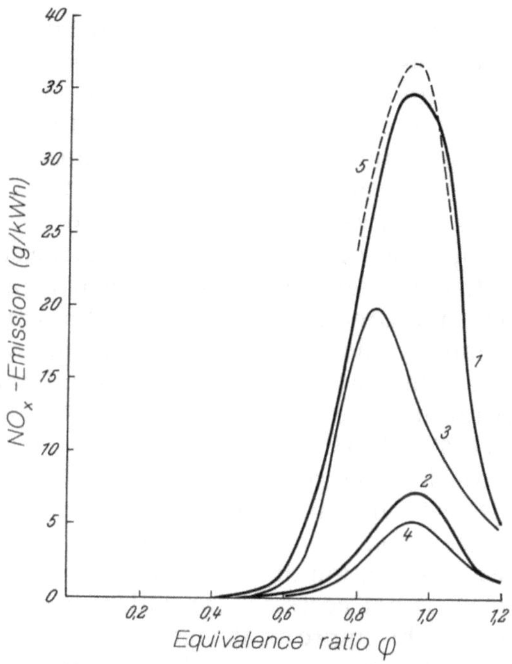

Fig. 68. Typical values for the NO_x emission of H_2 internal combustion engines as a function of the equivalence ratio. *1* external mixture formation, volumetric control (from [47]); *2* external mixture formation, quality governing, water injection (from [47]); *3* internal mixture formation (from [44]); *4* internal mixture formation (from [84]) and the best obtainable results (compare LIRIAM, from [43]); *5* typical data for conventional Otto and Diesel engines. The enormous reduction of NO_x emission for the operating range of $\varphi < 0.7$ possible with hydrogen is noticeable

a very low nitric oxide emission is obtained, which sharply increases near the stoichiometric mixture ratio and then quickly drops off again [44–48], (Fig. 68).

Engine operation with high excess air is possible when using hydrogen as a result of the wide range of flammability of 5 vol. % to 75 vol. %, i.e., the use of a lean mixture even with external mixture formation. In contrast to gasoline which has a flammability range of 2 vol. % to 8 vol. %, an Otto-cycle engine can be operated using hydrogen with both every lean and very rich mixtures. This enables quality governing by controlling the amount of fuel in the mixture as in the conventional Diesel engine. For this reason, a throttle valve (volumetric control) is theoretically unnecessary so that especially under lower partial load (urban traffic) the throttling losses are reduced and the efficiency and fuel economy can be perceptibly improved. The optimal efficiency of a hydrogen fueled motor is achieved earlier in the part load range than a gasoline engine at an equivalence ratio of about $\varphi = 0.4$, which is particularly advantageous in urban traffic. This is due to the fact that with lean mixture operation the influence of a change in the specific heat of the combustion gas with its undesirable effect on the thermal efficiency is only minimal [49, 50]. Moreover, hydrogen operation considerably improves the efficiency. Compared to gasoline its actual thermodynamic process more closely approximates the equal-volume process, which serves as a comparative process for the Otto-cycle-engine. A requirement of this equal volume process is the total combustion of the cylinder charge at top dead center where the flame speed must approach infinity. The considerably higher flame speed of hydrogen compared to gasoline under the same conditions is therefore the cause for improved conformity to the equal-volume process which in this respect is superior to any other process with slow combustion. The total potential fuel saved with hydrogen operation, according to different data, ranges between 25 and 40% [10, 28, 47, 48, 51–53, 55].

Since gaseous hydrogen mixes very well with air[12] as a result of its very small kinematic viscosity and the strong turbulence associated with it, it far outweighs all other gaseous and liquid fuels in this respect. For this reason and because evaporation and recondensation problems are ruled out, fuel enhancement is not necessary for hydrogen engines at cold start or during acceleration. This improves the fuel economy, provides improved idling with multicylinder engines as a result of the uniform distribution of the fuel to the cylinders and helps reduce cyclic variations.

In contrast to gasoline and Diesel oil, hydrogen does not cause thinning of the lubrication oil and ensures a permanent clean combustion chamber because there is no solid residue left over from combustion. Disadvantages with ambient hydrogen operation compared to gasoline are the lower heating value per unit volume, the potentially higher flame speed of hydrogen–air mixtures as well as the lower ignition energy. With external mixture formation, i.e., adding hydrogen to the air in the intake manifold, the lower heating value per volume leads to reduced engine output whereas the higher flame speed together with the lower ignition energy can cause backfiring of the mixture from the cylinder into the intake manifold. The flammable hydrogen–air mixture flowing into an open intake valve comes into contact with hot spots in the combustion chamber and with hot residual gas in the cylinder. This can lead to ignition and together with the high flame speed can cause backfire through the open intake valve into the intake manifold. Suppression of this backfire and uncontrolled preignition with external mixture formation can basically be accomplished through heat reduction measures. This includes the use of an aluminum cylinder head, the avoidance of acute angles in the combustion chamber, the construction of the combustion chamber to provide for as homogeneous

[12] Hydrogen's diffusion properties are often inappropriately cited as the cause.

a temperature distribution as possible, the use of cold spark plugs as well as exhaust gas recirculation or direct injection of water into the intake manifold [28, 51–53, 163].

In this process exhaust gas recirculation and water injection, on one hand considerably reduce the nitric oxide emissions but can also cause lower power output. Furthermore, the electronically or mechanically timed injection of hydrogen into the intake port, in which the injector valves only open during the intake stroke, has been successfully implemented so that the intake air alone can have a cooling effect on the residual gas and the walls of the combustion chamber [51].

The problems of backfiring and deterioration of power output can be considerably reduced with four-stroke and two-stroke engines through direct injection of hydrogen into the combustion chamber early during the compression stroke (internal mixture formation) [38, 41, 43, 47, 48, 54]. However, this process requires injection pressures of 10 to 20 bar which must be provided by the fuel conditioning system. Further development of hydrogen engines will tread this path. Another possibility is Diesel-like operation with the injection starting just before top dead center [43, 44] which however requires pressures of more than 100 bar. Furthermore, the rotary piston engine must be mentioned here. It offers a certain development potential for hydrogen as a result of its combustion chambers which are cooler than those of the reciprocating engine [10].

Several problems can occur with hydrogen engines as a result of its less effective anti-knocking properties compared to hydrocarbons and methanol [39, 57–59]. Whereas gasoline can have an octane number of between 90 and 100 or more, depending on quality, hydrogen has a considerably lower equivalent octane number of only about 60 according to [56].

Based on various other tests [60, 61] it has been established that hydrogen has an octane number much lower than that of gasoline. Knocking can be avoided by reducing the compression ratio. However, uncontrolled preignition with the hot residual gas is more likely due to the larger volume of residual gas. An indirect increase in the mixture octane number can be achieved by mixture cooling, for example by water injection or cryogenic mixture formation (see pages 155, 217).

The methane number (MZ) usually used to characterize gaseous fuel engines cannot be used to measure hydrogen's anti-knock properties because it is defined with hydrogen as the lower limit (H_2: $MZ = 0$, Methane: $MZ = 100$).

The measurement of anti-knock properties of hydrogen is further hindered because no clear definition of knocking and the required measurable verification associated with it, can be derived from the literature. According to [39] there are two types of knocking, where for an engine under test in one case a decrease of the spark advance leads to a decrease in power output—"detonation knock"—and in the other case an increase of power

output—"combustion knock". According to [40, 58] the hydrogen engine should exhibit a tendency for combustion knock.
Moreover, in addition to the dependency of the knocking behavior on the compression ratio and the intake air temperature there is a strong dependency on the fuel–air equivalence ratio φ and the engine speed. When approaching a stoichiometric mixture ($\varphi = 1$) the combustion cycle is negatively effected by the high flame speed of the hydrogen–air mixture which leads to loud combustion noise as well as high thermal and mechanical load on the engine [38].
Furthermore, depending on the engine speed the influence of the gas flow and turbulence in the combustion chamber and at the intake openings is important. In the range of a stoichiometric mixture an increase in the flame speed can also be observed with hydrogen operation, which can be ten times higher than the results obtained in the combustion of the mixture in a combustion bomb with no turbulence [49].
Since a high transitory pressure increase and high peak pressure can occur as a result of the high flame speed the exact start of the knock is difficult to pinpoint precisely. Furthermore, knocking cannot be considered completely independent from nitric oxide emissions because, depending on the type of mixture formation, measures which are taken to suppress knocking can also have an impact on nitric oxide formation. For example, low mixture temperature before and during the compression stroke, homogeneous mixing and low equivalence ratio (lean mixture) have a positive effect on knocking behavior as well as on the level of nitric oxide emissions. With external mixture formation, which is characterized by uncontrolled preignition as well as backfiring into the intake manifold, only insignificant nitric oxide formation takes place as a result of an homogeneous mixture of hydrogen and air below an equivalence ratio of $\varphi \sim 0.5$. NO_x formation reaches its maximum at $\varphi \sim 0.8$ and drops rapidly above it [45, 47].
In this process the nitric oxide content below $\varphi \sim 0.8$ and particularly at $\varphi \sim 0.5$ is principally limited by the reaction kinetics during formation whereas above $\varphi \sim 0.8$ it is essentially determined by the thermodynamic equilibrium between nitrogen and oxygen frozen at the expansion stroke [43].
A backfire into the intake manifold is no longer possible with internal mixture formation and the knock behavior can be improved or suppressed by appropriate measures like for example continuous Diesel-like fuel injection during the combustion phase, starting at about 6° before top dead center (b.t.d.c.) and ignition at about 5° b.t.d.c. Since the combustion pressure in the cylinder can be effected by the hydrogen injection, specifically the injection pressure and injector nozzle geometry as well as by multiple injection, "softer combustion" is essentially possible as opposed to external mixture formation.

With stratified and heterogeneous charges as can be typical with internal mixture formation, the interaction of the influences of nitrogen oxide formation and decomposition in the range of about $\varphi \sim 0.9$ cause a decrease in the nitric oxide content as compared to external mixture formation.
However, due to charge inhomogeneities, disproportionately high nitric oxide formation occurs in the intermediate range of $0.5 \leqslant \varphi \leqslant 0.8$ because a considerable amount of the combustion takes place in the range of φ where the reaction kinetics favor the formation of nitric oxides [42–44, 47, 48, 52, 53]. In this process a strong dependence of the nitric oxide emission on the start of the hydrogen injection can be ascertained. With injection start earlier than about 40° b.t.d.c. there is a corresponding long mixing time for fuel and air, which results in reduced charge stratification and thus lower nitric oxide formation. With late fuel injection start at about 30° b.t.d.c. as a result of a largely heterogeneous charge hydrogen burns predominantly with a local equivalence ratio of $\varphi > 0.8$ and thus reduced nitric oxide emission.
While a hydrogen injection start of 30° b.t.d.c. does not result in complete suppression of uncontrolled preignition and combustion knock, this suppression is possible with a very late injection start (6° b.t.d.c.) but as previously mentioned causes problems with combustion and nitric oxide emission as a result of the largely heterogeneous charge. Essentially, only a procedure which ensures thorough mixing in a short amount of time can alleviate these problems. An appropriate procedure, Late Injection Rapid Ignition and Mixing (LIRIAM) is described in [43] with which, however, the problem of rapid mixing could not yet be satisfactorily eliminated. Hybrid mixture formation is an improved implementation of the LIRIAM procedure, i.e., it is a combination of external and internal mixture formation (DLR patent DE 37 31 986 A1). Part of the hydrogen is mixed with the intake air during the intake stroke in which hydrogen precharging or air precharging in appropriate combinations can take place within the intake manifold. Thus, a lean H_2–air mixture ($\varphi \leqslant 0.6$) is formed which does not yet exhibit a tendency for uncontrolled preignition and is ignited at about 20–40° b.t.d.c. by a spark plug. At about 5–8° b.t.d.c. cryogenic hydrogen is injected under high pressure (200 bar) into the highly turbulent flame front which is formed either as a single or multiple pulses. With an appropriate arrangement of H_2 jets, the flame front is ruptured resulting in the formation of a large number of local combustion nuclei distributed three dimensionally, in which combustion with a local excess of hydrogen ($\varphi > 1$) takes place, although the effective combustion over the entire combustion chamber volume is stoichiometric ($\varphi = 1$). In this manner combustion takes place to the left or right side of the maximum of the NO_x curve (see Fig. 68) with appropriately low or decreasing NO_x emission.
Hydrogen engines with internal mixture formation and late start fuel injection apparently have several problems in common with conventional

Diesel engines like increased nitric oxide emission and insufficient homogeneity of the fuel–air mixture for example. With a conventional Diesel engine this requires lean combustion operation in order to reduce nitric oxide and soot emission and associated pollutants, whereas with a hydrogen engine operation in the stochiometric mixture range seems to be possible if additional measures can be taken to reduce the nitric oxide emission.

Flow conditions with respect to the boundary conditions of the fuel–air mixture in the cylinder including microturbulence have to be investigated more intensively. An interesting fact is obtained by the comparison of change in moles of different fuels during combustion:

Isooctane $2C_8H_{18} + 25O_2 \rightarrow 16CO_2 + 18H_2O$
27 moles 34 moles

Methane $CH_4 + 2O_2 \rightarrow CO_2 + 2H_2O$
3 moles 3 moles

Hydrogen $2H_2 + O_2 \rightarrow 2H_2O$
3 moles 2 moles

The first two equations are characterized by an increasing or constant number of moles. Initially disregarding the increase in temperature, the volume of the chemical reactants increases (first case) or remains constant (second case). In the case of hydrogen a decrease of moles and as a result a decrease in volume takes place. Taking into account the increase in temperature during combustion, an increase in volume is observed in every case due to thermal expansion of the gaseous reactants. Nevertheless, the microstructure and reaction kinetics of the hydrogen–oxygen reaction is quite different from the first and second case. The oxygen molecules act as if they are sinks for the hydrogen molecules. Therefore a kinetic pressure gradient pushes the hydrogen molecules towards the oxygen molecules and the reaction may be represented theoretically by a two-fluid model analogous to the two-fluid model in plasma physics. In fully ionized plasmas the two fluids are represented by the positive ions and the negative electrons. If the density of the electrons differs appreciably from the corresponding density of positive ions within a particular volume, the electrostatic forces act to restore electrical neutrality. As a result a discontinuous microstructure is maintained, characterized by a parameter known as the Debye shielding distance. It is a measure of the amount the density of electrons may perceptibly deviate from that of the ions.

In the case of the hydrogen–oxygen reaction the two-fluid model is represented by the hydrogen and oxygen molecules or atoms where the electrostatic forces in the plasma are replaced by the kinetic pressure gradient. Therefore a granulation exists in the microstructure of the gaseous mixture. This may deteriorate the flame front propagation as well as the flame front stability because of the hydrogen depleted zones within the granulated microstructure. Further investigation has to include the granulated structure and its connection to microturbulence and the influence on the macroscopic properties of the gaseous mixture.

Aside from exhaust gas recirculation and water injection, cryogenic mixture formation is a very effective measure for the reduction of nitric oxide emission [62, 63]. With external mixture formation this consists of adding cold gaseous hydrogen to the intake air and with internal mixture formation injection of cold hydrogen into the cylinder. By adding cold gaseous hydrogen the temperature of the mixture is perceptibly reduced primarily in the stoichiometric range which leads to a higher volumetric heat value of the mixture (charge specific heat) and, in principle also reduces the tendency to knock.

With external mixture formation the following applies for the charge specific heat value h_v (kWh/L) of the mixture

$$h_v = h_u \cdot \varphi \cdot \frac{r_{st} \cdot \rho_k}{1 + \varphi \cdot r_{st}} \tag{123}$$

where h_u is the lower heating value of the fuel (kWh/kg).

Table 25 shows the lower heating value h_u for gasoline and hydrogen, the density ρ_k at ambient temperature and gaseous fuel as well as the volumetric fuel–air ratio r_{st} with a stoichiometric mixture under standard conditions. With a stoichiometric mixture hydrogen reaches 42% of the air volume. According to Eq. (123) this corresponds to a fraction of hydrogen of

$$\frac{\varphi \cdot r_{st}}{1 + \varphi \cdot r_{st}} = \frac{r_{st}}{1 + r_{st}} = 0.3 \tag{124}$$

i.e., about 30% of the mixture volume.

Whereas with external mixture formation the large percentage of gaseous hydrogen results in a reduction in the air available for combustion and thus a corresponding deterioration in output power, with internal mixture formation pure air is sucked into the engine and hydrogen is not injected until the compression stroke. Thus the charge specific heat value increases to

$$h_v = h_u \cdot \varphi \cdot r_{st} \cdot \rho_k . \tag{125}$$

Table 25. Characteristic data

	h_u (Wh/g)	ρ_k (g/L)	r_{st}
Gasoline	12.7	3.5	0.026
Hydrogen	33.3	0.09	0.42

Table 26. Comparison of the charge specific heat value for gasoline and hydrogen with cryogenic mixture formation

		h_v max (Wh/L)				h_v at η_{max} (Wh/L)				h_v at φ_{min} (Wh/L)
		T_M (K)[g]				T_M (K)[g]				T_M (K)[g]
	φ	300	228[a,d]	256[b]	φ	300	266[a,d]	300[b]	φ_{min}	300[a,d]
Gasoline[c]	1.0	1.05	—	—	0.9	0.96[h]	—	—	0.83	0.88
		1.89[c]	—	—		1.25[h]	—	—		1.14[h]
Hydrogen	1.0	0.89[c]	1.77	1.04	0.4	0.43[c]	0.49	0.43	0.2	0.23[c,d]
Hydrogen[e]	1.0	1.14	1.50[h]	1.38[h]	0.4	0.56	0.64	0.58	0.2	0.30
Hydrogen[f]	1.0	1.26	1.95	1.79	0.4	0.50	0.83[h]	0.75[h]	0.2	0.25[f]
										0.32[h]

[a] Dry air.
[b] Hydrogen, saturated air.
[c] External mixture formation.
[d] Cryogenic mixture formation.
[e] Cryogenic external mixture formation and adiabatic supercharging at about 0.5 bar.
[f] Internal mixture formation.
[g] Charge temperature in K.
[h] Adiabatic supercharging at about 0.5 bar.

Therefore, in the case of a stoichiometric mixture, with internal mixture formation the charge specific heat increases by a factor of 1.42 as compared to external mixture formation under the same conditions, and can be further increased by supercharging the engine.

Table 26 provides typical values of the charge specific heat value for gasoline and cryogenic hydrogen at characteristic equivalence ratios φ at maximum charge heat value, maximum efficiency η_{max} as well as minimum usable charge specific heat value. The temperature of the gaseous hydrogen was assumed to be 40 K. Therefore, correspondingly higher values of h_v would have been obtained with liquid hydrogen.

From this table it becomes apparent that at maximum power output ($\varphi = 1$) cryogenic external mixture formation increases the charge specific heat value from 0.89 Wh/L to 1.17 or 1.04 Wh/L which approximates that of gasoline (1.05 Wh/L) and internal mixture formation with hydrogen (1.26 Wh/L). With adiabatic supercharging of the engine at 0.5 bar and external mixture formation with hydrogen it increases to 1.5 and 1.38 Wh/L. A further increase can basically be achieved with cryogenic internal mixture formation and supercharging so given that uncontrolled preignition and knocking can be prevented the hydrogen engine is superior to the gasoline engine.

Furthermore, it must be pointed out that at about $\varphi = 1$ the cooling effect of the cold hydrogen–air mixture corresponds quantitatively to that which can be obtained with external mixture formation through water injection without perceptible deterioration in output power. It is expected that initial test results will show that water injection, which is always complicated, is not necessary under certain conditions [62, 63, 66, 134, 139].
In addition to its use in racing engines, water injection, due to its cooling effect, has recently been introduced into gasoline and Diesel engines in order to reduce nitric oxide emissions despite its undesirable characteristics like for example corrosion and freezing at low temperatures.

6.2.3 Comparison of Hydrogen-Fueled Vehicles Regarding Propulsion and Storage

The applicability of hydrogen-fueled vehicles is as much dependent on the means of storage for the transported hydrogen as on the operating procedure itself [152]. In the following, a comparison is made between hydrogen fueled vehicles with LH_2 storage and hydride storage as well as those propelled by I.C. engines and electric motors with hydrogen-air fuel cells. Urban traffic and short distance traffic were selected because their particular requirements for economical and energy saving vehicle operation are extremely important [67]. The Europe cycle (Fig. 69) and the SAE-J-227D cycle for electrical vehicles (Fig. 70) are used as a basis. Typical data for these cycles are shown in Table 27 (see [65, 66]).
Disregarding idling time the required propulsion energy with a given distance is about the same for both cycles (see Table 27, gasoline consumption). Including idling for driving the same distance under the assumptions made here, total fuel consumption is slightly different for both cycles. Nevertheless both driving cycles can be said to be roughly equivalent from an energy standpoint with tolerable approximation and can be used for traffic in and around urban areas.

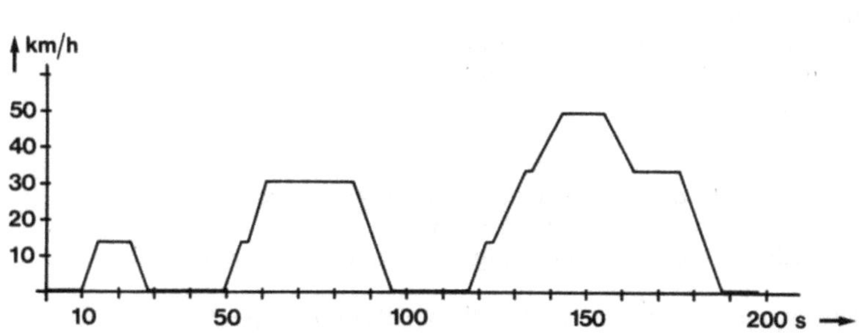

Fig. 69. "ECE Cycle"

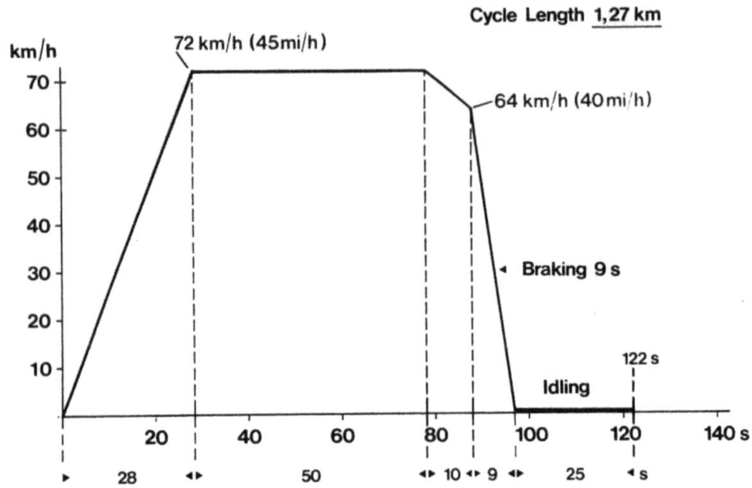

Fig. 70. SAE-J 227 D Cycle

Table 27. Comparison of the ECE cycle and SAE-J-227D cycle for a gasoline Otto-cycle engine

	ECE	SAE227
Cycle time (s)	195	122
Cycle length (km)	1.013	1.27
Average speed[a] (km/h)	18.7	47.13
Maximum speed (km/h)	50.0	70.0
Total idle time (s)	60.45	25
Propulsion energy per cycle[b] (kWh/1000 kg)	0.081	0.10
Gasoline consumption[c] (L/100 km)/1000 kg)	5.4	5.28
Gasoline consumption[d] at idle (L/100 km)	4.9	2.34
Total consumption[e] (L/100 km)	11.4	8.7

[a] No idling.

[b] For the wheels, air drag coefficient c_w times cross section area A: $c_w A = 1\ \text{m}^2$, rolling friction coefficient $c_r = 0.012$.

[c] Conversion efficiency $\eta = 0.17$, including engine and power train, no idle time.

[d] Otto-cycle engine, 2 L displacement, idle at $800\ \text{min}^{-1}$, $\varphi = 0.8$, volumetric control by throttle flap.

[e] Vehicle mass 1,200 kg.

The amount of primary energy for covering a given distance at a given driving cycle is informative in the comparison of vehicles with different hydrogen storage methods and various means of propulsion. The following is based upon the production and liquefaction of hydrogen via electricity and electrolysis from heat as primary energy[13]. According to [64–66] a characteristic parameter is the additional amount of primary energy that is necessary for the storage of 1 kg of hydrogen onboard a vehicle for a given range per refueling. In the following, hydride storage as well as storage in a liquid form will be considered. The amount of hydrogen which has to be stored for a given range based on the SAE-J-227D cycle (Fig. 70 and Table 27) is determined from the energy equation:

Case A: Hydride Storage

$$\frac{m_0 + m_{sp}}{1{,}000} \cdot p \cdot \frac{Z}{\eta_M} = \beta \cdot m_{Hy} \cdot h_u \tag{126}$$

where m_0 is the vehicle mass without hydride storage (kg), m_{sp} the total mass of the hydride storage system (kg), m_{Hy} the hydride mass (kg), $\gamma = m_{sp}/m_{Hy}$, p the specific energy consumption per cycle at the wheels per 1,000 kg vehicle mass (kWh/1000 kg), Z the number of cycles based on range per tank filling, $\beta = \frac{m_{H_2}}{m_{Hy}}$ the mass fraction of the hydrogen stored in the hydride, h_u the lower heating value of hydrogen (kWh/kg), η_M the conversion efficiency including I.C. engine and power train.
The following is obtained from (126)

$$m_{H_2} = \beta \cdot m_{Hy} = \frac{\beta}{\gamma} \cdot m_{sp} = \frac{\beta}{\gamma} \cdot \frac{m_0}{\frac{1000\eta_M \cdot \beta \cdot h_u}{p \cdot Z \cdot \gamma} - 1} \tag{127}$$

The following is obtained from (127) for the amount of primary energy E_{pr} based on the amount of hydrogen which has to be stored

$$\frac{E_{pr}}{m_{H_2}} = \frac{p}{1000} \cdot \frac{m_{sp}}{m_{H_2}} \cdot \frac{1}{\eta_M} \cdot \frac{1}{\eta_1 \cdot \eta_2} \cdot Z \tag{128}$$

where η_1 is the efficiency of the conversion from heat into electricity and η_2 is the electrolyzer efficiency.

---

[13] It is interesting to note that under the assumptions outlined above 30% of the required primary energy is contained in the liquid hydrogen, whereas for synthetic hydrocarbon fuels this value ranges between 30% and 40%. If the improved fuel economy of hydrogen I.C. engines (page 150) is considered, LH_2 compares favorably to synfuels from an energy standpoint, without their environmentally detrimental effects.

Table 28. Required primary energy related to the amount of stored hydrogen for a specific range and varying means of propulsion and storage methods (SAE J-227-D driving cycle. From [66]

Cruising range	Storage	Propulsion	Mass[a] (kg) m_{H_2}	m_{Hy}	m_{sp}	Primary energy required (kWh/kgH$_2$)			Total required (kWh/kg)	Total H$_2$ (kg)
150 km	FeTiH$_{1.95}$	Otto engine	2.84	162	260	2.6[b]	22.75[d]	9.5[e]	34.8	1180
	MgNiH		2.56	81.2	130	2.6[b]	12.6[d]	10.5[e]	25.7	1065
	LH$_2$	Otto engine	2.33	—	15	31.4[c]	1.85[d]	0.25[e]	33.5	970
		Electric engine and fuel cell	2.64	1800[h]	15	31.4[c]	0.72[g]	10.5[e]	42	1098
500 km	FeTiH$_{1.95}$[a]	Otto engine	21.6	1233	1974	2.6[b]	75.65[d]	9.5[e]	87.75	8985
	FeTiH$_{1.95}$[f]		13.6	835	1000	2.6[b]	56.76[e]	7.7[e]	67.0	5657
	MgNiH		11.8	374	598	2.6[f]	41.9[d]	10.5[e]	55.0	4908
	LH$_2$	Otto engine	8.1	—	50	31.4[c]	5.96[d]	0.25[e]	37.6	3370
		Electric engine and fuel cell	8.8	1800[h]	50	31.4[c]	2.41[g]	10.5[e]	44.3	3600

[a] m_{H_2} Hydrogen mass, m_{Hy} hydride mass, m_{sp} total mass of the hydride storage ($m_{sp} = 1.6\ m_{Hy}$).
[b] Primary energy required to compress hydrogen to 30 bar for storage.
[c] Primary energy required for liquefaction with: liquefaction work 11 kWh/kg, power station efficiency 0.31, electrolyzer efficiency 0.8.
[d] Engine efficiency including power train 0.17. The improved efficiency of H_2 engines under partial load is not considered.
[e] Primary energy required for the production of the storage system and the fuel cell, based on weekly refueling over 8 years (refueled 416 times).
[f] $m_{sp} = 1.2\ m_{Hy}$.
[g] Average efficiency of the fuel cell and electrical propulsion motor including the power train $\eta_M = 0.6 \cdot 0.65 = 0.39$.
[h] Fuel cell mass 1800 kg (15 kg/kW). Operating power output 60 kW, peak power output 120 kW. It is assumed that the mass of the propulsion motor and its accessories is equivalent to the mass of the 100 kW Otto engine. In the best case a total vehicle weight of 1100 kg + 1800 kg = 2900 kg is assumed as opposed to 1100 kg (without storage) in the normal case.

Case B: Storage of Liquid Hydrogen

The energy equation is

$$\frac{m_0 + m_{sp} + m_{H_2}}{1000} \cdot p \cdot Z = m_{H_2} \cdot h_u \cdot \eta_M \tag{129}$$

where m_{sp} is the mass of the storage tank.
The following is obtained from (129)

$$m_{H_2} = \frac{m_0 + m_{sp}}{\dfrac{1000\eta_M h_u}{pZ} - 1}. \tag{129}$$

With (128) the amount of primary energy per unit of mass storage can be obtained

$$\frac{E_{pr}}{m_{H_2}} = \frac{p}{1000} \cdot \frac{m_{H_2} + m_{sp}}{m_{H_2}} \cdot \frac{Z}{\eta_M \cdot \eta_1 \cdot \eta_2}. \tag{131}$$

Table 28 provides a comparison of both storage methods and is based on a vehicle with a mass of 1100 kg without the storage mass or the mass of the fuel cell aggregate. The mass of the propulsion engine is contained in the 1100 kg in all cases. Additionally it was assumed the mass of the electrical engine (60 kW continuous rating, 120 kW peak capacity) was considered to be equivalent to that of the I.C. engine (100 kW) which was advantageous for the fuel-cell vehicle. Vehicles with hydride storage or liquid hydrogen storage and I.C. engines as well as vehicles with liquid hydrogen, fuel cells and electric engines are compared. The assumed ranges per refueling are 150 km and 500 km. In particular the total amount of primary energy required for the storage of hydrogen (data column 7) is of interest. It is about the same at the 150 km cruising range for iron–titanium hydride and liquid hydrogen for this driving cycle whereas at the 500 km cruising range per refueling, liquid hydrogen (37.6 kWh/kg) is clearly superior to iron–titanium hydride (87.75 kWh/kg) with respect to energy. For the driving cycles considered here, a vehicle with fuel cells, electrical propulsion and liquid hydrogen storage behaves similarly for both distances. With respect to the total energy required in carrying out the driving cycle it is comparable to the combination of the I.C. engine with liquid hydrogen, but is clearly inferior with respect to the total mass of the system (see Table 28, note[h]).

The storage of hydrogen in the vehicle with magnesium–nickel hydride is cited here for comparison. However, it can only be used in combination with iron–titanium hydride due to the high decomposition temperature of the nickel hydride (see Table 23). The iron–titanium hydride storage provides hydrogen until the magnesium–nickel hydride storage has been sufficiently heated by the exhaust gas (combined hydride storage). A large reduction in the total storage weight does not seem to be possible here. Furthermore, the high heat of decomposition of the magnesium–nickel hydride, amounting to 27% of the lower heat value of hydrogen together with the necessarily high decomposition temperature of about 250 °C (Table 23) is a problem. The average I.C. engine exhaust gas temperature of about 550 °C can result in an insufficient amount of usable waste heat to liberate the hydrogen that is needed for the combustion in the engine under all possible operating conditions due to the low temperature difference available. This is especially applicable when the engine is driven at optimum combustion efficiency under part load, i.e., with minimum exhaust heat [59].

The distribution of the various ways that primary energy is used, depending upon kind of hydrogen storage, is informative. With hydrides, most of the primary energy (column 5, Primary energy required) serves to accelerate the heavy storage system in performing the driving cycle. For the liquid hydrogen storage the liquefaction process requires the most primary energy. While in this respect liquid hydrogen is equivalent to either the simple or the combined hydride storage at an unrefueled range below 150 km, without exception liquid hydrogen is superior at an unrefueled range of 500 km, and corresponds to gasoline with respect to the mass of the tank and fuel. In inter-city traffic with a correspondingly long unrefueled range and nearly constant driving speed, i.e., rarely accelerating, the hydride storage is more favorable with respect to the amount of primary energy consumed. Even on long graded road sections the large weight of the required storage system results in an unreasonable reduction in driving speed as well as maneuverability in city traffic, and a passenger car would have the road performance of a loaded pickup truck. Conceptual advantages of hydride storage lie more in stationary applications than in mobile ones [17, 18].

Electrical propulsion via fuel cells and stored liquid hydrogen can roughly be compared with the I.C. engine regarding range and amount of primary energy for hydrogen storage, but regarding the high weight of the currently available fuel cells and electrical propulsion it is inferior in the same manner as hydride storage [10, 67]. This shortcoming could be improved by the use of a hybrid propulsion system, i.e., the coupling of a fuel cell with low power, and thus low weight, with an efficient storage battery, which handles the acceleration peaks in city traffic and is charged by the fuel cell [68]. Apart from the price it seems doubtful that future development will lead to light weight fuel cells that can handle the higher continuous power demand of the vehicle propulsion system in inter-city traffic and its corresponding driving speeds. The fuel cell electrical vehicle could be important primarily for the traffic in and around the city. Liquid hydrogen could also represent the best energy solution when using appropriate light weight fuel cells—propulsion and H_2-storage aggregates.

Liquid hydrogen storage represents the most favorable compromise between driving characteristics in urban traffic and cross-country traffic with regard to vehicle operation and required amounts of primary energy. From this point of view it is nearly equivalent to hydrocarbons and gasoline [67].

6.2.4 Experimental Liquid Hydrogen Vehicles

Electrical Propulsion via Fuel Cells: Although many test vehicles have been demonstrated with hydrogen–oxygen and hydrogen–air fuel cells, no vehicle has yet been demonstrated in which hydrogen was stored in its liquid state.

The General Motors and Union Carbide test vehicle which was used in 1965–66 was initially designed for liquid hydrogen and liquid oxygen fuel components. It was a minibus for six persons which was equipped with fuel cells and an electrical motor instead of a gasoline engine [67, 70]. The normal output of the 1,500 kg fuel cell aggregate, which consists of 32 modules, was 32 kW under constant load and could be increased to 160 kW for a short time. A cruising range of about 250 km was planned for the liquid hydrogen and liquid oxygen storage. Under realistic test conditions, however, pressure bottles were used because the storage of cryogenic hydrogen and particularly of cryogenic oxygen in the vehicle caused technical safety problems.

Oxygen was used instead of air in order to achieve the power which is necessary for accelerating the vehicle (0–100 km in 30 s) as well as for maintaining a corresponding top speed (115 km/h) with the fuel cells that were available at that time. Operation of the vehicle revealed that, at that time, fuel cell technology had reached a development stage which for the first time provided the power necessary for the vehicle's propulsion (at 100% greater weight, compared to gasoline operation) for electrical propulsion.

Vehicles with Internal Combustion Engine: As the following Table 29 shows, this type of vehicles was developed and tested between 1973 and 1988 by different groups in the U.S., Japan and Germany (see [162]).

The following discussion briefly describes characteristics of these test vehicles with emphasis on the cryogenic fuel system. In the interest of coverage of recent developments in this field, the DFVLR test vehicles and test vehicles which were designed by LASL and DFVLR together as well as those which were developed in cooperation with BMW are described in more detail. The referenced literature gives further details, especially concerning engine behavior, exhaust gas emissions and further test results. (Also see [85, 86]).

1. The Perris Smogless Automobile Association Test Vehicle (1971): According to the author's knowledge this vehicle, a Ford F250 pickup

Table 29. LH_2-vehicles

1. 1971 Perris Smogless Autom. Association [71]
2. 1973 Billings Energy Corp. (BEC) [42, 73]
3. 1973/74 Los Alamos Scientific Laboratory (LASL) [74, 162] (see footnote 16 on page 176)
4. 1973/74 University of California Los Angeles (UCLA) [75, 76, 162]
5. 1975 Mushasi Institute of Technology, Tokyo [62]
6. 1978/79 DFVLR-Stuttgart and University Stuttgart FKFS [77] (se footnote 16 on page 176)
7. 1979–1982 LASL-DFVLR joint project [78–80, 162]
8. 1980 Musashi Institute of Technology [54, 81, 83]
9. 1981 DFVLR-Stuttgart [66, 82, 162]
10. 1982 Musashi Institute to Technology [84, 132, 162]
11. 1984 DFVLR-BMW joint project, 745i [113, 120, 140, 125]
12. 1984 Musashi Institute of Technology (Musashi VI) [118, 121]
13. 1985/86 BMW-DFVLR joint project, 745i internal mixture formation [121, 122, 113]
14. 1986 Musashi Institute of Technology (Musashi VII) [151, 154, 155]
15. 1988 BMW, modification of the 1984 vehicle with a new model 735i [141, 142]

truck with an 8-cylinder engine, was the first automotive vehicle ever driven with stored liquid hydrogen. Liquid hydrogen and liquid oxygen were stored in Union Carbide–Linde laboratory storage containers [71, 85]. Evidently, there are no other publications of test programs and corresponding test results since 1971[14].

2. The Billings Energy Corporation Test Vehicle (1974): After a camper (Winnebago Motorhome, 1972) with a 7.2 liter 8-cylinder engine, including additional use of oven, refrigerator and electrical generator was converted to operate with liquid hydrogen fuel [42, 73, 85]. A Chevrolet Monte Carlo V8 with an 8-cylinder engine and 5.7 liter displacement and external mixture formation was introduced in 1974, a Beech-Aircraft Corporation LNG-tank was used for the storage of liquid hydrogen and hydride storage was used reserve. In order to reduce NO_x emissions and to avoid backfiring into the intake manifold, water injection was evidently applied for the first time.

3. The Los Alamos Scientific Laboratory Test Vehicle (1973/74): This vehicle, a Dodge 1/2Ton Pick-up Truck with four-speed transmission, 5.2 liter V8 engine was converted to hydrogen operation with external mixture formation and filling control with the following modifications (see [74, 85]):

The carburetor was replaced by an IMPCO gas mixer designed for use with LPG and propane. In order to avoid uncontrolled preignition, and backfiring into the intake manifold and knocking, exhaust gas recirculation (EGR) was applied.

[14] Hydrogen–oxygen engines, see [72].

Fig. 71. Los Alamos National Laboratory liquid hydrogen test vehicle (1973) (Source: Los Alamos National Laboratory)

A 190 liter spherical aluminum storage dewar was designed and fabricated by the Minnesota Valley Engineering Company (MVE) (see Fig. 71). It was mounted in the back of the pickup truck. With respect to capacity at 42.6 kg it is the lightest tank used in vehicles operated by liquid hydrogen to date. The maximum boiloff rate of the container was less than 1% per day but amounted to about 3.3% per day when the necessary provision for instrumentation and filling and discharge of hydrogen were used. Since this vehicle (Fig. 71) was not constructed in the scope of a sponsored research project it only underwent limited testing. It was driven with hydrogen over a cumulative distance of 500 km on roads in and around Los Alamos (Approx. 2200 m sea level) which had long inclines of up to 7%. For hydrogen it had a specific energy consumption of 1.11 kWh/km which corresponds to 67% of the specific energy consumption of gasoline operation (1.65 kWh/km). Therefore, this vehicle was mainly driven under part load which on one hand was due to the high elevation and on the other hand due to the exhaust gas recirculation. The maximum engine power obtained with hydrogen was estimated at about 50% of that with gasoline operation. The liquid hydrogen in the dewar was under an overpressure of about 0.5 bar and reached the IMPCO gas mixer via a heat exchanger after evaporation and warming.

The experience with this vehicle, especially with the cryogenic components, pointed the way for the subsequent development in this field in the following years.

4. The University of California Los Angeles (UCLA) Test Vehicle 1973/74: After dynamometer testing in 1973 on a Chevrolet V8 engine (5.7 liter displacement) which was fueled with stored liquid hydrogen, an AMC-Jeep (provided by the U.S. Post Office) with 3.8 liter 6-cylinder engine, was converted to hydrogen operation with external mixture formation and was widely demonstrated [75, 76, 85].

A dewar of the same type as in the Los Alamos test vehicle was used as storage for liquid hydrogen. It was installed in the rear of the vehicle which usually served to store the outgoing mail.

The modified carburetor of the engine was used for water injection and supplied gaseous hydrogen at ambient temperature via two IMPCO gas mixers one of which worked under low partial load with a throttling flap (volumetric control) while the second was switched on under high partial load to full load where quality governing was carried out. The hydrogen–air equivalence ratio was kept constant under low partial load at about $\varphi = 0.45$ and then increased continuously under high partial load and at full load to $\varphi = 1$. More specific information about this remarkable vehicle is provided in publications [76, 77].

5. The Musashi-Institute of Technology Test Vehicle, Tokyo (1975): This vehicle (Musashi III), a Datsun B210 with 1.4 liter, 4-cylinder, Otto-engine was converted to hydrogen operation with liquid hydrogen storage to participate in the "Student Engineered Economy Design Rallye" (SEED-Rallye) which took place in California in August 1975 [62, 85–87].

For this purpose, a conventional storage vessel (inner vessel connected to the outer vessel at the neck) with a capacity of 230 L was installed in order to avoid refueling during the 400 miles rally (Fig. 72). The tank weighed 120 kg, the inner vessel was made of stainless steel with a wall thickness of 3 mm whereas the outer vessel was made of aluminum with a wall thickness of 5 mm (1,000 mm, approximately 800 mm height). The maximum operational pressure was 5 bar, the boil off rate amounted to about 2.5% per day. In order to achieve rapid pressure buildup in the tank after refueling, cold gaseous hydrogen from the tank was heated via a heat exchanger and was pumped back into the liquid phase of the tank.

Cold gaseous hydrogen was removed from the tank via a vacuum-insulated line which was routed inside of the vehicle. It was supplied to the engine at a temperature of about 140 K through a throttle valve which was mechanically actuated by an accelerator pedal.

The engine was operated with cryogenic external mixture formation, in which the cold gaseous hydrogen was injected into the intake ports of each cylinder at an overpressure of about 1 bar through valves which were mechanically actuated by the camshaft. The hydrogen intake valves opened simultaneously with the cylinder intake valves of the corresponding cylinder

Fig. 72. View of the Musashi 3 test vehicle. The rear construction served for covering the upper part of the liquid hydrogen tank (Source: Musashi Institute of Technology)

and closed after reaching a crankshaft angle of 90 degrees. Since the opening period of the hydrogen intake valve decreases inversely proportional to the engine speed the inlet pressure at the valves should increase at a greater rate than the rate of revolutions in order to achieve linear behavior of the accelerator pedal.

Linear behavior of the accelerator pedal was achieved by means of a second valve in series with the engine throttle valve which increased the inlet pressure at the hydrogen intake valves with increasing engine speed.

In the interest of obtaining low NO_x emissions, the engine was operated under middle to full load with quality governing (throttle flap open) and below an equivalence ratio of about $\varphi = 0.6$ with intake air throttling and increasing equivalence ratio (volumetric control).

With this engine, to the author's knowledge the first that operated with cryogenic external mixture formation (see page 156), the performance of the gasoline engine was attained with hydrogen and the advantages of cryogenic mixture formation were demonstrated.

The specific energy consumption of the vehicle at the SEED-Rally, in which most of the 2,781 km were driven at nearly constant speed (80–88 km/h), amounted to 0.6 kWh/km according to official results given by the organizer which, calculated for gasoline, corresponds to a value of 0.57 kWh/km (15.3 km/L; 6.5 L/100 km).

6. and 9. The German Aerospace Research Establishment (DFVLR) Test Vehicle (1978/79): This vehicle, a BMW 518, was originally designed in a joint program between the DFVLR (now DLR) in Stuttgart and the

Forschungsinstitut für Kraftfahrwesen und Fahrzeugmotoren (FKFS) at the University of Stuttgart as the first LH_2-fueled car in Europe in 1978–79 and was demonstrated for the first time in 1979. At this time, the main emphasis was placed on the demonstration of total engine-LH_2-storage system. Therefore, the FKFS only used minimum effort to convert the engine to hydrogen, whereas more effort was placed on the cryogenics of the fuel storage system. This vehicle was completely taken over by the DFVLR in 1980 and modified several times according to the test results [66, 63, 82, 87].

The following brief description represents the state of development since April 1982 [63]:

The 4-cylinder BMW 520 engine with a 2 liter displacement is used in the U.S. version (reduced compression ratio 1: 8.3). This vehicle is equipped with an exhaust gas turbocharger which starts at 2,500 r.p.m. and delivers a boost pressure of 0.3 bar. The dual-fuel operation concept has been maintained[15], so that operation with gasoline as well as hydrogen is

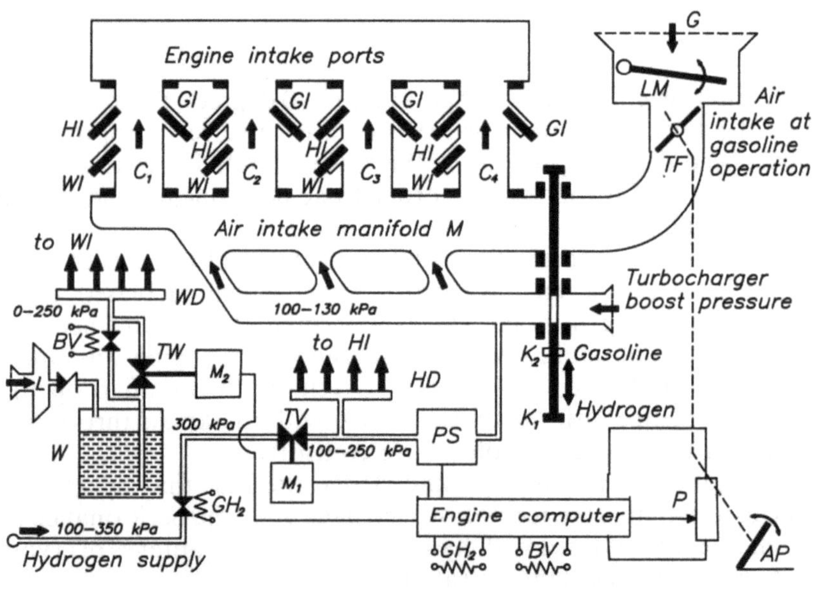

Fig. 73. Functional diagram of the DFVLR test vehicle (BMW 520) for the gasoline (L-Jetronic) and hydrogen operation. *P* potentiometer; *TV* throttle valve, *PS* pressure sensor for quality governing with H_2 operation (cap K_1 mounted, K_2 removed); *HI* H_2 injection; *WI* water injection at the cylinder intake ports C_1 to C_4; M_1 servo motor for throttle valve operation; *TW* throttle valve for water injection with BV bypass valve; M_2 servo motor; *AP* accelerator; *D* throttle flap; *LM* air flow meter of the L-Jetronic with gasoline operation (cap K_2 put on); *G* air intake at gasoline operation; *GI* gasoline injectors of the L-Jetronic for gasoline

[15] Not to be confused with mixed hydrogen–gasoline operation, compare [88].

possible. A Bosch L-Jetronic gasoline injection system without the turbocharger is used for gasoline operation with filling control.
Hydrogen operation was carried out with electronically controlled external mixture formation and water injection. Water injection starts only in the acceleration phase under high partial load to full load. A diagram of the principle of operation is shown in Fig. 73 while Fig. 74 shows a view of the engine compartment.
Above the cylinder head is an air intake manifold *M* (Fig. 73) which is supplied with air by the turbocharger under hydrogen operation. The air

Fig. 74. Engine compartment of the DFVLR test vehicle (BMW 520). (See also Fig. 73). *W* storage container for water, *L* membrane compressor for pressurization, *P* potentiometer, *TV* throttle valve for power control, *PS* pressure sensor, GH_2 gaseous hydrogen main solenoid valve; *LT* connecting pipe to the turbocharger, *M* air intake manifold. *D* throttle flap for gasoline operation; *GI* fuel rail for gasoline injection; *LM* air flow meter of the L-Jetronic; *LF* air filter, *V* vacuum pump for power brake amplifier with hydrogen operation

mass flow-meter (*LM*) of the Bosch L-Jetronic injection system is separated from the air intake manifold *M* which is in turn closed at K_1. The air from the intake manifold *M* reaches the intake ports of cylinder 1 to 4 via U-shaped connectors C_1 to C_4. After warming to ambient temperature in its gaseous state (GH_2) at an overpressure of 1.5–3.5 bar, hydrogen from the LH_2-tank reaches a throttle valve *TV* which is electronically controlled and regulated by the accelerator pedal (*AP*). It is then carried to the intake ports of the four cylinders via a distributor (*HD*) and continuously injected into the intake air (continuous individual port injection). The volume of the four U-shaped connectors is kept sufficiently large to avoid the formation of a flammable hydrogen air mixture upstream in the air intake manifold *M*. The water injection is accomplished with specially designed injection nozzles *WI* in which part of the hydrogen, needed for idling the engine, is swirled through a rifled block and then mixed with water. Thus, at an injection pressure of only 1.8 bar, proper water atomization within a cone formed at about a 20° aperture angle aimed at the cylinder intake ports could be achieved. The injection pressure was provided by a cushion of compressed air in the water tank *W* while the injection flow rate and injection start was regulated via an electronically controlled throttle valve TW actuated by a servo motor M_2 dependent on the accelerator pedal position and during the acceleration process via bypass valve *BV* which can be manually shut off. The water supply (16 liter) is sufficient for about 400 km. For the urban driving cycle no water is needed under partial load. This distance is reduced to about 150 to 200 km when driving for longer periods of time at full power.

When regulating the engine, a nominal value is determined for the hydrogen pressure according to the position of throttle valve *TV* by means of a potentiometer *P* which is connected to the accelerator pedal. By comparison of the actual value provided by the pressure sensor PS and the nominal value, a control signal is conducted via the electronic unit which insures the appropriate position of the throttle valve via the servo motor M_1, with the pressure sensor *PS* given signal. The transition time of the throttle valve *TV* from idle to full power was set to about 1 second in order to prevent over-shooting which would result in oversaturation of the hydrogen–air mixture during acceleration. It has to be pointed out that this arrangement works over a wide range independent of inlet pressure at the throttle valve *TV* so that it is unnecessary to install the normally required voluminous pressure regulator.

The ignition timing is carried out by a custom-made centrifugal governor (initial operating point about 1,500 r.p.m., timing control to 8° b.t.d.c. at 4,000 r.p.m.) and due to lack of a vacuum within the air intake manifold (quality governing, no throttle flap) via a load-controlled electromechanical timer (signal from the potentiometer *P*). When the engine is started, the main solenoid valve GH_2 (Fig. 74) is initially opened with a half second delay to avoid oversaturation of the mixture. For the same reason, pressing the accelerator pedal during start procedure has no effect. Additionally, if the engine speed drops below 100 rpm during operation, the main solenoid valve is closed automatically in order to avoid the discharge of hydrogen into the idling engine.

When operating with gasoline the connection to the turbocharger is closed and the air intake port to the air intake manifold is opened at K_1 (Fig. 73). The engine is then regulated as usual via the throttle flap and the ignition timing is carried out via centrifugal and vacuum governors.

A tank, which was developed and tested in several programs in 1974, with a capacity of about 120 liters is used for the storage of liquid hydrogen [82, 86, 89]. It has a boil off rate of 8% per day and is installed in the trunk next to a small gasoline tank with a capacity of about 18 liters. This hydrogen tank (Figs. 75 and 76) was initially developed solely for the use in a motor vehicle. In contrast to the conventional storage dewars used in hydrogen vehicles it can be placed in any position without being damaged. Discharge and filling of liquid hydrogen is possible up to a deviation of about a 20° angle from horizontal position. The maximum operating pressure is about 4.5 bar (activation of the safety valves), the maximum permissible operating pressure is about 8.5 bar, the burst pressure about 15 bar. The tank is vacuum insulated and was made of stainless steel

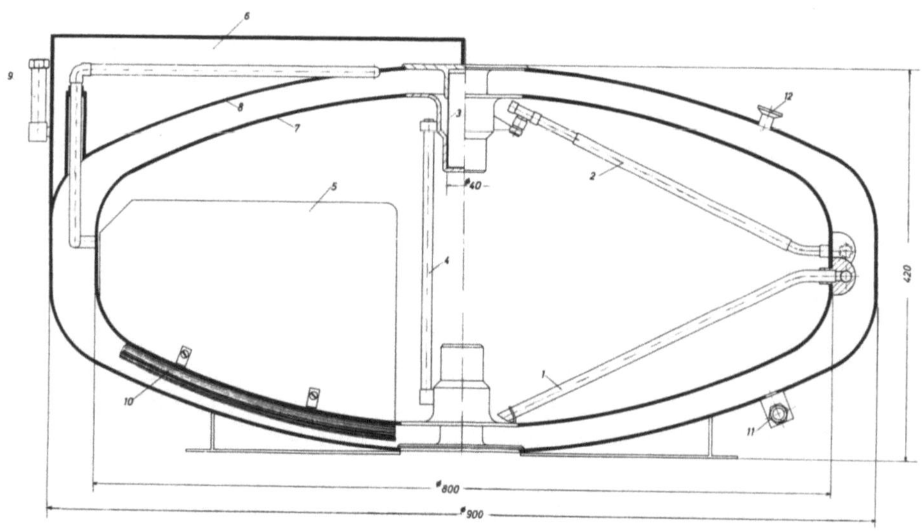

Fig. 75. Cross section of the tank of the DFVLR test vehicle. *1* and *2* line for liquid and gaseous hydrogen respectively; *3* upper bracing tube of the inner vessel (glass-fibre epoxy); *4* level sensors; *5* anti-splashing walls; *6* space for instrumentation and heat exchanger; *7* inner vessel; *8* outer vessel; *9* safety valves; *10* multiple layer insulation; *11* safety valves of the vacuum space; *12* evacuating nozzle

Fig. 76. View of the 120 L liquid hydrogen tank installed in the trunk of the DFVLR test vehicle. On the left a gasoline tank of approx. 16 L capacity for gasoline operation of the test vehicles (Photo: DFVLR)

for functional efficiency and weighs about 90 kg. Lighter tanks built according to the same principle, were made of aluminum during later developments [77, 82, 86, 87].

During operation, hydrogen is initially vented from the gaseous phase if the pressure in the tank exceeds a selectable pressure level in the range of 1.5 to 4.5 bar (Fig. 77). In this process, the pressure decreases until it falls below the preselected pressure level and an electronic control releases hydrogen from the liquid phase in order to maintain constant pressure in the tank. The liquid or gaseous hydrogen is evaporated and warmed to ambient temperature via a heat exchanger (Fig. 78) which is integrated into the tank with heat supplied by the engine coolant. It then reaches the engine through a transfer line which is run underneath the vehicle's floor via the main solenoid valve GH_2 and throttle valve *TV* (see Fig. 74). Thus, tank boil off losses due to pressure build-up during inactive periods can be avoided for about a maximum of two days because the resulting increased pressure level is reduced during the follow-on operation by the release of hydrogen from the gaseous phase.

Ten carbon resistors were used for fuel liquid level indication in the tank. When fed with constant current, they give a signal as a result of the change in resistance when the liquid level falls below the resistor which provides a signal relative to the number of carbon resistors.

The gaseous hydrogen vented from the tank after longer periods of inactivity via safety valves, is either emitted directly to the open air where it either

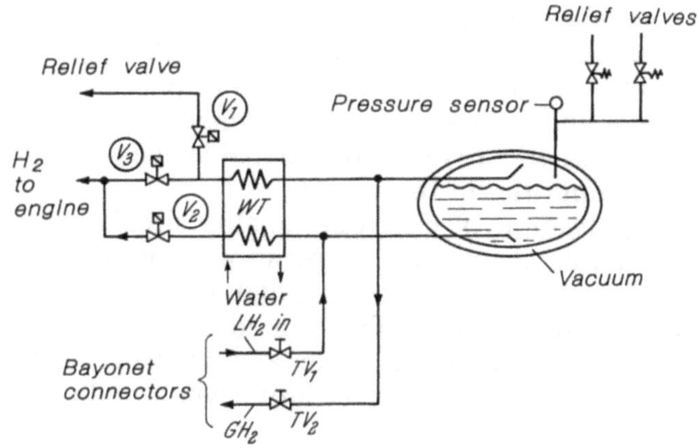

Fig. 77. Schematic of the hydrogen discharge from the tank of the DFVLR test vehicle. Hydrogen warmed to ambient temperature by the heat exchanger WT is discharged from the liquid and gas phase via the solenoid valves V_2 and V_3 dependent upon the pressure in the tank and is supplied to the engine.

Fig. 78. Compact heat exchanger for evaporation and warming of the hydrogen (Tube outer diameter 12 mm). Used in the vehicles in Figs. 79 and 86

is diluted or it is ignited and combusted by means of a pilot flame located underneath the vehicle. When the vehicle is parked in closed or poorly ventilated rooms the burner is always used and its pilot flame is monitored by a thermocouple and is automatically reignited when it goes out. The original "catalytic burner" proved to be unreliable due to the susceptibility of the catalyst to impurities in the air [66].

The consistent application of electronic analog circuits for the mixture formation, electronic accelerator pedal operation, ignition timing and water injection as a preliminary stage for using digital motor electronics, made test operation easier due to the flexibility in adjustment. It fully confirmed the initial expectations placed on them and verified the fact that the development of hydrogen vehicles in the future will be influenced to a large degree by the electronics.

In 1982, the test vehicle was driven on public streets in the U.S. with liquid hydrogen and underwent test programs in and around Los Alamos (New Mexico) including high altitude tests (2,300–3,000 m above sea level). The total distance driven amounted to about 3,200 km (Fig. 79). Table 30 shows typical data regarding acceleration and fuel consumption (see [63, 66]).

It can be seen from the acceleration response that under the conditions in Los Alamos the engine output is about the same for either gasoline or hydrogen operation. The turbocharger, which was primarily installed to

Fig. 79. View of the DFVLR test vehicle (BMW 520) on the occasion of high altitude testing in Los Almos in 1982

Table 30. Acceleration[1] and specific fuel consumption of the DFVLR experimental vehicle

Acceleration[a]	Hydrogen operation		Gasoline operation
0–90 km/h[a]	17 s (16.2 s)[c]		11 s (10.5 s)[c]
0–100 km/h[b]	19.5 s[k] (14 s)[c]		19 s (13.75 s)[c]
Empty weight (operational)	1,350 kg		1,250 kg
Total weight	1,510[g] kg		1,510[g] kg
	1,630[h] kg		1,630[h] kg
Fuel consumption			
A: DFVLR test track[d]	about.	60 L/100 km	—
		1.7 km/L	—
		1.4 kWh/km	—
Unrefueled range	approx.	200 km	—
B: ECE cycle[e]	approx.	40 L/100 km	—
		2.5 km/L	—
		0.944 kWh/km	—
Unrefueled range	approx.	300 km	
C: Test program in the U.S.[f]	approx.	32 L/100 km[k]	9.16 L/100 km[i]
		3.13 km/L	10.9 km/L
		0.755 kWh/km	0.96 kWh/km
Unrefueled range	approx.	400 km	approx. 180 km

[a] On the DFVLR test track. Air temperature 20 °C, atmospheric pressure 735 mm Hg.
[b] In Los Alamos. Engine speed in all the gears up to 4500 r.p.m., air temperature 26 °C, atmospheric pressure 588 mm Hg.
[c] Reduced to sea level and 20 °C.
[d] "DFVLR-cycle". 0.12 kWh/kg 1,000 kg, no idle, average engine efficiency including power train $\eta_M = 0.18$.
[e] 0.0814 kWh/km, no idle, B calculated from A.
[f] Average value.
[g] Including driver and passenger.
[h] Including luggage, repair parts and reserve gasoline.
[i] Regular, unleaded.
[k] With turbocharger.
[l] The speed data was corrected as opposed to data published earlier because an examination of the speedometer at questionable speeds reveiled an error in calibration of about 10 km/h.

avoid an increased nitric oxide emission was adjusted for a lean mixture without considerable loss in efficiency. With hydrogen operation it could therefore make up the loss in power caused by the high altitude. This was confirmed by the fact that with gasoline as well as with hydrogen operation, a speed of more than 130 km/h could be maintained on long stretches of approximately 4% gradient where the knock limit was not reached. For this purpose about 20 Vol. % tertiary butylalcohol was added to the water as an anti knock additive in order to avoid possible damage to the engine. It was noticed that through this measure the required amount of water injected under full load could be considerably reduced.

The gasoline acceleration data from Los Alamos and the DFVLR test track are not compatible even if the different total weights of the vehicle are used for comparison. This can probably be attributed to the L-Jetronic which could not be optimally adjusted due to the low atmospheric pressure (588 mm Hg) or by improper spark advance with respect to the fuel used (regular unleaded).

Since the application of the turbocharger in the test vehicle resulted in quite satisfactory driving performance, the application of timed cryogenic mixture formation (hydrogen temperature 40–60 K) could provide further improvement which, according to previous test results, reduced uncontrolled preignition and NO_x emissions and made water injection indispensable [63, 139] (Fig. 80).

7. Test Vehicle Designed and Constructed in a LASL-NMEI-DFVLR Joint Project (1979/80): During a project sponsored by the US Department of Energy, a test vehicle modification was designed in a joint effort by the Los Alamos Scientific Laboratory[16a], the New Mexico Energy Institute (NMEI) and the Deutsche Forschungs- und Versuchsanstalt für Luft- und Raumfahrt (DFVLR)[16b]. On one hand, this vehicle demonstrates the state of the art in automotive application with liquid hydrogen, including fuel handling and on the other hand it should provide guidelines and recommendations for further developments based on test results [79, 90–93, 135].

The LASL was responsible for the coordination of the project including management, engine conversion to hydrogen fuel, test operation and methodology. The test vehicle, a 1979 Buick-Century with a 3.9 L, 6-cylinder engine with a turbocharger and automatic transmission was provided by NMEI. The DFVLR initially provided an aluminum tank for liquid hydrogen (Fig. 81)[17], which was originally planned for another project, as well as a semiautomatic refilling station (see Fig. 119) including the required electronic equipment [77, 80].

The required engine conversion to external mixture formation with volumetric control was done by the Billings Energy Corporation (Independence, Missouri) included the installation of an IMPCO gas mixer on the carburetor, the adaptation of the spark advance as well as the tuning and dynamometer testing. The gas mixer and the original gasoline fuel pump were modified for the water injection. The vehicle was put into operation in May 1980 and after surmounting initial problems it was operated until July 1982 in New Mexico, mostly around Los Alamos, however, it was also successfully demonstrated in various cities in the U.S.

[16a] Since 1981 Los Alamos National Laboratory, LANL; [16b] DFVLR: now DLR.

[17] The author wishes to express his gratitude to the Los Alamos National Laboratory and especially to W. F. Stewart for the permission to publish Figs. 71, 81–83.

Fig. 80. Four-cylinder BMW-M20 engine with cryogenic external mixture formation on the test stand. Continuous individual intake port injection of hydrogen at 80 K. The air intake manifold was removed for the sake of clarity

Fig. 81. View of a liquid hydrogen tank installed in the trunk of the Los Alamos test vehicle. (Photo: Los Alamos National Laboratory)

Fig. 82. View of an experimental tank for liquid hydrogen installed in the trunk of the Los Alamos test vehicle. It was developed in a mutual program between the DFVLR and the Los Alamos National Laboratory (Photo: Los Alamos National Laboratory)

It must be pointed out that due to Los Alamos' altitude, the vehicle acceleration was not completely satisfactory. The turbocharger, which normally compressed the fuel–air mixture in the gasoline version and was not designed for the hydrogen–air mixture could only partially eliminate the loss of power output due to altitude and external mixture formation. Furthermore, during the test program, a cylindrical tank for liquid hydrogen was used (Figs. 82, 83) which had been designed and constructed in a joint effort between the LASL and the DFVLR. This tank made better use of the space in the trunk of the test vehicle [90, 91]. The aluminum outer vessel of this experimental tank is fitted with flanges for access purposes. The inner vessel was made of stainless steel and had a capacity of 155 L liquid hydrogen. It is surrounded by two vapor-cooled, superinsulated, copper radiation shields [80, 93]. This design explains the relatively large mass of the tank of 156 kg, including valves and tubing. The boil off rate was approximately 4% per day. Pressure buildup as well as continuous pressure control while driving could be carried out by means of an installed electric heater (100 W).
Some refueling of the vehicle (refueled total of 65 times) was carried out by the DLR-filling station. Outside of Los Alamos the refueling was carried out by means of a transportable liquid hydrogen storage vessel [79, 87,

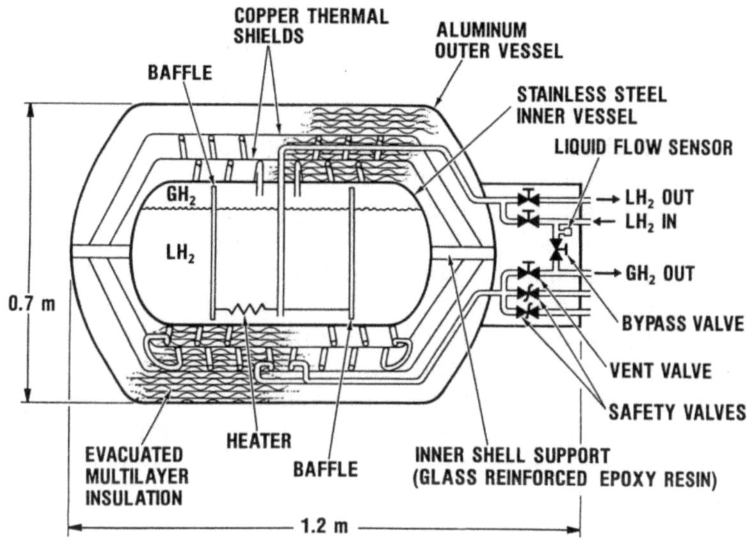

Fig. 83. Schematic of the LH_2 fuel tank of the Los Alamos test vehicle (Fig. 82) (Source: Los Alamos National Laboratory)

Table 31. Typical data of the LASL experimental vehicle (see [90, 91])

		Hydrogen	Gasoline
Acceleration from 0–80 km/h	(s)	16.8[a]; 24.8[b]	14.4[c]
Weight, operational	(kg)	1,610; 1,715	1,520
Fuel consumption	L/100 km	41.3[d]	13.8[e]
	km/L	2.42	7.23
	kWh/km	0.84	1.21
Nitric oxide emission	g/km	0.04; 3.7[f]	0.37–0.5[h]
Cruising range	(km)	274[d]; 362[d,f]	483
Total cumulative driven distance	(km)	3633	—

[a] In Independence, Missouri, altitude 229 m.
[b] Los Alamos, New Mexico, altitude 2,225 m.
[c] Around Los Alamos determined by a similar Buick Regal.
[d] Cross country around Los Alamos, to Santa Fe and Albuquerque.
[e] Environmental Protection Agency estimate.
[f] Idling and at full load, not an average value.
[g] Second tank.
[h] Data provided by manufacturer for the average achieved in operation over a standard driving cycle.

90, 91]. With a cold tank refueling times of 9 minutes could be achieved (14.5 LH_2/s) with an evaporation of 9.1 liter LH_2 caused by the cooldown of the transfer line. This corresponds to a volumetric efficiency of 92% without filling station and 88% with filling station. The loss of 9.1 liter LH_2 without refueling station is caused by the cooldown procedure of the transfer line from the storage tank to the DFVLR filling station. This transfer line is sized for the considerably larger flow rates to the 50,000 liter LH_2 main storage tank which served for the refueling in Los Alamos. Improved refueling efficiencies could have been obtained with transfer lines designed for lower flow rates.

The goal of the project were the operational test of a vehicle with liquid hydrogen under normal traffic and road conditions including the collection of technical data and measurements, the demonstration of the vehicle and its refueling, and the determination of recommendations for further research and development. All of this goals were achieved. Vehicle and refilling station were later sent to the "Institute for Hydrogen Systems" (IHS) in Toronto for further testing.

8. and 10. (see Table 23) Musashi Institute of Technology Test Vehicles with Internal Mixture Formation (1980–82); In 1980, a vehicle with a 3-cylinder two-stroke engine and electronic ignition (Musashi IV) [81] was exhibited at the 3rd World Hydrogen Conference in Tokyo. A similar vehicle with a 3-cylinder two-stroke glow-plug-assisted Diesel engine was exhibited at the World Hydrogen Conference in Pasadena in 1982 (Musashi V) [84] (Fig. 84).

With the knowledge that external mixture formation is practically excluded with two-stroke engines because a more or less intensive mixture of intake gas and exhaust gas takes place during charge exchange, which leads to uncontrolled preignition with a hydrogen–air mixture, an attempt was made to operate two-stroke engines with pure air scavenging and internal mixture formation [54, 83, 87].

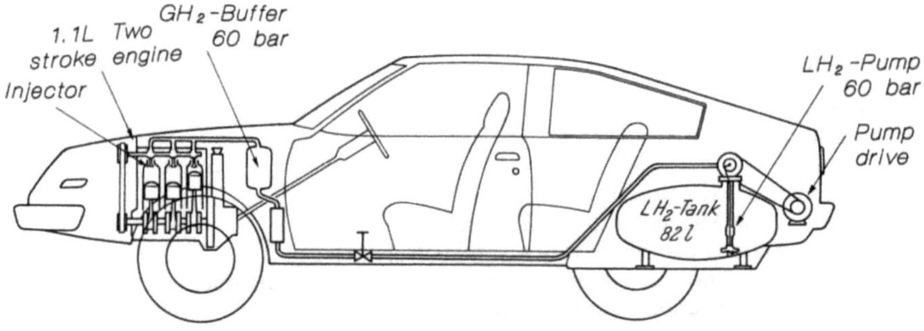

Fig. 84. Schematic of the construction and cryogenic system of the Musashi 5 test vehicle (Photo: Musashi Institute of Technology)

The results show that the residual gas from the charge exchange of the two-stroke engine corresponds to the exhaust gas recirculation of the four-stroke engine so that without additional measures, a nitric oxide emission is obtained which is considerably lower than that obtained with the four-stroke engine with internal mixture formation. Knock and especially uncontrolled preignition can be reduced by injection of cold gaseous hydrogen. However, with "early injection", i.e., directly after closing the inlet port by the piston, efficiency and power output of the engine being tested (1,100 cm^3) decreased to a large degree with a hydrogen intake temperature below about 210 K, as a consequence of low flame speed of the cold gaseous hydrogen and the accompanying reduction of the peak pressure as well as cyclic variations in the pressure. This was avoided in the Musashi III (539 cm^3 3-cylinder two-stroke engine of the Suzuki CERVO) and the Musashi IV (1,100 cm^3) by the injection of cold gaseous hydrogen at a temperature above 150 K [81]. Furthermore, as a result of the injection of cold gaseous hydrogen it was also possible to increase the compression ratio from about 6.9:1 with gasoline operation to 8:1 with hydrogen operation. Thus, results for power output and efficiency were obtained which were higher than those with gasoline operation. Moreover, the nitric oxide emission could be reduced to about 60% of that for gasoline operation.

The hydrogen injection took place via one injection valve per cylinder which was mechanically actuated by a camshaft. Injection valves and camshaft are installed on the cylinder head.

An important problem is the production of the required injection pressure of about 10 bar onboard the vehicle which would require a larger compressor for ambient gaseous hydrogen with appropriately large shaft power. This problem could be solved by the development of a small, compact reciprocating pump for liquid hydrogen [81]. The cold side of this pump, driven by a 70 W electrical motor, is located in the tank in liquid hydrogen. The Musashi III vehicle used a cylindrical tank whereas the Musashi IV vehicle uses a flat tank made of stainless steel similar to that of the DLR-test vehicle. This pump arrangement was chosen because proper pump operation is only possible close to 20 K as a result of the clearance between piston and cylinder wall which had to be kept small. In this manner the critical phase of the pump cooldown with a cold tank can be avoided. However, due to this arrangement, a considerably higher boiloff rate of the hydrogen in the storage vessel has to be accepted.

The further development of this pump for liquid hydrogen finally provided an injection pressure of 60 bar and thus permitted "late injection" close to top dead center. The Musashi V vehicle uses a 3-cylinder two-stroke engine with 1.1 liter displacement and a compression ratio of 12:1 which, equipped with platinum glow plugs (1300 K) and catalytic ignition of the

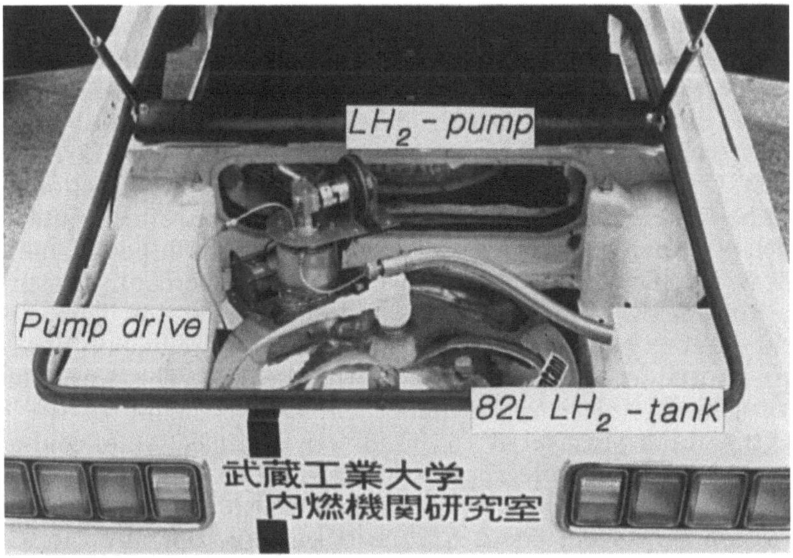

Fig. 85. View of a LH_2 tank with LH_2 high pressure pump in the rear of the Musashi 5 test vehicle (Photo: Musashi Institute of Technology)

hydrogen–air mixture[18], demonstrated Diesel-like operation of a vehicle with a hydrogen operated engine for the first time [84] (Fig. 85).

11. DFVLR-BMW Test Vehicle 745i. The test vehicle, a BMW 735i with a 745i motor, was designed in 1983/84 in a joint effort between the DFVLR and BMW[19], Table 32, (Fig. 86). It demonstrates the state of the art in cryogenic technology regarding hydrogen storage and fuel conditioning systems, as well as digital motor electronics for external mixture formation with quality governing and control of the engine characteristics under hydrogen operation. It was the first LH_2-vehicles in Europe licensed for liquid hydrogen operation. The fuel is stored in a liquid hydrogen tank manufactured by Messer Griesheim GmbH using DFVLR design and experience (Figs. 87, 88, 89).

It has a 130 liter capacity and a mass of 65 kg with an evaporation rate of less than 1.8% per day (Figs. 88, 89).

All tank refueling components, with the exception of the safety valves, are contained in a compact valve block located in a vacuum box (Fig. 90). All

[18] The ignition temperature of the hydrogen–air mixture is about 100 K higher than for a Dieselfuel–air mixture. Tests showed that Diesel operation with hydrogen is not possible without appropriate auxiliary devices [94–98, 124].

[19] Demonstrated at the 5th World Hydrogen Conference 1984 in Toronto and in the U.S. with the support of the Bayerische Motorenwerke AG. Afterwards, it was driven to Los Angeles partially under hydrogen operation.

Fig. 86. DFVLR-BMW 745i test vehicle on the chassis dynamometer. Rear view with tank and valve combination

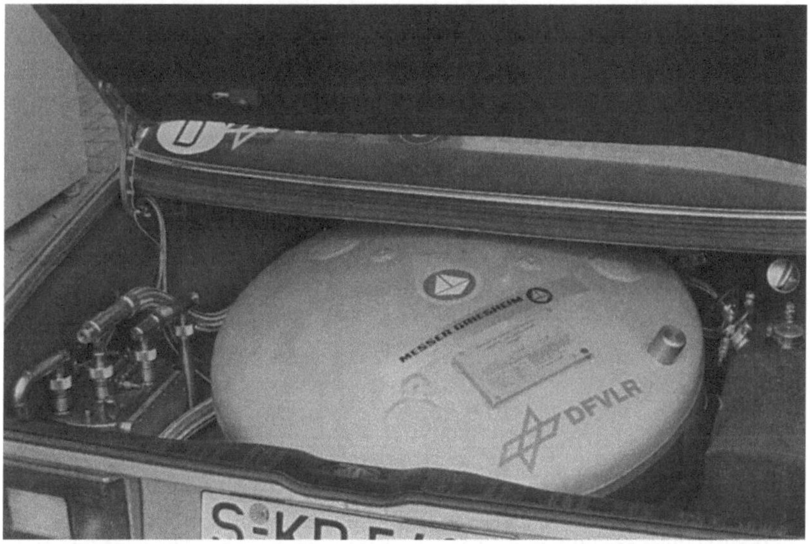

Fig. 87. DFVLR-BMW test vehicle. Rear view with tank and valve combination

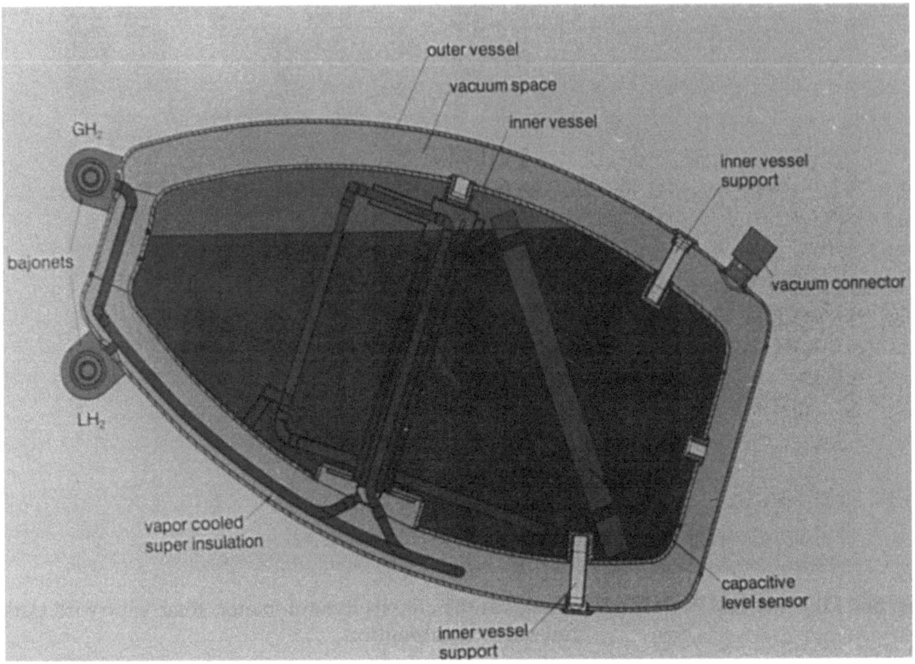

Fig. 88. Schematic of the LH_2 tank of the DFVLR-BMW 745i test vehicle

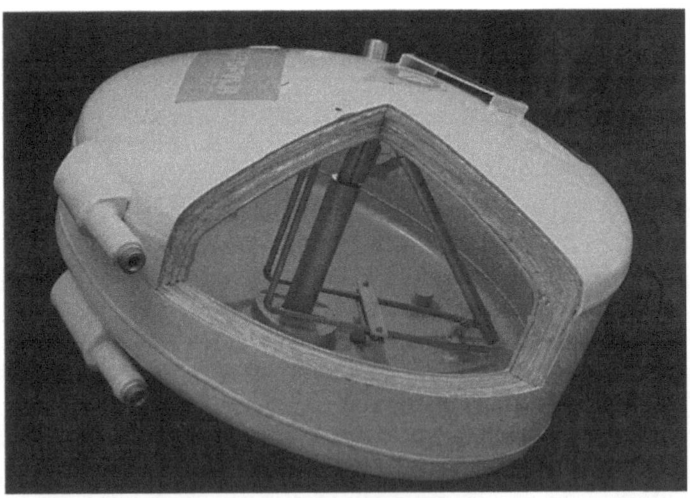

Fig. 89. Original LH_2 tank of the vehicle in Fig. 86 cut open. The capacitive level sensor as well as the LH_2 line for filling and LH_2 discharge can be seen. The vent line ends in the upper part of the vertical construction.

Fig. 90. Valve block and vacuum box of the vehicle in Fig. 86. All but one of the solenoids of the welded magnetic valves have been removed for the sake of clarity

processes such as tank filling, and releasing hydrogen from the liquid and gaseous phase are controlled by a central electronic unit by means of the valve block, which contains solenoid valves especially developed for this purpose. Hydrogen reaches the engine compartment in a liquid state via vacuum-insulated line. There, it is vaporized and warmed up to ambient temperature in the mixture formation system. Then it is injected individually into the intake ports of the six cylinders of the supercharged 3.4 L engine via valves which are actuated by the camshaft (Figs. 91, 92, 93).

Air precharging is used, i.e., during the intake stroke the relevant cylinder is initially supplied only with air while the corresponding hydrogen valve remains closed. The hydrogen induction, timed by the hydrogen valve was started at about 40° after top dead center (a.t.d.c.) and stopped at about 140° a.t.d.c. (Fig. 93).

For the test vehicle this basic approach proved to be the most favorable compromise up to an engine speed of about 4,300 r.p.m. under load which corresponds to a speed of 170 km/h. Using this approach, engine operation under load was not possible even with an increased water injection rate, at an engine speed from 4,300 to 6,000 r.p.m., especially with load transitions caused by the backwash of the exhaust gas at the turbine of the turbocharger and the accompanying uncontrolled preignition with the hot residual gas. With the test stand version of the vehicle engine, an additional hydrogen precharging could be obtained by mechanical modifications (sequence H_2–air–H_2 during the intake stroke) and thus an engine speed in the range of about 4,300 r.p.m. to about 6,000 r.p.m. could be achieved under load. In the vehicle it was planned to use electromagnetic solenoid injectors

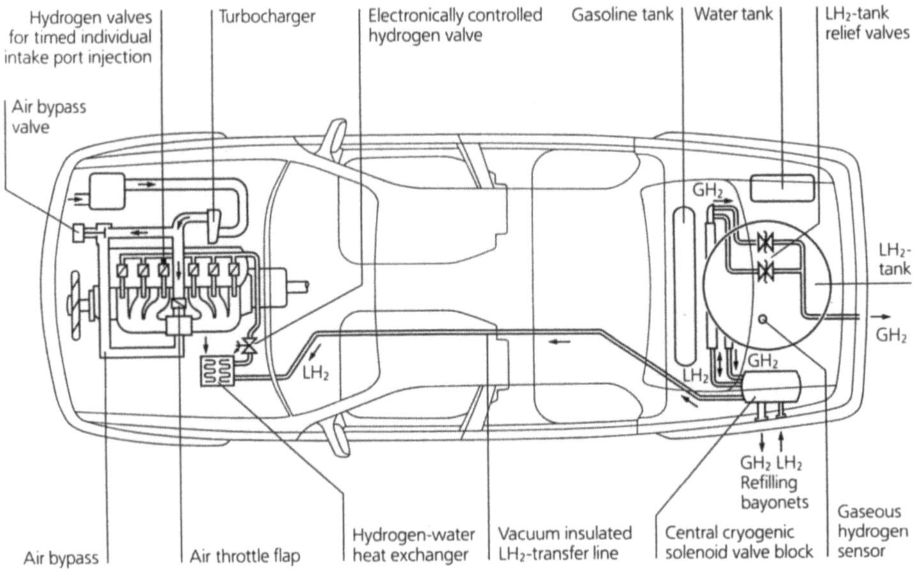

Fig. 91. Schematic of the LH_2 fuel system of the vehicle in Fig. 86

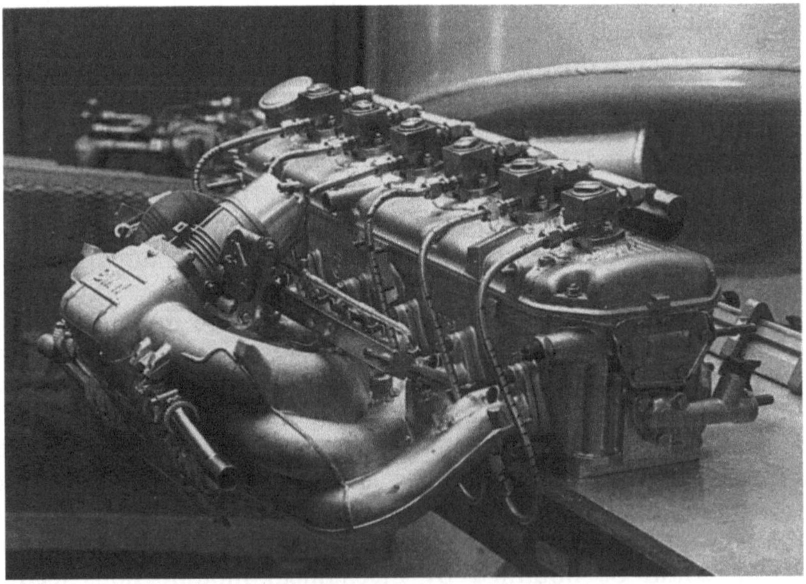

Fig. 92. Test stand version of the cylinder head for the vehicle in Fig. 86. Timed individual intake port injection by the timing valves mounted on the valve cover which are actuated by the rocker arms of the camshaft

which are sequentially controlled by the digital engine electronics [100, 102] however this was not carried out. This process has been recently adopted again by an automobile manufacturer.

In order to suppress uncontrolled preignition as well as reduce NO_x under full load, electronically controlled injection of water into the individual intake ports (see Figs. 92, 94) is used in parallel with a lean mixture concept and the exhaust gas turbocharger.

Due to the lean mixture concept and improved efficiency of the engine under partial load with hydrogen operation (maximum exhaust gas temperature 650 °C compared to 850 °C with gasoline operation) the regular turbine cylinder (16 cm^2 intake cross-section) had to be replaced by one with a considerably smaller intake cross-section (8 cm^2) in order to have sufficient boost pressure under partial load with an engine speed above at 2500 r.p.m.—although with an increased backwash of the exhaust gas and increased fraction of residual gas in the cylinder.

The engine is controlled by a Motronic microcomputer and a digital turbocharger boost pressure control. Through the use of special software and stored tables for ignition timing and fuel injection rates the optimum operating conditions for the engine are achieved dependent on engine load and engine speed. An additional EPROM was provided in the Motronic microcomputer for storing the hydrogen ignition table. Conversion from hydrogen to gasoline operation and vice versa is accomplished purely through electronics without any additional manual steps. For quality governing, the throttling flap was bypassed. Because the air flow rate does

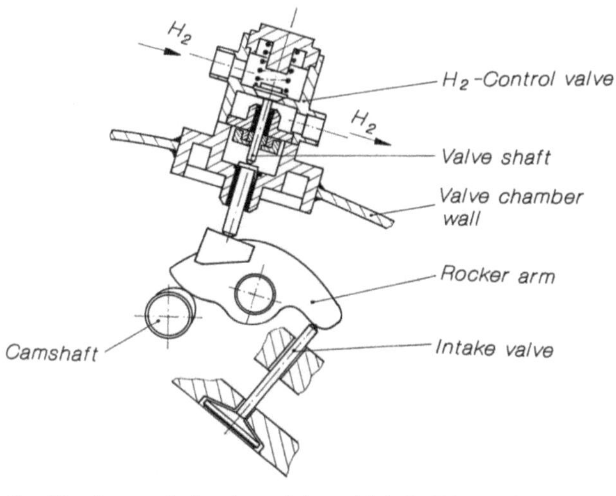

Fig. 93. Schematic of hydrogen induction of the vehicle in Fig. 86 by timing valves operated by the camshaft

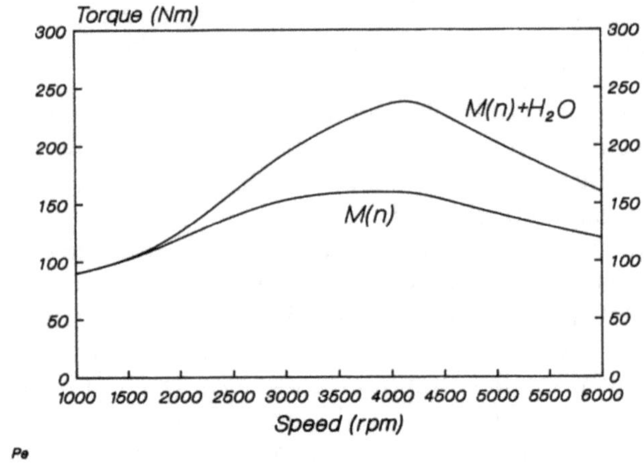

Fig. 94. Engine torque of the M30 engine under hydrogen operation with timed individual port injection and supercharging. The break in the torque curve at about 4,300 rpm caused by uncontrolled preignition can be seen

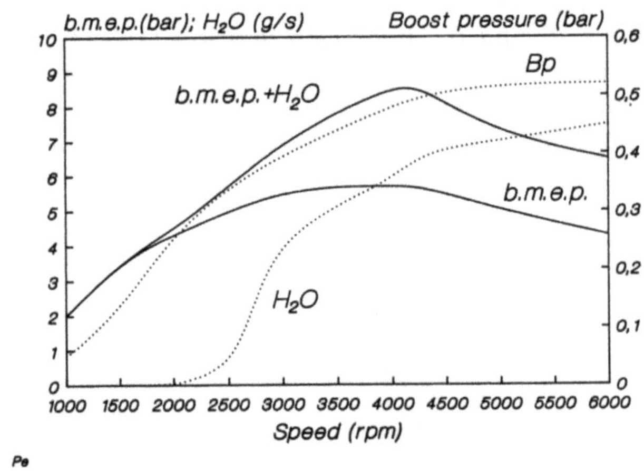

Fig. 95. M30 engine for the vehicle in Fig. 86. Brake mean effective pressure, Boost pressure and water injection rate

not depend on the engine power output at constant engine speed, it no longer represents a load signal. Therefore a true load signal is derived from the accelerator pedal position. For external mixture formation and timed individual intake port injection six, hydrogen timing valves actuated by the rocker arms of the intake valves were located on the top of the valve chamber.

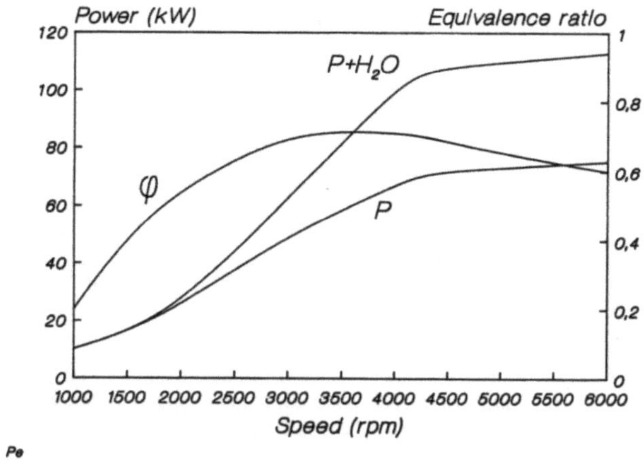

Fig. 96. M30 engine for the vehicle in Fig. 86, power output with and without water injection as well as equivalence ratio

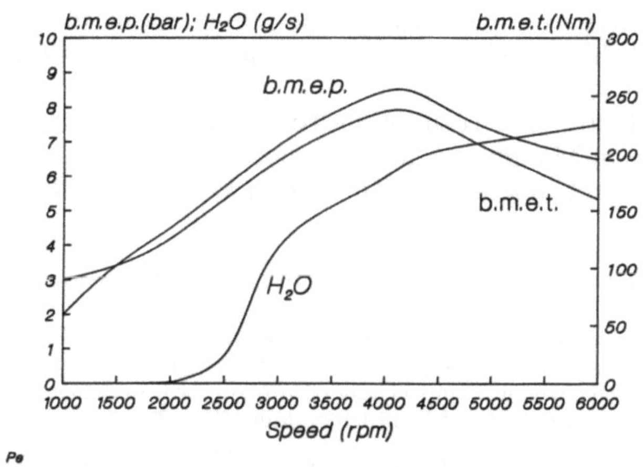

Fig. 97. M30 engine for the vehicle in Fig. 86. Brake mean effective pressure and brake mean effective torque at optimum water injection rate

The amount of hydrogen injected is dependent upon the engine load and is regulated by a central electronically controlled throttle valve. With this arrangement, the mixture formation system can balance out irregularities of the hydrogen supply pressure (Pressure in the LH_2 tank) in the range of 1.9–3 bar. Therefore, the LH_2 tank does not require an electrical heater to maintain constant pressure while driving.

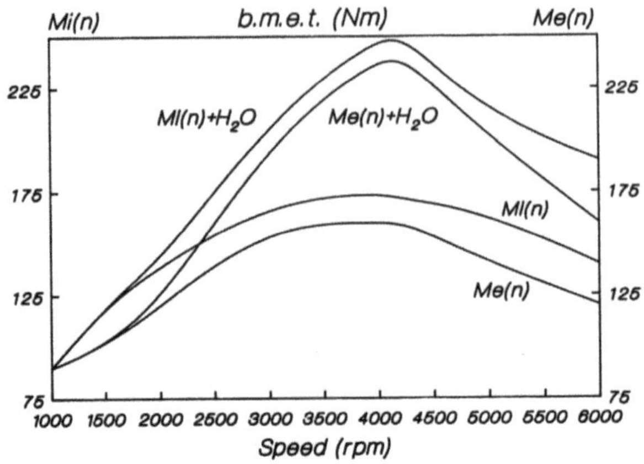

Fig. 98. M30 engine. Comparison of brake mean effective torque for external mixture formation $M_e(n)$ (see Fig. 98) and internal mixture formation $M_i(n)$ with early injection start. The improvements due to additional water injection can be seen

Six specially designed water injection nozzles with a clear spray path on each intake valve were mounted on the intake manifold. A more detailed description of this vehicle may be found in [113, 120, 140, 125].
(see further Figs. 94–98).
12. Musashi Institute of Technology Test Vehicle (Musashi VI) (1894): This vehicle, a Nissan Maxima Sedan, was successfully demonstrated at the 5th World Hydrogen Conference in Toronto in 1984. The 2 liter 4-stroke turbo

Table 32. Typical technical data from the DFVLR-BMW 745i experimental vehicle

Storage:	LH_2-tank 130 L (gasoline equivalent 40 L) boil off rate 1.8% per day compact solenoid valve block (Fig. 90) for refilling and LH_2 feed to the engine, compact water/hydrogen heat exchanger for warming hydrogen to ambient temperature (Fig. 78).
Engine:	BMW M-30 Otto-cycle engine, 3.5 L, external mixture formation by timed individual port injection, lean mixture concept ($\varphi \leqslant 0.7$) exhaust gas turbocharger, variable water injection, engine control by a digital motor electronics system, dual fuel operation (H_2 or gasoline). 150 kW at 4,300 r.p.m. (top speed 170 km/h[20]), 170 kW at 5,500 r.p.m. if run with gasoline.
Emissions:	Water vapor, NO_x: 0.27 gram per ECE cycle[20] vs 3.91 gram per ECE 1984 and a target of 2.5 gram in 1987 for gasoline.
Range per tank filling:	Hydrogen about 400 km at 100 km/h Gasoline about 160 km at 100 km/h (20 liter gasoline backup tank)

[20] Data made available by the BMW test center in Ismaning/Munich.

Diesel engine was converted to internal mixture formation (late injection) with hydrogen and catalytic ignition on a ceramic glow plug which was heated up to 900 °C [160]. To pressurize the fuel a LH_2 piston pump was developed with an injection pressure in the range of about 8–10 MPa and was installed inside the 82 liter LH_2 tank (cruising range 350–400 km). Further hydraulically actuated high pressure injectors for ambient hydrogen were developed (Fig. 99).

A special direct injection combustion chamber with a piston cavity was designed. The compression ratio was decreased from 21:1 to 13.5:1 and the hydrogen injection nozzle was relocated to be near the center of cylinder bore. Hydrogen was injected from eight holes located around the circumference of a circle. One of the jets was directed at the hot surface igniter.

Fig. 99. Engine of the Musashi VI vehicle

The NO_x emissions decreased when injection start approached t.d.c. It was considerably less than that of a premixture type engine using external mixture formation. The engine performance deteriorated with increasing engine speed, caused by slow flame propagation from the initially ignited H_2 jet to the other jets.
This vehicle demonstrates the continuously consistent and successful development of the internal mixture formation with late-start injection by Musashi since 1978. It is remarkable that currently no vehicles with late injection from other R&D groups or the automobile industry have been successfully demonstrated or further developed, which shows the pioneer achievement of Musashi.

Fig. 100. Hydrogen version of the M30-engine for internal mixture formation of the BMW-DFVLR test vehicle on test stand.

13. BMW-DFVLR Joint Project Test Vehicle (1985/86): This vehicle, a BMW 745i, was exhibited at the 6th World Hydrogen Conference in Vienna [125, 142]. It was the first LH_2 test vehicle in Europe with direct cylinder injection of cryogenic hydrogen (Fig. 100). In this process, cold hydrogen (40–100 K) is injected directly into the combustion chamber by hydraulically actuated cryogenic injectors at a maximum pressure of 2 MPa and electrically ignited. (Cryogenic internal mixture formation with early-start injection at approximately 150° b.t.d.c.). The development of the cryogenic injectors (see Figs. 114, 115), as well as the LH_2 intermediate-pressure pump, installed outside the LH_2 tank and driven via hydraulic motor, were the essential DFVLR contributions to this vehicle (see Fig. 101).

14. Musashi Institute of Technology Test Vehicle (Musashi VII) (1986): The experience obtained with Musashi VI led to Musashi VII, a 3.5 ton medium-size truck from Hino Motors Limited [151, 154, 155, 116]. The 3.8 liter, four-cylinder turbo-charged Diesel engine from Hino Motors was converted to hydrogen in a manner similar to the engine from Musashi VI. The glow plugs, normally used to start the Diesel engine, are used for catalytic ignition with H_2 operation. The same type of injectors and LH_2

Fig. 101. 50 liter LH_2 tank and LH_2 fuel pump (see Fig. 107) in the BMW-DFVLR test vehicle for internal mixture formation with early injection start (Photo: BMW)

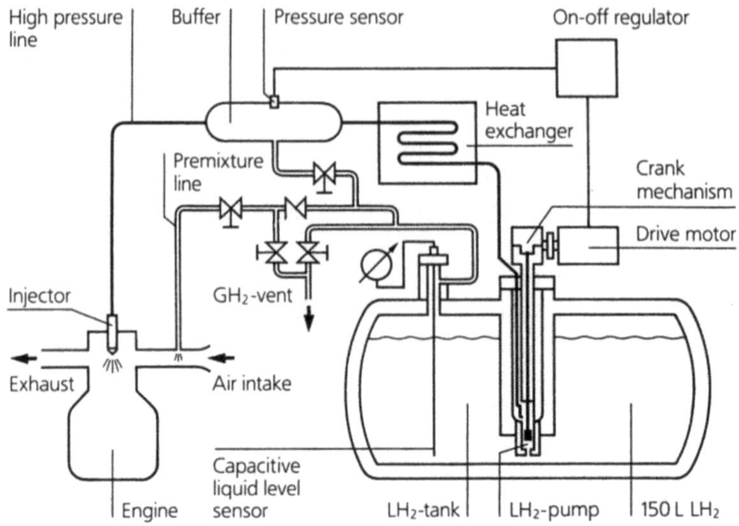

Fig. 102. Fuel conditioning system of the Musashi VII test vehicle (Source: Musashi)

high-pressure pumps as in Musashi VI were used. The pump is installed in a cylindrical 150 liter LH_2 (Fig. 102, 103). Gaseous hydrogen from pump losses during pump operation was mixed with the engine intake air and led to reduction or elimination of the pressure oscillations which usually occur with normal direct late injection during the combustion stroke. At

Fig. 103. Musashi VII test vehicle fuel tank (Source: Musashi)

IVDC in Vancouver it was demonstrated that at a top speed of 120 km/h the cruising range was 240 km per refueling which corresponds to a gasoline fuel economy of 6 km/L. The laboratory engine test showed a maximum output power of 107 kW at about 3,500 r.p.m. compared with the 96 kW of the base engine. The maximum brake thermal efficiency (b.t.e.) was 0.32 compared with 0.38 of the base engine. The truck has demonstrated that large-size hydrogen engines have excellent prospects for use in trucks by applying high-pressure, direct-cylinder hydrogen injection assisted by a small percentage of hydrogen forming a premixture with the intake air.

15. BMW 735i Test Vehicle (1988): Up until 1988, vehicle (9) and (11) were the only two operational liquid hydrogen test vehicles in Europe. A BMW test vehicle (long version of the new 750i) started undergoing tests in 1988. It has a cylindrical 100 liter liquid hydrogen storage tank supplied by Messer Griesheim GmbH. The complete cryogenic fuel conditioning system, including a heat exchanger for warming liquid hydrogen up to ambient temperature, was designed and manufactured in accordance with the successfully proven LH_2 fuel conditioning system of the DLR test vehicle (11) also by Messer Griesheim GmbH [142].

Mixture formation is carried out via external continuous individual intake port injection of ambient hydrogen as in the DFVLR vehicle 6 and is controlled via a central timing valve. Supercharging of the engine is carried out via a centrifugal compressor (max. 90,000 r.p.m) which is driven by a geartrain[21] and variable speed drive (Fig. 104). Thus, the problem of uncontrolled preignition with hot residual gas while driving under constant load associated with the exhaust gas turbocharger and exhaust gas backwash is mitigated considerably. Variable load and acceleration phase problems have been reported [143]. Due to the lean mixture concept, water injection is not necessary. Engine management including storage of the ignition and mixture performance curves is accomplished with digital electronics. The top speed of the vehicle is set at 170 km/h. Further concrete engine power data has not yet been published[22], but at best might correspond to that of vehicle (11).

6.2.5 LH_2 Fuel System Components

High-Pressure Liquid Hydrogen Pumps. The development of compact LH_2-reciprocating pumps suitable for the operation of vehicles poses a central problem of the internal mixture formation in hydrogen fueled

[21] Developed by the Zahnradfabrik Friedrichshafen.

[22] Meanwhile BMW has developed a vehicle of this type, without charging compressor. Engine power data has also not been released.

Fig. 104. View into the engine compartment of the BMW 735i test vehicle from 1988 with individual continous port injection. In the front the charging compressor driven by the engine (Source: BMW)

internal combustion engines. Whereas technically sound solutions for stationary pumps for cryogenic liquids like LH_2, LO_2, LN_2 or LCH_4 [143–146], have existed for more than two decades this does not apply to automotive applications for several reasons.

As opposed to the substantial flow rate of current stationary piston pumps (30–100 L/min), the typical flow rate for automobiles is more than one order of magnitude smaller. The following problems arise from the required compact construction:

1. Piston–cylinder leakage has an increased effect. The standard sealing methods for large pistons such as piston rings with piston slots, or labyrinth seals are hardly applicable here due to the small piston diameter (approx. 20 mm) and the low flow rate because the leakage rate can be larger than the flow rate.
2. A piston–cylinder seal that functions between ambient and cryogenic temperatures can be implemented either via gas film bearing pistons over this temperature range permitting radial clearance of a maximum

of one micrometer or via piston sleeves or piston rings with dry lubrication (PTFE, PTFE carbon, PTFE bronze, fiberglass—PTFE graphite, etc.).

Gas film bearing pistons result in a minimum of friction loss and are state of the art in expansion engines [147, 148].

Problems are caused by the very narrow tolerance of radial clearance which is necessary for radial stabilization of the piston movement and the resulting requirement to maintain a radial pressure gradient in the boundary layer in the piston cylinder gap. Materials with defined thermal expansion properties that can maintain this piston clearance have to be further developed.

A prototype of this type of pump was constructed at the DFVLR for pressures up to 50 bar in 1980/81 and tested with LN_2. Difficulties in priming the pump can be alleviated by the addition of a supercharger [149] developed by Cryomec AG. Since this pump proved to be very sensitive to impurities and irregularities, as a result of the very small radial piston clearance this development was not continued because as at this time small low-loss, compact pumps with piston ring sealing or piston sleeves are already available for this medium-pressure range.

Piston rings with gaps have the advantage of sealing by making contact with the cylinder wall and the ring groove as a result of the pressure. In order to reduce the gas leakage of the gap a larger number of piston rings and increased friction cannot be avoided. Piston sleeves do not have gaps but must be carefully designed and installed.

The total thermal expansion of piston sleeves (for materials suitable within the range of 20–300 K) is about seven times more than the thermal expansion of the austenitic steel used in cylinders and pistons. This results in the fact that a sleeve fitted for ambient temperature with minimal or negligible leakage shows intolerable leakage after cooldown.

A compensation through initial stress of the sleeve is not possible because of the viscous flow of the synthetic materials and the increased production of frictional heat caused by the initial stress at the inner cylinder wall.

With increasing stroke rate, this heat cannot be transferred fast enough from the cylinder wall and therefore leads to a premature partial evaporation of cylinder contents during the intake phase.

3. Material cross sections and wall thicknesses cannot be reduced in the same proportion as the other geometrical dimensions of the pump for reasons of strength and inherent stability. That is why thermal loss has a greater effect the more the system is reduced in size. Locating the pump within the LH_2 tank avoids the cooldown procedure at start up but leads to increased thermal loss caused by the coupling of the cold pump sections with the mechanical drive at ambient temperature.

4. Vapor formation during the intake phase of the piston pump also has a greater effect with a decrease in system size. Primarily, the pressure drop at the intake valve cannot be reduced to the required extent. The flow and pressure loss experienced during the opening period of the intake valve has an increasing negative effect with decreasing density of the fluids and achieves the worst results with liquid hydrogen. Due to the small construction size, a direct mechanical positive control of the valve could be achieved best through electromagnetic actuation. This applies as well to the installation of auxiliary pistons or asymmetrical displacers acting as superchargers for priming the pump [149, 150].
5. The thermal loss of the pump feed line in combination with the reduction in geometrical size causes increased vapor formation, an effect, which can only be controlled by a phase separator in the intake region of the pump.

In contrast to the LH_2 high-pressure pumps (6–10 MPa) [81, 121, 152] of the Musashi Institute of Technology which have been successfully demonstrated in automotive applications, the CRYOMEC LH_2 high-pressure pump and the DFVLR intermediate-pressure pump (2.5 MPa max.) are not installed inside the LH_2 fuel tank. The large boiloff rate (up to 10% per day and more), resulting from the required access to the pump, is thus avoided [117, 125, 113]. In this manner, the low boiloff rates of the LH_2 fuel tanks (1.5–1.7% per day) currently found in operation are not deterimentally effected.
Figure 105 shows a LH_2 high-pressure pump of this type under development by CRYOMEC AG in Basel which represents a miniaturization of the relatively large LH_2 pumps shown in Chap. 4.4, p. 96. The flow rate is about 4 L/min at a pressure up to about 300 bar and a stroke rate between 200–600 per min. This pump, which is equipped with a supercharger to allow operation with NPSH = 0, is currently under test at the DFVLR (now DLR) for automotive applications.
Continuous development work with the DLR intermediate-pressure pump includes increasing the injection pressure to more than 100 bar, increasing the reliability and lifetime as well as determining an optimal location of the pump within the fuel storage tank and mixture formation system [63, 82]. Placing the pump as a separate unit outside the tank causes more problems with respect to warming up the pump of a parked vehicle, cooling down the pump upon starting and maintaining a minimum cooling requirement in order to provide a short cooldown period, i.e., minimum fuel expenditure on the cold side of the pump.
The design and operation of the pump is shown in Fig. 106. The polished and honed 1.4541 (321 AISI) cylinder *Z* (bore 20 mm, stroke 26 mm) is attached to the top of the closing flange *F* of the evacuated pump vessel *P*. Inside the cylinder *Z* (Fig. 106) is the piston *K* with compensated radial

Fig. 105. View of a Cryomec LH_2 fuel pump. Pressure 250 bar, 240 L/h

thermal expansion which carries the two completely sealed PTFE graphite piston sleeves *M*1 and *M*2 (Germ. Pat. DE 3621 726). The piston is hinge-mounted to the hydraulic drive which is mounted on top of the closure flange *F*, via a piston rod *KS*. The upper guidance of the piston rod *KF* can be removed from the flange for the purpose of disassembly of the piston and access to the spring loaded intake valve *E* (mass approx. 0.5 g). For cooling purposes, the lower cylinder section is surrounded by a precooling chamber *VK* located in a vacuum acting as a phase separator which is connected to the LH_2 feed line of the LH_2 tank and a line for venting the gaseous phase. The exhaust valve *A* is accessible externally via openings in the thermal radiation shield *S* (copper) and in the bottom of the pump vessel.
The pumped LH_2 cools the middle part of the cylinder to about 40 K via the heat exchanger *WT*, dissipates the heat flowing in from the upper part of the cylinder and keeps the temperature of the pressure sealed thermal radiation shield *S* at about 60–70 K. Thus, the inflow of ambient heat to the lower cylinder section is reduced by approximately 95%, effectively the remaining piston rod heat conduction *KS*. During cooldown, LH_2 flows through the pump from the tank via the LH_2 line, the prechamber *VK*,

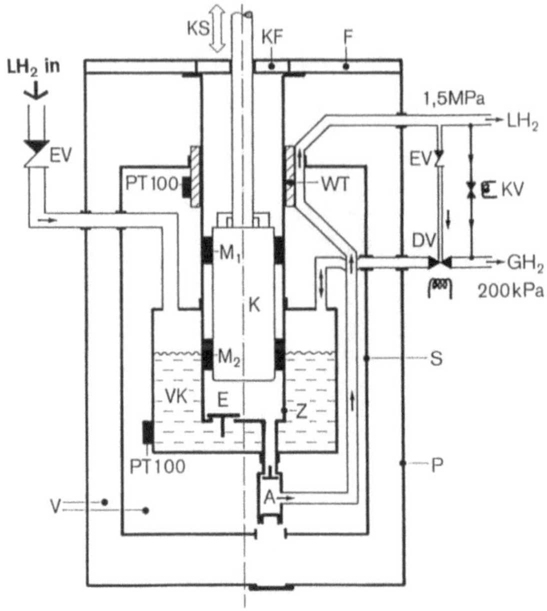

Fig. 106. Cross section of the DFVLR intermediate-pressure LH_2 pump. Pressure 25 bar, 180 L/h

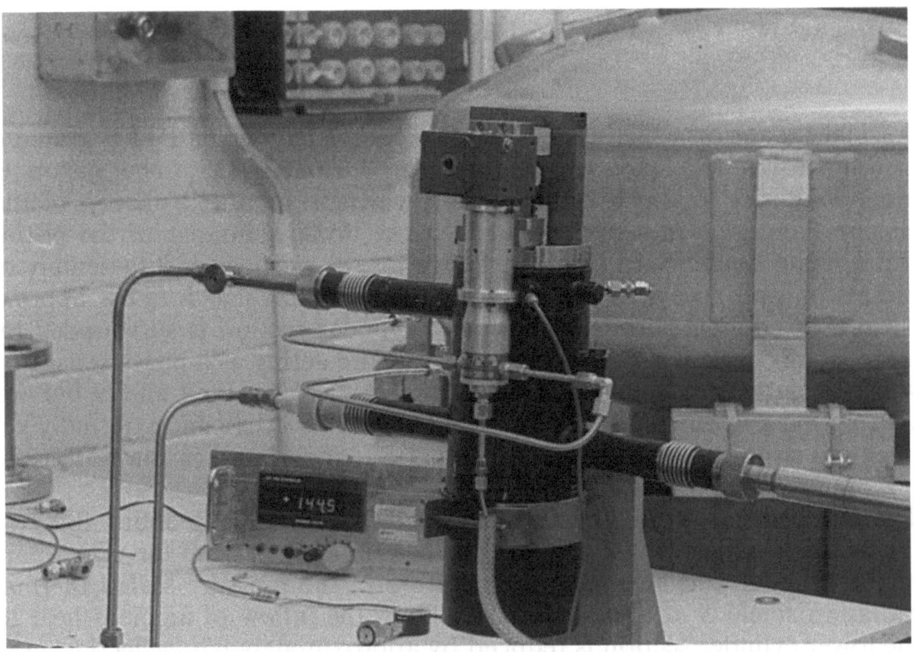

Fig. 107. DFVLR intermediate-pressure pump on the test stand. The vertically arranged hydraulic motor with miter gear and eccenter drive can be seen

the intake valve *E*, the exhaust valve *A*, the heat exchanger *WT* and the cooldown valve *KV*. As a result of the relatively small mass which has to be cooled, the cooldown period from ambient temperature to operating temperature is only about 3 minutes in duration and requires approximately 1 liter of hydrogen.

Due to the combination of the heat exchanger *WT*, thermal radiation shield *S* and prechamber *VK*, the warmup rate of the prechamber and of the cold cylinder section is reduced to the low value of approximately 15 K/h. For this reason, the pump remains precooled for several hours after shutdown and a steady operational condition can be achieved in approximately 20–30 seconds after restart.

Complete filling of the pump cylinder *Z* during the intake stroke is achieved through a cyclic pressurization (*EV*, *DV*) of the prechamber *VK* functioning as a phase separator via a pulsating cushion of cold gaseous hydrogen, which is removed from the delivery side of the pump. Figure 107 shows the overall view of the pump on the test stand and Fig. 101 shows the pump (Germ. Pat. DE 37 10 363) installed in the test vehicle 745i.

The pump has proven its basic suitability as a prototype LH_2 pump for automotive application. While cooldown and warmup periods approach nearly ideal conditions, developments and improvements concerning controller action and feeding of the pump have to be carried out:

- Reduction of the friction between piston sleeves and cylinder wall through improved compensation of the radial thermal expansion.
- Reduction of the pressure drop at the intake valve of the pump.
- Reduction of the heat leaks in the feed line to the pump, if necessary through counter current heat exchange with the discharged hydrogen.

Instead of cyclic pressurization at the intake section or a supercharger system, the LH_2 intermediate-pressure pump mounted outside of the tank can be fed with subcooled liquid supplied by a low pressure pump mounted inside the tank. With a pressure buildup in the range of only about 0.2–0.3 MPa, this type of pump can be designed with the required reliability as well as with a "cold drive". Furthermore, this design can be compactly constructed, enables access to the pump via the wall of the outer vessel, and does not lead to an increase in the static boil off rate.

Table 33. Liquid hydrogen test results

Pressure *P*: average value 2.0 MPa, peak value 2.4 MPa
Cooldown of the pump from 300 K: 2 min 30 s ($\dot{m}_{H_2} = 1.2$ g/s)
Bypass flow rate at steady operation: $\dot{m}_{H_2} = 75$ g/s
Maximum flow rate at 160 strokes/min: 75 L/h (approx. 5.25 kg/h)
Average warmup rate after shutdown: $dT_1/dt = 15$ K/h
Cooldown period from $T_1 = 135$ K to 20 K: 25 s

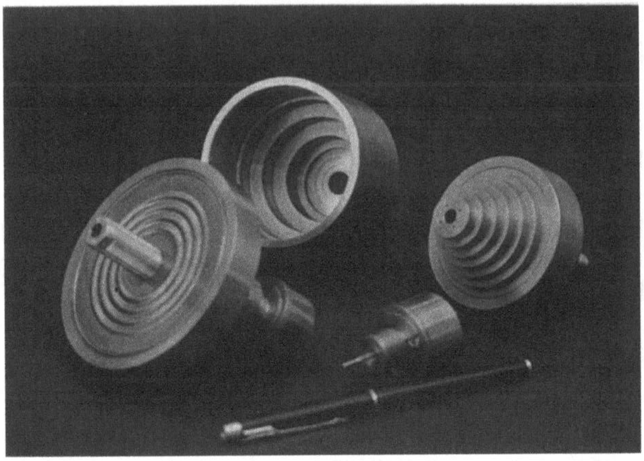

Fig. 108. LH_2 membrane pump. 600 strokes per minute, 3 bar, 80 L/h

In addition to magnetic bearing centrifugal pumps with cold electrical induction motors (frequency 1,000–2,000 Hz, 60,000–120,000 rpm), membrane pumps with electromagnetic linear drives can also be considered. Figure 108 shows this type of pump with a welded membrane made of 1.4541 (321 AISI) high alloy steel and a "COLENOID" drive [153], which supplies a stroke of about 2.5 mm. With a stroke rate of 300–400 strokes per minute this pump supplies about 200 L/h at a pressure of about 0.25 MPa. The average electrical output is about 25 W with power peaks during the working stroke of about 300 W.

Although a supercharger was provided, considerable problems were experienced filling the pump with liquid, especially at the beginning of the intake stroke. Both of the original spring loaded intake flapper valves, which were actuated by inflowing liquid, were substituted with electromagnetically actuated ones. Thus at the start of the intake phase (reversal of the membrane movement), complete opening of the intake valves, thus minimal pressure drop, as well as optimum operation of the supercharger were immediately achieved.

Figure 109 shows the primary location of such a pump on the cold wall of a vacuum-insulated LH_2 tank with multilayer insulation. Access to the pump is obtained via a flange in the wall of the outer vessel and an opening in the superinsulation. The pump's pressure connection as well as the electrical connections are flexibly attached to the inner tank. The vacuum-tight, cold-seal flange on the wall of the inner tank carries the connections for the single layers of the vapor-cooled, superinsulated, hinged shields made of fiberglass epoxy. Thus, large temperature differences between the different groups of layers and the hinged shields can be avoided and the related

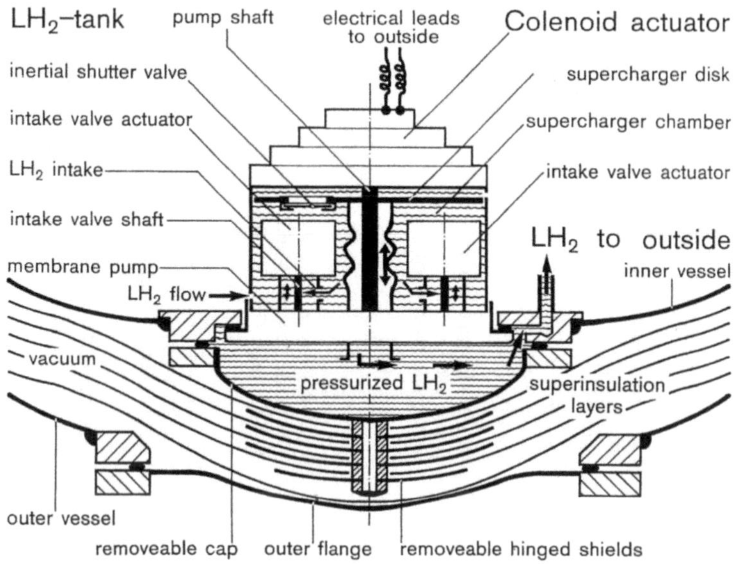

Fig. 109. Location of the LH_2 membrane pump on the cold LH_2 inner tank with optimal obtainable access

negative effect of heat leakage from a gap in the superinsulation can be mitigated.

LH_2 Fuel Tank. While in commercial vehicles with a centralized refueling system and nearly 24 hour operation, relatively simple LH_2 tanks with an acceptably high evaporation rate (5–8% per day) can definitely be used, stricter requirements have to be fulfilled for use in passenger vehicles.

Development of LH_2 fuel tanks has been going on at the DFVLR [113, 114] since 1976 (see page 204). Now tanks suitable for automotive application, with a capacity of up to approximately 150 liter are currently available for use in test vehicles (see Table 34). This is equivalent to up to 40 liter of gasoline. Tanks made of aluminum alloy and equipped with vapor-cooled shields[23], weighing about 50–70 kg, have an evaporation rate of approximately 1.8% per day when using fiberglass or carbon fiber composites for the inner vessel suspension.

At about 10% full, this corresponds to a pressure buildup rate of about 0.05 bar/h (20 h/bar) and 0.03 bar/h (30 h/bar) at about 85% full[24].

These tanks supply either cryogenic gaseous or liquid hydrogen using available automatically controlled instrumentation, valves, safety valves,

[23] Joint project between DFVLR (now DLR) and Messer Griesheim GmbH.

[24] Despite this state of the art technology and about 15 years of experience, newcomers tend to identify the storage onboard the vehicle as an essential problem which still has to be solved.

Table 34. Data for DFVLR liquid hydrogen fuel tanks for motor vehicles [115]

Tank/type manufactured	LH_2 (liter)	Capacity (kg)	Shape	Geom. data (mm)	Material	Weight (kg)	Boil off rate (%d^{-1})	Project	Ref.
					1st generation				
1974	120	8.5	flat cyl.	O.D. 900 length 560	stainless steel	90	8	DFVLR experimental vehicle BMW 520, in operation since 1979	Figs. 75, 76
1978 (VLHD-A)	110	7.8	flat cyl.	O.D. 900 length 660	$AlMg_{3.5}Mn$	83	10	LASL-DFVLR joint vehicle project. Since 1982 at the IHS	Fig. 81 [90, 91]
1981 (VLHD-B)[a]	155	11.0	cylinder	O.D. 712 length 1,220	outer vessel: $AlMg_{3.5}Mn$ inner vessel: stainless steel	156	3.4[b]	LASL-DFVLR joint vehicle project 1981, 1982–1988 at IHS, now DFVLR (DLR)	Figs. 82, 83 [90, 91]
					2nd generation				
1984	128	9.1	flat cyl. tapered	O.D. 900 max. length 400	$AlMg_{3.5}Mn$	65	1.8[b]	DFVLR-BMW coop. 1983–1986; 735i vehicle: tank design DLR Manuf.: Messer Griesheim GmbH	Figs. 87, 88, 89 [121, 126, 141]
1985	130	9.2	flat cyl.	O.D. 900 length 400	$AlMg_{3.5}Mn$	65	1.8[b]	Daimler Benz-DFVLR coop. 3 tanks, design: DLR Manuf.: Messer Griesheim GmbH	Fig. 110 [115, 126]
					3rd generation[c]				
1985/86	120	8.5	asymm. non circular shape	length 1,200	outer vessel: fiberglass epoxy inner vessel: stainless steel	65	1.5[b]	Daimler Benz-DFVLR coop. tank design: DFVLR	Fig. 111 [115]

[a] Experimental tank, designed and constructed in a joint effort of the DLR and the Los Alamos National Laboratory (LANL).
[b] Vapor-cooled shields.
[c] Final design.

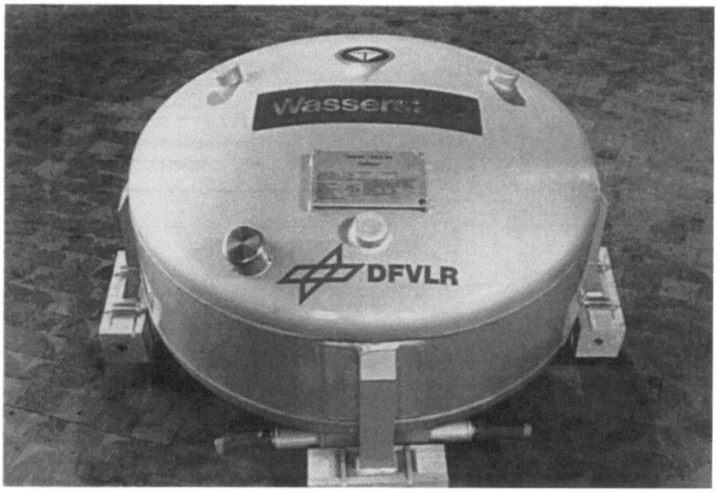

Fig. 110. LH_2 tank, 140 L capacity, operating pressure 3.5 bar for a Daimler-Benz project see (Table 34)

tubes systems, and heat exchangers, suitable for automotive application. They also supply hydrogen at ambient temperature, up to a pressure of approximately 0.5 MPa, by means of compact heat exchangers. The external design of these flat tanks has either a rotational symmetry, i.e., cylindrical with convex ends [114] and very rudimentary cylindrical part (Fig. 110), or they may be tapered using rotationally symmetric convex ends (Fig. 89). Tanks of this design are installed inside the trunk or near of the vehicle, which can only be tolerated with test vehicles. Based on current experience with hydrogen storage on board vehicles, a tank with nearly prismatic shape, installed behind the back seat, is an attractive solution with regard to safety aspects. This leaves about 60% of the trunk space available in the vehicle (Fig. 111). The first development steps have started at the DLR for tanks similar in shape to flat prisms, which are required for conventional vehicles [117].

With the currently available tanks the pressure difference of about 2 bar between minimum and maximum operating pressures (activation of the safety valves at about 3.5 bar over-pressure) corresponds to a time period of about 40–60 hours (pressure buildup time). Within this period no losses occur caused by release of the safety valve with the vehicle at rest. Since after this time gaseous hydrogen is emitted to the air in accordance with the boiloff rate of the tank, parking this type of vehicle in closed and poorly ventilated places (garage) could cause technical safety problems.

Therefore, a further increase in the pressure buildup time and a reduction in the heat leakage is desirable. Since laboratory storage containers for

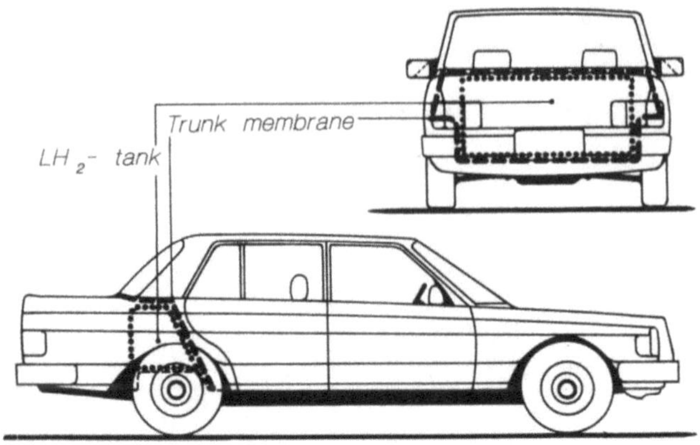

Fig. 111. The location of a LH_2 tank with flat geometry in the vehicle with a minimum use of space

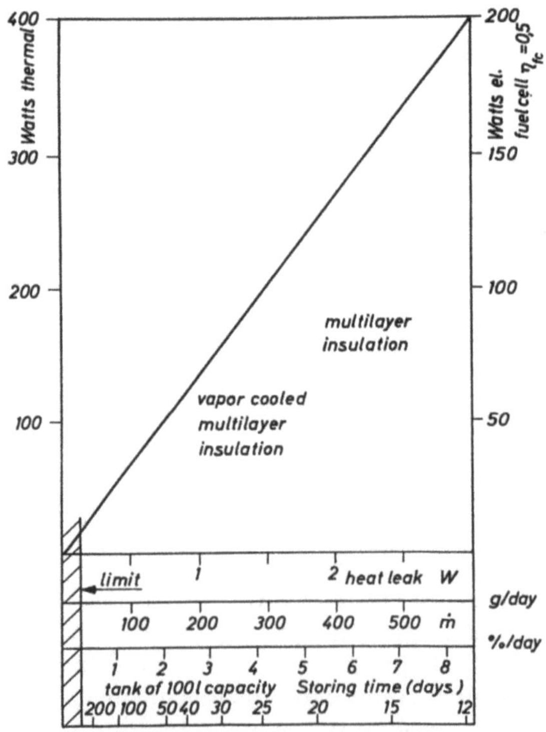

Fig. 112. Boil off rates of LH_2 storage tanks and the energy in the vented hydrogen

liquid hydrogen are currently available with vapor cooled multilayer insulation and a boil off rate of 0.5% per day at a capacity between 100 and 200 liter (see Fig. 36), further improvements are very possible. Tanks for vehicles with a boil off rate of about 1% per day are available now from Messer Griesheim GmbH. Since the suspension of the inner vessel contributes considerably to the total heat leakage as a result of heat conduction, further reduction of the boiloff rate of the inner vessel [99] to about 0.25% per day can be achieved by magnetic suspension. This will probably be the limit of technically feasible improvements. The hydrogen, which still has to evaporate can feed a small fuel cell battery [6, 65, 66, 77], which charges the starter battery of a vehicle at rest and could totally replace the alternator when the vehicle is operating (Fig. 112). This possibility is advantageous from an energy perspective because the total efficiency of the electrical energy production related to fuel is currently very low for a vehicle in urban traffic (Fig. 113). As a result of the higher efficiency of the fuel cell, a relatively small but perceptible amount of fuel

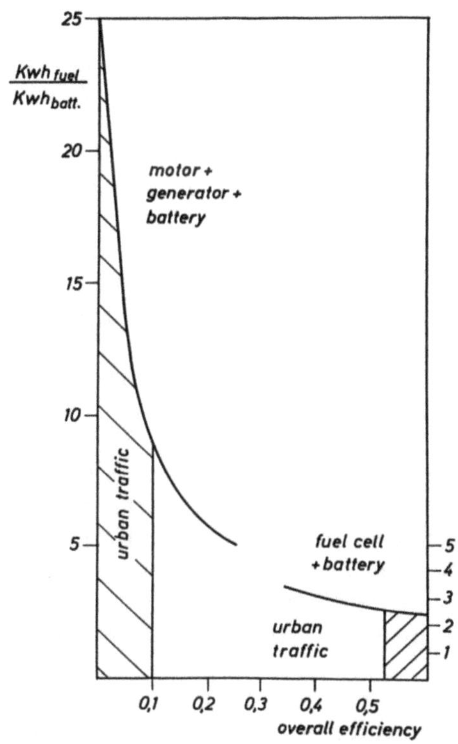

Fig. 113. Amount of fuel (kWh_f) for the production of electrical energy onboard vehicles related to the electrical energy stored in the battery (kWh_{Batt}). Comparison of the total efficiency between conventional processes and fuel cells

can be saved. The small fuel cells, suitable for automobile operation and which are necessary for this process, are not yet available.
Another means of avoiding the emission of evaporated hydrogen into the atmosphere is to install a small auxiliary hydride storage reservoir. In the case of a boil off rate for the fuel tank of 0.5% per day (about 100 g hydrogen per day) it would require only 2 kg of storage material in the form of magnesium in order to absorb the evaporated hydrogen for over 48 hours. While driving, this stored hydrogen is released by heat supplied from the exhaust gas and delivered to the engine in addition to the fuel from the LH_2 tank. Thus, energy problems which are encountered with pure hydride storage can be avoided. If this concept is further developed, auxiliary storage of this type could serve as an emergency reserve when the LH_2 tank is empty.
LH_2 Fuel Tank Level Indication. Only a few of the various methods for measuring the level of the LH_2 are usable for an LH_2 fuel tank.
From all the methods which could be considered for liquefied gas level indication such as float level indicators, hydro-static pressure manometers (measuring the liquid column), superconducing sensors, weighing devices, ultrasonic sensors, capacitive sensors and discrete level sensors, only the last two have proved to be suitable for automotive applications.
Sensors in the form of Allen–Bradley carbon resistors, supplied with a constant current which can detect 10 discrete levels were used for LH_2 tank developed in 1974 and have been used in the DFVLR BMW 520 test vehicle since 1979 [89, 141] (Fig. 76), as well as for the tank used in a LASL and DFVLR test vehicle in 1979 [80, 90, 122]. As a result of the difference in heat transfer between the liquid phase and gaseous phase, the electrical resistance of these sensors charges with a change from liquid to gaseous phase which can be used to indicate a particular level. It became evident, however, that a reliable level indication at all vessel pressure levels (pressure remaining in thermodynamic equilibrium with the temperature of the liquid) can only be carried out by the use of resistors with considerably complicated electronics because the influence of the warming of the liquid must be compensated. Further, the electrical leads to the level sensors increase the heat leak which can increase the boil off rate of the vessel. Despite positive operational experience of reproducible and accurate liquid level sensing the following applications have used capacitive level sensors which do not cause a perceptible increase in the heat leak due to the single lead. Moreover, in principle they can also provide continuous level indication.
Capacitive LH_2-level indicators can either be implemented as plates of a condenser with a flat surface or as cylinder condensers which are situated vertically in the vessel. With liquids that are not electrically conductive but that have an appropriately large dielectric susceptibility, the change in capacitance related to the capacitance of the empty container is propor-

tional to the level. The zero capacitance corresponds to the capacitance measured with an empty tank. It can be calculated from the vacuum capacitance of the sensor by use of the dielectric constant of the gaseous phase of the stored liquid.

In many cases, the zero capacitance can effectively be considered as equivalent to the vacuum capacitance without unacceptably large measurement errors. This is applicable when the dielectric constant of the liquid phase e_F is large compared to that of the gaseous phase e_g which partially applies to hydrogen or nitrogen for example, but not for helium. Furthermore, the variation in e_F with the temperature of the liquid is important. This dependency should be as small as possible in order to provide a liquid level indication which is independent from thermal stratification and tank pressure with a minimum amount of effort required for the measuring technique. Table 35 shows e_F and e_g for LH_2 indicated as a function of the temperature T of the liquid. Also given in Table 35 are the vapor pressure p as well as the density of the liquid phase ρ_F and the gaseous phase ρ_g (see [127]) is shown.

Table 35 shows that the assumptions made above are correct up to $T = 24$ K, i.e., up to a pressure of about 2.5 bar. There is a tolerable measurement error of less than 5% compared to the measurement of the liquid level at the normal boiling point. If the measurement is extended to larger pressures in the tank, a correction of the indication is necessary in order to avoid unacceptably large liquid level measurement errors.

In order to measure the capacitance basically all known methods can be applied but only two are of interest due to their simplicity. These are measurement of a square wave pulse width of a monoflop that is controlled by the sensor's capacitance as well as frequency-voltage conversion by an integrated circuit that is also controlled by the sensor's capacitance.

In the first process, square-wave pulses are produced by a triggered monoflop with a pulse width proportional to the sensor's capacitance.

Table 35. Density and dielectric constant of liquid and paseous hydrogen

T (K)	p (MPa)	ρ_F (kg/m^3)	ρ_g (kg/m^3)	e_F	e_g
20.4[a]	0.1013	70.7864	1.3378	1.22978	1.00404
22	0.1585	68.7200	2.0711	1.22261	1.00625
23	0.2039	67.4149	2.6119	1.21810	1.00789
24	0.2579	66.0112	3.2548	1.21325	1.00984
27	0.4800	61.0065	5.9999	1.19608	1.01820
30	0.8116	53.9303	10.8866	1.17208	1.03319
33.18[b]	1.3130	31.4285	31.4285	1.09795	1.09795

T temperature, p pressure, ρ density, index F: liquid phase, index g: gaseous phase.
[a] Boiling point (NPT).
[b] Critical point.

Simultaneously, a second monoflop is triggered and produces square wave pulses by means of a reference capacitance corresponding to the zero capacitance of the sensor, including the capacitance of the leads to the sensor. The square-wave pulse width from the second monoflop is proportional to the sensor's zero capacitance plus the lead capacitance. A square-wave pulse is derived from both impulses. The resulting pulse width is proportional to the difference of the output pulses, that is proportional to the actual sensor capacitance and thus to the liquid level.

In the frequency-voltage conversion method [128] the characteristic capacitance (one shot capacitor), which is necessary for the operation of a frequency-voltage converter I.C. is represented by the sensor's capacitance including lead capacitance. In addition, a quartz controlled frequency signal is supplied to the I.C. This way an output signal is obtained which is proportional to the actual sensor capacitance plus lead capacitance. A replication of the sensor's zero capacitance of the sensor including lead capacitance produces a second signal from a second I.C. of the same type. The difference of both of the signals is proportional to the liquid level.

Both processes have been successfully used and provided a satisfactory level indication under specific conditions, i.e., sensor zero capacitance about 100 pF, maximum capacitance at full level of about 120 pF and lead capacitance of about 400 pF.

While it is possible to record the empty capacitance including lead capacitance by means of a reference capacitance it is considerably more difficult to replicate the variation with temperature of the lead capacitance with the reference capacitance which is the result of the temperature dependence of the leads' (mini-coaxial cable) insulating material.

A design for the reference capacitance where the leads to the sensor are duplicated and run over the same route but not connected with the sensor, proved to be reliable. Mini-coaxial cables were used outside the tank (approx. 50 pF/m). The routing of the currently available mini-coaxial leads inside the vent line was not successful because of the small vent line diameter (vent line 10 mm $\phi \times 1$) and the risk of wire break and short circuiting when pulling. Moreover, the copper coating of the coax line causes a considerable heat leak.

Inside the tank therefore use was made of niobium-titanium wires (0.3 mm ϕ) insulated by a PTFE sheath, routed inside the approximately 3 m long coiled vent line leading from the inner vessel to the outer vessel. Niobium–titanium wire, usually used as a superconductor was chosen because of its high tensile strength. Thus, on one hand, problems with pulling the leads through the vent line can be avoided and on the other hand because of the low thermal conductivity of the wire (only 20 μ copper coating) heat leak caused by the leads can be disregarded.

Basically, the problems caused by the lead capacitance could have been solved by placing parts of the electronics system in the inner vessel. The "cold electronics" required for this are basically available and have been used in various applications [129]. In the case of an installation inside the LH_2 tank—this should be carried out during the construction of the tank—there is no access upon failure of a construction element so that standards have to be established regarding reliability and lifetime which, if at all possible, will probably be very difficult to fulfill.

In order to determine the tank content—defined as volume of the liquid based on the density at the normal boiling point (NBP)—the volume of the liquid phase has to be determined from the level. For this, the fill level function has to be known, which is defined as the volume relative to the liquid level (distance of the liquid level from a reference point). The fill liquid level function can either be determined by calculation or through successive filling and weighing of the vessel with simultaneous measurement of the sensor capacitance.

If the linear deviation is not too large it is possible to establish a linear relationship between liquid volume and sensor capacitance by shaping the level sensor appropriately (a flat plate level capacitor with appropriately shaped plates). This was achieved in a cylindrical tank which had been installed in the test vehicle designed and constructed in a joint effort between the LASL and the DFVLR in 1980 [91]. Four level sensors were installed in this tank. These sensors were installed in the vehicle tank diagonal to the travel direction, which as a result of their arrangement also compensated for the influence of the banking of the vehicles indicated by the level indicator. The influence of the liquid splashing during driving was balanced out by a correspondingly slow indication (response time approximately 2 sec.).

While at low tank pressure, i.e., with relatively little warming of the liquid phase, the liquid volume can be determined with sufficient accuracy by directly measuring the sensor capacitance. This is no longer possible with increasing pressure and when approaching the critical point because under certain circumstances a considerable fraction of the fuel can be in the vapor phase depending upon the respective liquid level. This has to be taken into consideration particularly with LH_2 or also with LHe because the critical point is very close to the normal boiling point (NBP).

If, for example, the tank is partially filled with LH_2 at boiling conditions (20.4 K, 1 bar) and V is the volume of the liquid phase in the tank then a volume $\bar{V}$ is obtained after warming of the liquid phase to a temperature T which corresponds to the vapor pressure p, resulting from partial evaporation of the liquid, and taking the increase in volume of the remaining liquid phase due to its warming into consideration. Since in this process, the total mass of the liquid plus vapor phase remains constant the following is obtained:

$$V \cdot \rho_F + (V_G - V) \cdot \rho_D = \bar{V} \cdot \bar{\rho}_F + (V_G - \bar{V}) \cdot \bar{\rho}_D \qquad (132)$$

where V is the volume of the liquid phase at NBP
ρ_F density of the liquid phase at boiling point (NPT)
ρ_D density of vapor phase at boiling point (NPT)
V_G is the geometric tank volume

values indicated with an overbar correspond to values of liquid and vapor along the vapor–pressure vurve.
Thus the following is obtained:

$$\frac{\bar{V}}{V_G} = \frac{(\rho_F - \rho_D) \cdot \dfrac{V}{V_G} - (\bar{\rho}_D - \rho_D)}{\bar{\rho}_F - \bar{\rho}_D}. \qquad (133)$$

Table 36 shows $\dfrac{\bar{V}}{V_G}$ as a function of $\dfrac{V}{V_G}$.

If a tank is filled with liquid volume V at boiling conditions (20.4 K; 1 bar) after warming to the temperature T the volume $\bar{V}$ of the liquid phase is obtained.
The volumes of V where $\bar{V} = 0$ are just sufficient to fill the geometric tank volume V_G with vapor at the corresponding pressure as a result of evaporation and warming, therefore $\bar{V}$ is exactly zero at this point.
A small fill ratio V/V_G causes a decrease in volume ($\bar{V} < V$) because the required liquid volume, which has to be evaporated to fill the vapor space, is larger than the increase in volume of the liquid as a result of warming. With increasing fill ratio this behavior reverses and results in an increase in volume of the liquid phase ($\bar{V} > V$). With the values of the fill ratio which correspond to $\bar{V}/V_G = 1$ there is no more volume of the vapor phase in the tank after warming. Therefore, a further increase in the fill ratio V/V_G is not advisible because liquid would have to be vented out of the tank (actuation of the safety valves) in order to avoid an unacceptable increase in pressure in the tank. The maximum fill ratio is, for example, only 43% of the geometric volume if warming to the critical point should occur.
The value for V which is necessary for measuring the tank volume can be determined from $\bar{V}$ dependent upon the pressure p along the vapor pressure curve by means of a microprocessor and an EPROM [156]. The corresponding values for $V(\bar{V})$ dependent upon the measured tank pressure as well as for the fill level function $\bar{V}(\bar{C})$ are stored in an appropriately large number of reference points in the EPROM. ($\bar{C}$ is the capacitance signal of the liquid level sensor which corresponds to each level.) The tank volume is determined by the microprocessor where values between the reference points are calculated through interpolation. The influence of the

Table 36. Liquid hydrogen volume at temperatures between 23 K and critical temperature (33, 18 K) as a function of the liquid volume V at NBP. Volumes are related to the geometrical tank volume V_G

$\frac{V}{V_G}$	23 K 2,0 bar	24 K 2,6 bar	26 K 3,9 bar	27 K 4,8 bar	28 K 5,7 bar	29 K 6,9 bar	30 K 8,1 bar	33, 18 13, 1 bar
0.018	0							
0.028		0						
0.052			0					
0.067				0				
0.086					0			
0.1	0.087	0.080	0.058	0.043	0.019			
0.109						0		
0.138							0	
0.2	0.195	0.191	0.178	0.169	0.153	0.133	0.101	
0.3	0.302	0.301	0.298	0.259	0.288	0.278	0.262	
0.4	0.409	0.412	0.418	0.422	0.422	0.424	0.423	
0.433								1.000
0.5	0.516	0.523	0.537	0.548	0.557	0.569	0.585	
0.6	0.623	0.633	0.656	0.675	0.691	0.714	0.746	
0.7	0.730	0.744	0.777	0.801	0.825	0.860	0.907	
0.757							1.000	
0.796						1.000		
0.8	0.834	0.855	0.897	0.927	0.960			
0.83					1.000			
0.85				1.000				
0.886			1.000					
0.9	0.945	0.965						
0.931		1.000						
0.952	1.000							

lead capacitance as well as the sensor empty capacitance can be determined and taken into consideration via periodic calibration by the microprocessor. Tests have shown that in this manner very good tank level indicators can be developed for LH_2-fuel tanks as well as for stationary storage tanks.

High-Pressure Hydrogen Injectors. An essential prerequisite for timed external mixture formation as well as for internal mixture formation is reliable injectors with suitably long lifetimes and minimum leakage rates. While for external mixture formation a major challenge is a valve opening which is large enough to supply the fuel needed for combustion during the typical opening periods of 1–5 ms (typical value of 100 cm^3 at ambient temperature), with internal mixture formation the challenges lie in coping with the large forces for valve actuation and in the lifetime of the valve seat and valve cone which remain in contact with the combustion chamber [100].

For external mixture formation, several proposals to resolve injector actuation and experiences have been reported which range from injectors mechanically actuated by the camshaft and pure electromagnetic actuation to combined electro-hydraulic functions [101–103, 125]. According to experience to date, only mechanical actuation by the camshaft, hydraulic and electro-hydraulic actuation and recently even electromagnetic actuation [131, 133, 138, 158] can be considered for internal mixture formation. In the past few years several injectors were developed and tested in various locations. These injectors are designed for hydrogen at, or close to, ambient temperature. In this process, mechanical actuation by the camshaft or hydraulic actuation was used. In two cases [84, 118] cold hydrogen was injected at 130–200 K where the injectors had a design that was only partially consistent with good cryogenic design.

According to results to date and contrary to common opinion, considerable problems still exist in achieving a suitable mixture homogeneity with respect to the distribution of hydrogen in general and in particular with late injection, which above all requires the development of suitable cryogenic injectors for hydrogen. Figures 114 and 115 show, respectively, the principle and a view of a cryogenic injector developed at the DFVLR for the BMW/DFVLR test vehicle with direct cylinder injection. The hollow injector needle which supplies the hydrogen, is closed by means of spring tension against the cylinder pressure and is opened via a hydraulic linear actuator which admits pressure pulses. The pressure pulses, with changing pulse width, can be supplied via diesel fuel injector pump or via an

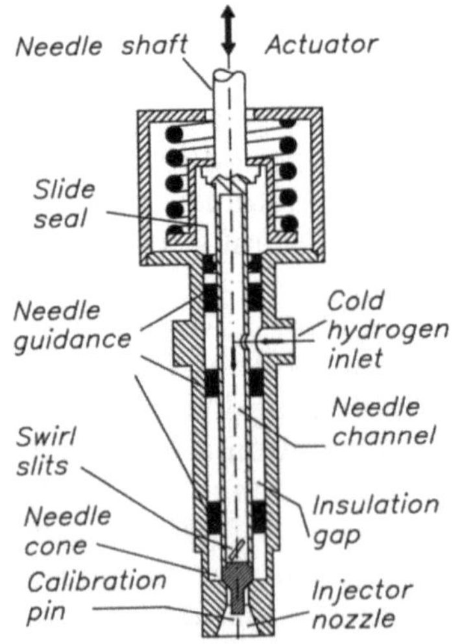

Fig. 114. Schematic of the DFVLR high-pressure injector for cryogenic hydrogen.

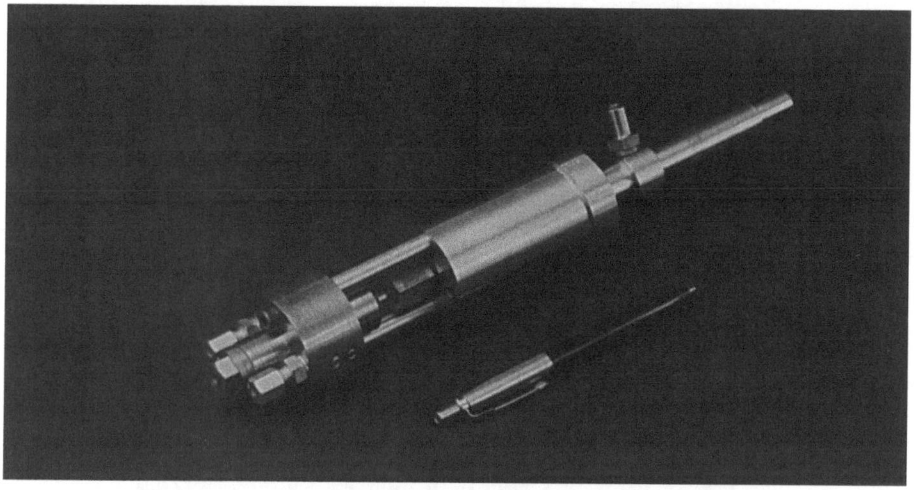

Fig. 115. View of a DFVLR high-pressure injector for cryogenic hydrogen with hydraulic actuator (on left)

electro-hydraulic device as is currently done with test vehicles. Thus, injection pressures for liquid or supercritical hydrogen in the range of 10–15 MPa and above can be controlled safely [118, 151, 113].

In contrast to conventional diesel injectors, with cryogenic hydrogen injectors, a cold gas with a certain volume has to be injected into the combustion space and has to be distributed accordingly. Regarding jet formation and gas mixing with air in the combustion space, totally different conditions have to be fulfilled compared with diesel fuel. This different conditions result from the gaseous nature of the jet medium, the large density difference between fuel and air, the considerably higher combustion rate as well as completely different turbulence characteristics which influence mixing. Evidently, hydrogen engines with internal mixture formation and late-start injection share several problems with conventional diesel engines. Insufficient homogeneity of the fuel–air mixture has a tendency to increase nitric oxide formation, even with hydrogen air mixtures that may be lean in the average. The reason is, that combustion may take place locally at fuel air ratios close to the stochiometric mixture, which leads to highly increased NO_x emissions. To keep the NO_x emissions low, very lean combustion in the average, or improved mixture homogeneity must be achieved by increasing the turbulence of the cylinder charge. Basically, stoichiometric combustion with hydrogen appears possible, if additional measures, in the form of injection into an already existing flame front for example could be taken to reduce nitric oxide emission (hybrid mixture formation, see page 217).

Fig. 116. View of the M30 engine with hybrid mixture formation (DFVLR procedure). Beside the high-pressure injectors, the low-pressure injectors for hydrogen designed for the timed individual port injection can be seen. They are marked by the black connection hoses with low-pressure hydrogen rail

Figure 116 shows the BMW M30 engine converted to hybrid mixture formation on the DLR hydrogen engine dynamometer. The low-pressure injectors for external mixture formation as well as cryogenic, hydraulically actuated cryogenic high-pressure injectors for late-start injection can be seen.

This experience can be modified by developments in cryogenic mixture formation in which, as a result of the substantially higher density of the cold gaseous hydrogen, considerably smaller orifices and thus smaller volumes are tolerated [63, 102].

6.2.6 Concept for Vehicles Fueled with Liquid Hydrogen

Based on experience to date with test vehicles and test engines, the importance of the following concepts becomes evident when taking advantage of the cryogenic properties of liquid hydrogen in order to improve engine operation while reducing NO_x emissions:

- Cryogenic external mixture formation.
- Cryogenic internal mixture formation and electronic ignition as well as early- and late-start injection.
- Combination of external with internal mixture formation (hybrid system).

The cryogenic internal mixture formation with late-start injection corresponds to conventional diesel operation and, due to controlled combustion, offers the largest potential regarding low-pollution or perhaps pollution-free combustion. Whereas, with cryogenic external mixture formation, the pressure level in the fuel tank is sufficient for engine operation, a pressurization with reciprocating pumps is necessary for cryogenic hydrogen with internal mixture formation, as previously indicated. With early-start injection, this can be effectively achieved with an electrically or hydraulically driven one stage liquid hydrogen pump installed in the immediate vicinity of, but separated from, the fuel tank. In order to avoid two-phase flow during the intake stroke of the pump an adequate pressure level in the fuel tank (approximately 0.1–0.2 MPa) and a low stroke rate (3000–4000 per min) are necessary in addition to supercooling of the cryogenic liquid.
The vapor formation, caused by heat leaks from cryogenic lines and components, as well as two-phase flow to the pump, can be avoided by choosing a pressure level above the critical pressure of hydrogen (1.3 MPa) and by warming up to the critical temperature (33 K). Under these conditions the density of the gaseous cryogenic hydrogen is still approximately 50% of that of the liquid at NBP and therefore small orifices in lines, valves and injectors are adequate for the given mass flow rates (1–10 g/s). Pressure oscillations of the pump are eliminated via a buffer volume which is installed cose to the engine. The fuel control for the engine is effectively implemented via variation of the opening period of the cryogenic fuel injectors. Application of digital motor electronics enables dual-fuel operation, i.e., the use of hydrogen or gasoline as a fuel without a large amount of transition effort. Furthermore, it must be emphasized that dual-fuel operation is basically necessary in order to enable adequate flexibility in selecting available fuels. DLR test vehicles for instance, permit gasoline operation as well as hydrogen operation using digital motor electronics.
The following vehicle concept results from internal mixture formation and late-start injection:

Contrary to current experience, pressurization up to the necessary injection pressure of over 10 MPa is achieved in a minimum of 2 stages (Fig. 117). The electrically or hydraulically driven LH_2 pump is maintained next to the LH_2 tank for the purpose of pressurization up to approximately 1.5–2.5 MPa and warming of hydrogen to the critical temperature (33 K) while avoiding two phase conditions [117, 125, 161]. At this point, the fluid cryogenic hydrogen reaches the engine compartment via the cryogenic feed line and there it is brought to the necessary injection pressure via a multicylinder high pressure cold compressor which is driven by the vehicle engine. The design of this cold compressor is not critical with regard to flow orifices. Furthermore, most of the compression work (1–2 KW) is done directly by the vehicle's engine. This system demonstrates considerably high redundancy despite a considerably sophisticated design because, for example, upon failure of the cold compressor or intermediate pressure pump, engine operation is then still possible with reduced power output. Upon simultaneous failure of the intermediate and high pressure pump emergency operation can still be carried out via external mixture formation by means of the LH_2 tank pressure. The same procedure can be applied

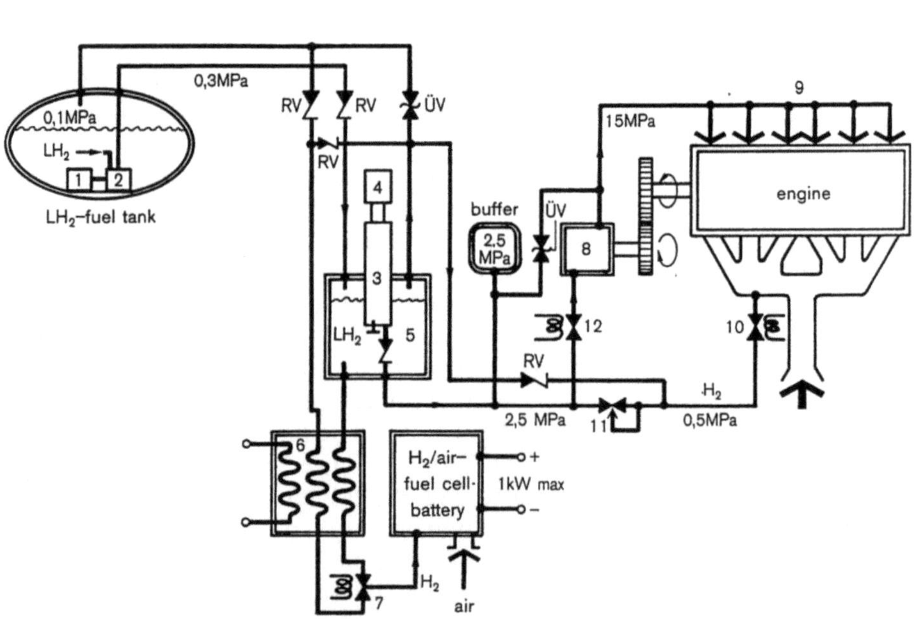

Fig. 117. Fuel conditioning system for hybrid mixture formation. *1*; *2* boost pump and cold electrical drive, *3–5* intermediate pressure LH_2-pump, pump drive and phase separator, *6* heat exchanger, *7* fuel cell control valve, *8* cold compressor, *9* high pressure cryogenic fuel rail, *10* low pressure hydrogen injector on air intake manifold, *11* pressure regulator, *12* cold compressor control valve

to the start and cooldown phase of warmed up pumps after longer periods of non-operation.
While cryogenic internal mixture formation can be considered for engines with a large piston displacement in passenger vehicles, commercial vehicles etc., due to its considerable complexity as well as the increased need for space on the cylinder head, electronically controlled cryogenic internal mixture formation with early-start injection, possibly in combination with timed external mixture formation, seems to be especially attractive for use in engines with medium piston displacement and medium power output. Here it is assumed that the internal combustion engine will still be the optimal solution for automotive vehicle power for a long time.

6.2.7 *Vehicle Refueling*

Although handling large amounts of liquid hydrogen during refueling and discharge of tankers, large containers, or rocket tanks is state of the art and is routinely carried out by specially trained personnel, the problems of refueling automotive vehicles are completely different. In contrast to the handling of hydrogen on a large scale, refueling cannot be limited to technically trained personnel. Furthermore, since the amounts of hydrogen and mass flow rates used in refueling a vehicle are magnitudes smaller than in the large-scale industry, components and procedures developed there are generally not easily transferred. For example, the heat capacity of transfer lines, valves, etc. is relatively insignificant with the large amount and high mass flow rates used for refueling transporters and containers. In the refueling of automotive vehicles where periodically about 120 to 150 liters have to be transferred in a maximum period of about 5 minutes, the heat capacity of the refilling station including valves, flexible hoses, transfer lines as well as the flow losses have a major impact. For lack of an appropriate demand, suitable components are not on the market and have to be developed when needed.
These problems described here were solved by the development of a fully automatic refilling station controlled by a microprocessor. Here, the user must only establish a connection to the vehicle's tank, start the refueling process and detach the connection from the vehicle after completion of the process.
Although in some publications it has often been stated that the storage of LH_2 onboard vehicles and refueling are still problems which have to be solved, it can be stated that practical solutions have been demonstrated for both problems. Further developments of LH_2 tanks and LH_2 filling stations for use in larger numbers and the corresponding questions about construction and standardization are still required. These requirements can

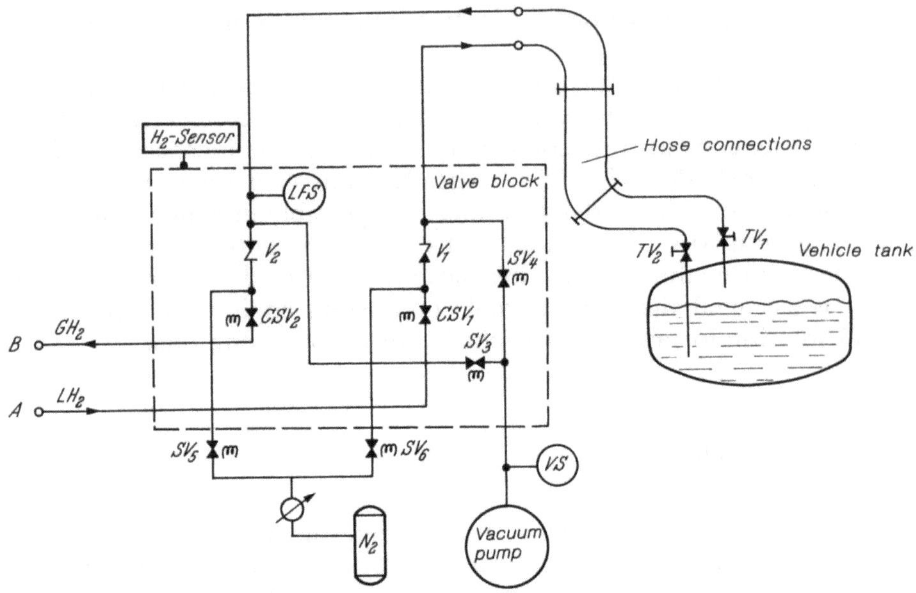

Fig. 118. Schematic of the liquid hydrogen prototype gas station TV_1; TV_2 tank valves, CSV_1; CSV_2 cryogenic solenoid valves, $SV_3 - SV_6$ ambient solenoid valves, V_1; V_2 check valves, *LFS* liquid flow sensor, *VS* vacuum sensor

be fulfilled with the means and facilities available in the current stage of development of electronics.

Initial considerations in this direction were made by F. J. Edeskuty and W. F. Stewart in 1973 [6] but could not be pursued at that time. In 1978, the DFVLR demonstrated a semiautomatic refilling station which, with the exception of points 6 and 7, essentially met the prerequisites listed below [77, 82, 90, 91]. However, because two manual steps still have to be carried out during the refueling, this refilling station, whose basic principle is shown in Figs. 118 and 119 was not yet the fully automatic system which was sought.

A fully automatic refilling station which can be safely operated and installed without specially trained personal, was developed and tested (Fig. 120)[25]. Operational error is prevented by the microprocessor controlled program. Basically the experience aquired with liquid hydrogen is also applicable for the use of liquid methane (LNG).

The automatic system must meet the following requirements:

1. Operational error must not lead to a dangerous situation. In particular, the release of liquid hydrogen into the air must be avoided.

[25] Germ. Pat. DE 33 44 770 C2.

Fig. 119. View of a semiautomatic filling station for liquid hydrogen. The plant shown was used in the U.S. within the scope of a mutual program between the Los Alamos National Laboratory and the DFVLR from 1979 to 1982

2. The refueling process must stop automatically upon failure of the electronics or the hydrogen carrying lines.
3. Refueling must stop automatically when the vehicle tank is full.
4. It must be possible to stop the refueling process at any time by external command, i.e., partial refueling must be possible.
5. The interior of lines carrying hydrogen and valves must be protected from air and moisture.
6. The combination of starting the engine, and hence driving away the vehicle, must be prevented automatically as long as there is a connection between vehicle and refilling station.
7. Connections between the refueling station and the vehicle should not be able to be interrupted during the refueling process.

The following brief operational procedure for the prototype LH_2 refueling stations developed at the DFVLR (Fig. 120) resulted from extensive experience in the use of LH_2 in space research activities. Pressurized LH_2 (0.15 MPa) flows from the main storage tank of the refilling station through

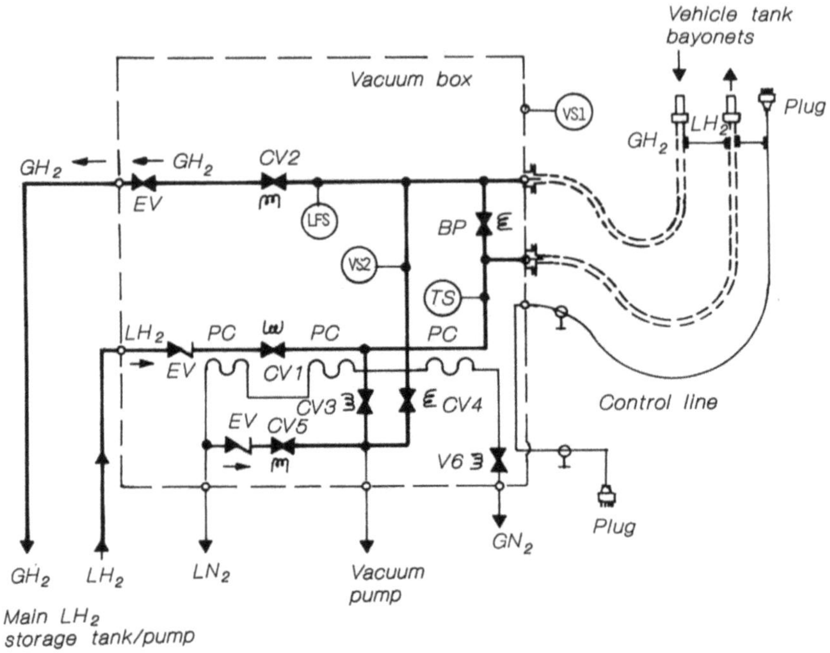

Fig. 120. Schematic of a fully automatic LH_2 filling station

a LH_2 line, a cryogenic check valve *EV* and a cryogenic solenoid valve CV_1 via a flexible vacuum-insulated transfer line LH_2 to the bayonet connection on the vehicle tank. The small amount of gaseous hydrogen produced by evaporation during the refueling process (a total of about 5% of the amount fueled) returns to a recovery system through the vacuum-insulated flexible recovery line GH_2 via a cryogenic solenoid valve CV_2 and another check valve *EV* via the line GH_2 The complete valve installation, designed as a valve block, is located in a vacuum (vacuum box) for thermal insulation. To avoid hydrogen release into the atmosphere due to operational error, the hydrogen supply and return system, up to the including the bayonets on the vehicle tank, are evacuated by a vacuum pump through solenoid valves CV_3 and CV_4. Only when the hoses are perfectly coupled, i.e., when a certain vacuum in the hydrogen line is reached, controlled via a vacuum sensor VS_2, can the refueling process be initiated by the press of a button. Than the valves on the vehicle tank (CSV_1, *BP*, CSV_2, see Fig. 121) open the tank. When the vehicle tank is full, liquid hydrogen flows into the recovery line, and the valves CV_1, CV_2 as well as valves CSV_1, *BP*, CSV_2 (Fig. 121) are closed by a liquid-phase sensor *LFS*.

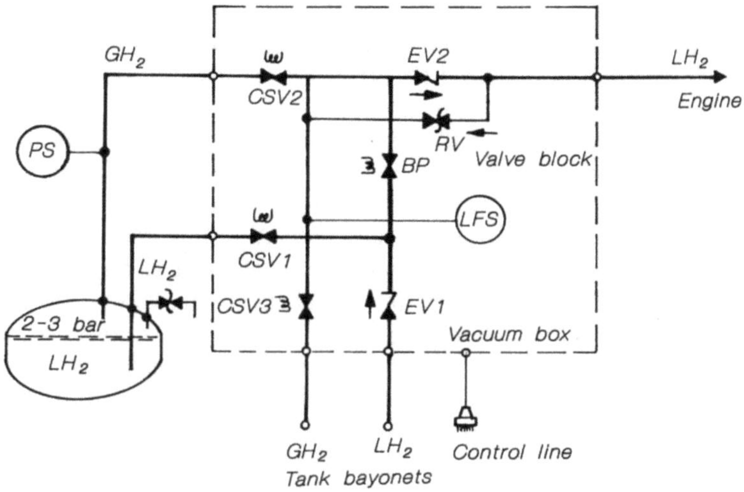

Fig. 121. Principle arrangement of the refueling equipment at the side of the vehicle for fully automatic refueling and discharge

Finally, the lines are evacuated again and filled with dry gaseous nitrogen. For partial refueling, the filling procedure is interrupted by pressing a button, instead of by the liquid phase sensor as in the full refueling process.
Through control of the permanent vacuum in the valve box used for thermal insulation, and the vacuum space of the flexible fill and recovery lines by the sensor VS_1, it is guaranteed that if a leak develops, the entire unit is shut down, and the release of hydrogen into the atmosphere is prevented.
The solenoid valves CSV_1, *BP*, CSV_2 on the vehicle tank, (see Fig. 121) are actuated by an electrical connection between the refilling station and the vehicle. Thus, starting the vehicle's engine during refueling is also simultaneously prevented. During periods of non-operation of the fill-station the valve V6, which is periodically actuated (see Fig. 120), keeps the refueling instrumentation cold with liquid nitrogen (LN_2) via the heat exchanger *PC*. The bypass valve *BP* is also opened for a short time when the refueling begins. In a similar manner, the bypass valve *BP* of the valve block on the tank (see Fig. 121, see also Fig. 90) serves to cool the flexible hoses between refilling station station and vehicle at the beginning of the refueling process. With systems working according to this principle and the vehicle-side valve block (see Fig. 90), a total refueling time (line cooldown included) for a cold 130L fuel tank of between 4 and 4.5 minutes was achieved (DFVLR-BMW 745i vehicle see page 224) which agrees well with the

required amount of time for refueling with coventional fuel. Fifteen minutes are required for the tank at ambient temperature including cooldown. The overpressure (pressure in the LH_2 storage tank or pressurization via a centrifugal pump) which is necessary for the refueling process is only 1–1.2 bar. If no pump is used for pressurization in the vehicle, in order to achieve the required pressure buildup in the vehicle's tank, an additional warming of the LH_2 tank to about 24 K is required. In addition to an electrical heater (Los Alamos vehicle, BMW 735i) requiring warm up periods of about 20–30 minutes, the induction of gaseous ambient hydrogen via the fill line into the liquid phase in the tank proved to be successful. In this process the gaseous hydrogen is cooled and condenses into the liquid phase which itself is warmed up. This process is carried out automatically by the refilling station and lasts about one minute.

A total of three refueling stations were built and tested within the scope of these projects, the first two in somewhat simplified versions. One was used for refueling test vehicles in a joint project between Los Alamos National Laboratory and the DFVLR (Fig. 119). The third refueling station (Fig. 122), which exhibited all of the previously described features in laboratory tests, was never put into the demonstration phase due to

Fig. 122. View of the fully automatic LH_2 filling station prototype

scheduling problems, but serves as a basis for further development by relevant industry[26].

Furthermore it is important that the DFVLR refueling stations, which respect to cryo-technology, were consistently designed for use under poor weather conditions (high humidity, low outside temperature) and also do not require helium as a scavenging gas.

Measuring the amount of hydrogen drawn is a problem, which has not yet been solved with satisfaction. Volume flow meters, turbine flow meters [104, 105], which are used in cryo-technology, lead to excessive measurement error for the small mass flow rate (up to about 30 g per second), which occurs when refueling a vehicle, where it is seemingly impossible to avoid vapor formation within the flow. A solution to this problem is expected to be a direct mass flow rate measurement which is either based on inertial effects (axial flow or gyroscopic transverse momentum mass flow meters), the anemometer principle (heat transfer) or proton spin resonance, for which considerable development work still has to be done [106, 107].

6.2.8 Behavior of the Fuel Storage and Conditioning System in Accidents

No reliable statements can currently be made about effects of accidents on the LH_2 fuel system and specific damage that can arise. On one hand, there are parallels to liquid propane (LPG) and liquid methane (LNG), but on the other hand these fuels cannot be directly compared to liquid hydrogen because, for example, their behavior when dissipating in the air is completely different. Furthermore, from NASA experience [108] in dealing with liquid hydrogen it can be inferred that, in accidents to date with LH_2 tank vehicles, the chain of events of accidents with other flammable liquids, like for example gasoline or liquid propane gas, in most cases would have had more dramatic consequences.

In order to obtain a reliable answer to the questions posed here, it is necessary to gather further experience and to carry out corresponding test programs [119, 123].

From carrying out spill experiments with large amounts of liquid hydrogen as well as from results of tests carried out with liquid hydrogen in aeronautics, statements can be made which support the opinion that liquid hydrogen would not be worse in this regard than gasoline or liquid propane gas when weighing its advantages and disadvantages [109, 110] (also see Chap. 8). A brief discussion of some typical accident situations follows.

---

[26] Messer Griesheim GmbH as well as the LINDE-AG are currently developing refueling stations based on this unit for LH_2 storage and transport containers as well as for commerical LH_2 test vehicles. The manually actuated container instrumentation is substituted by a combination of solenoid valves which are in the vacuum space, similar to Figs. 90 and 121.

The Formation of a Flammable Hydrogen–Air Mixture in Closed Volumes of the Vehicles, Especially the Passenger Compartment

This should be prevented by all means with all the three fuels (gasoline, LPG, and LH_2) because a source of ignition in the vehicle can never be ruled out and the over-pressure caused by deflagration after igniting a gaseous mixture can result in a detonation. Due to its wide range of flammability in air of 4 to 75 Vol.% compared to methane with 5 to 15 Vol.% and gasoline 1.0 to 7.6 Vol.%, hydrogen could be considered more dangerous than either of the two. However, of the three, gasoline has the smallest lower ignition limit. Furthermore, heating value (based on volume) at the lower ignition limit for a hydrogen–air mixture is only about one fifth of the value of gasoline–air or methane–air mixtures.
Therefore, in automotive vehicles the whole fuel system is generally installed outside the vehicle and the fuel tank should be located outside the crumple zone. In vehicles with LPG, the tank is generally installed in the trunk outside the crumple zone. Here the trunk must be sealed from the passenger compartment. In case of leaks, it is recommended that hydrogen sensors be installed in the passenger compartment of test vehicles in accordance with [111].

Tank Damage Without Liquid Hydrogen Release

The tanks used in the test vehicles in Los Alamos and the DLR represent a very rigid structure (double walled with vacuum and rugged outer vessel) which are considerably more resistant to accidents than the thin-walled tanks used for conventional fuels. Furthermore, the fiberglass epoxy suspensions of the inner vessels withstand shocks between 20 to 30 g[27]. This was confirmed by the experience with one of the DFVLR test vehicles which had a collision during a stay in the U.S. The collision pushed the vehicle about 7 m into an intersection. The backwall of the trunk, the floor of the trunk and left fender were considerably damaged. The drive shaft tunnel had cracks. The tank was filled with liquid hydrogen, but the tank as well as its instrumentation were not damaged in this accident.

Vacuum Loss Due to a Leak in the Outer Vessel. Tests show that the insulating property of the multilayer insulation, although considerably reduced, is still partially effective even with a total loss of the vacuum, so that unacceptably high pressure buildup in the tank will not occur with an appropriate installation of vent lines and safety valves. For example, the heat leak of the tank in the DFVLR test vehicles is about 3 W and, as

[27] $1\,g = 9.81\,m\,s^{-2}$.

tests confirmed in the case of a vacuum loss, increases initially to about 500 to 1,000 W. With a complete loss of the insulating property—tearing the insulation from the inner vessel—the heat leak would be about 40 W/m^2 K according to [5] (Chap. 4). This corresponds to a heat leak of 19 kW with about 2 m^2 surface of the inner vessel and would cause an evaporation rate of approximately 4 g LH_2/s. Since hydrogen flows through the vent line at a temperature slightly above boiling temperature its density is relatively high. At 25 K and 3.5 bar it is about 0.064 g/cm^3 for the liquid and about 0.004 g/cm^3 for the vapor compared to $9. 10^{-5}$ g/cm^3 under normal (NPT) conditions. Because of the very low kinematic viscosity of hydrogen at low temperatures the usual vent line of the inner vessel (12 $\phi \times 1$ mm) will cause a pressure drop in the order of 1 bar. Therefore, at an operating pressure of 3.5 bar, which is sufficiently below the critical pressure, no unacceptably high increase in pressure is expected with this type of accident [130, 157]. To complete the picture it has to be mentioned that the conditions are considerably more critical with liquid helium due to its low heat of evaporation (20.6 J/g compared to 443.5 J/g at LH_2).

Failure of the Suspensions of the Inner Vessel with and Without Vacuum Loss. In both cases it can be assumed that the heat leak cannot cause a great evaporation rate or increase in pressure than the previouly mentioned values.

Tank Damage with Liquid Hydrogen Release

This situation can occur with vacuum loss due to a leak in the outer vessel and simultaneous damage of the inner vessel. As a result of the rigidity of the tank structure, this is only possible in very severe accidents, for example if the tank is struck by a heavy vehicle, in multiple car collisions etc., where in general the severity of the damage is no longer characterized by the damage to the hydrogen tank.

In this case, liquid hydrogen is released but vaporizes within a short time (see page 141). Since hydrogen, as mentioned, is lighter than air at a temperature of about 30 K. i.e., 10 K above the boiling point, it mixes and rarefies in the air when it rises with strong turbulence so that in case of ignition the hydrogen fire essentially is limited to the area above the vehicle and will last only a few seconds. Here, it was assumed that hydrogen can directly emerge to the open air which is true if the tank is installed underneath the vehicle floor. In test vehicles, where the tank is in the trunk, the lid of the trunk usually springs open or is ripped off in this type of severe accident, so that the above assumption can also be applied. The behavior of hydrogen in accidents, described above, is completely different from that of conventional fuels. For example, neither gasoline vapor nor vaporized liquid propane gas is lighter than air so that at the scene of accident the gas clouds on the ground persist for long periods of time.

Upset of the Vehicle

If the tank is constructed according to the design shown in Fig. 75 there are no further consequences if the vehicle is turned upside down with an undamaged tank. The pressure buildup rate remains unchanged even though the vent line is filled with liquid hydrogen. With vacuum loss due to damage of the tank, liquid hydrogen would have to be vented via the safety valves as a result of the pressure buildup. However, if the safety valves are located outside under the floor of vehicle, the hydrogen can emerge unhindered into the open air, evaporate, mix with the air and rarefy. The tank can be emptied considerably faster than in the situation described in point 2 because of the considerably higher density of the liquid (0.07 g/cm^3). Wether this period is sufficient to avoid an unacceptably high pressure buildup remains to be determined.
Finally, an accident from 1975 should be mentioned which involved the UCLA vehicle with a full tank as it was being transported on a trailer [112]. As a result of a flat tire on the tractor, the trailer went out of control, broke the trailer coupling, went off the road and turned over. The liquid-hydrogen vehicle came to rest on its on roof underneath the trailer (Fig. 123). The mounting support as well as the hydrogen tank itself remained externally undamaged and—as was later shown—there was no leak in the vacuum jacket. Since the tank was constructed as a storage dewar with a vertical neck (see Fig. 36), liquid hydrogen came in contact with the neck section of the tank which was at ambient temperature. This

Fig. 123. Situation directly after the accident of the UCLA vehicle originally on top of the trailer (see [112])

Fig. 124. Situation shortly after the accident of the UCLA vehicle. The cloud caused by the vented cold hydrogen can clearly be seen on the left (see text) (from [112])

caused a quick increase in pressure in which liquid hydrogen was vented via the safety valve and quickly vaporized (Fig. 124[28]). Gaseous hydrogen was quickly vented by manually opening the valves of the liquid line, the inner end of which was in the gaseous phase, and the pressure was reduced to below the set level of the safety valves. Afterwards, the vehicle was set upright and was able to drive back to Los Angles under its own power despite considerable damage to the automobile body.

As events tend to repeat themselves, the DFVLR BMW 520 test vehicle was involved in an almost identical accident in 1984. The fuel tank which did not carry liquid hydrogen at that time remained undamaged.

The 745i DFVLR-BMW test vehicle which was demonstrated at the 6th World Hydrogen Conference in Vienna 1986 suffered a no-fault lateral collision with no damage to the fuel system.

References

[1] C. Marchetti: Air Transport Hydrogen, Master Key to the Energy Market. Euro-Spectra *10*, 117–129 (1971).

[2] C. Marchetti: Hydrogen and Energy. Chem. Econ. Eng. Rev. *5*, 7–25 (1973).

[3] C. Marchetti, N. Nakicenovic: The Dynamics of Energy Systems and the Logistic Substitution Model. Ar-78-1B, Int. Inst. for Appl. Systems Analysis (IASA) (1978).

---

[28] The author would like to express his gratitude to J. Finegold from the Solar Energy Research Institute in Golden, Colorado for providing Figs. 123 and 124 as well as for the permission for their publication.

[4] A. S. Manne, C. Marchetti: Hydrogen: Mechanisms and Strategies of Market Penetration. In: T. N. Veziroglu, "Hydrogen Energy", Part P. (ed.), pp. 1193–1209. New York Plenum Press (1974).
[5] C. Marchetti: The Evolution of the Energy Systems and the Aircraft Industry. In: Proc. Symp. Hydrogen in Air Transportation, DFVLR, Stuttgart (1979).
[6] L. W. Jones: Liquid Hydrogen as a Fuel for the Future. Science *174*, 367–370 (1971).
[7] A. Baine: LH_2-Spacecraft Experience Applicable to Aircraft Operations. In: Pro., Working Symp. on Liquid Hydrogen-Fueled Aircraft. NASA-Langley Res. Center, Hampton, VA, May 15–16 (1973).
[8] A. V. Cleaver: Cryogenic Fluids in the Aerospace Industry. Cryotech. *73*, 107–111 (1974); see also: Proc., Ind. Gases Conf. Guildford England, IPC Sci., Technol. Press (1974).
[9] I. I. Pinkel: Alternative Fuels for Aviation. Impacts Mil. Res. Dev., AGARD, Annual Meeting, pp. 31–36 (1974).
[10] W. J. D. Escher, G. D. Brewer: Hydrogen: Make Sense Fuel for an American Supersonic Transport. J. Aircr. *12*, 3–10 (1975).
[11] A. Silverstein, E. W. Hall: Liquid Hydrogen as a Jet Fuel for High Altitude Aircraft. RME 55C28a, NACA, pp. 1–2 (1955).
[12] B. E. Gammon: Preliminary Evaluation of the Air and Fuel Specific-Impulse Characteristics of Several Potential Ram-Jet Fuels, IV: Hydrogen, Methylnaphtalene and Carbon. RME 51Fo5, NACA (1951).
[13] W. T. Mikolowski, L. W. Noggle: The Potential of Liquid Hydrogen as a Military Aircraft Fuel. Int. J. Hydrogen Energy *3*, 449–460 (1978).
[14] L K. Carson, G. W. Davis, E. F. Versaw, et al.: Study of Methane Fuel for Subsonic Transport Aircraft. NASA-CR-159320, Lockheed Calif. Corp. (1980).
[15] G. D. Brewer: The Prospects for Liquid Hydrogen Fueled Aircraft. Int. J. Hydrogen Energy 7, 21–41 (1982), see also: Proc., 17th IECEC, Los Angeles (1982).
[16] R. D. Witkofski: Comparison of Alternate Fuels for Aircraft. NASA-TM-70155 (1979).
[17] R. D. Witkofski: Alternate Aircraft Fuels—Prospects and Operational Implications. NASA-TM-X-74030 (1977).
[18] J. F. Sloop: Liquid hydrogen as a Propulsion Fuel 1945–1959. NASA-SP-44-4, Stock-No. 033-000-00707-8, U.S. Governm. Printing Office, Washington, D.C. (1978).
[19] G. D. Brewer: Hydrogen Usage in Air Transportation. Int. J. Hydrogen Energy *3*, 217–229 (1978).
[20] W. J. D. Escher: Prospects for Liquid Hydrogen Fueled Commercial Aircraft. Escher—Foster Technol. Ass. Rep. PR-37 (1973).
[21] W. J. D. Escher: Liquid Hydrogen Future Aircraft Fuel: Background, Payoff and Cryogenic Engineering Challenge. In: Adv. Cryog. Eng. *20*, 70–81. New York, Plenum Press (1974).
[22] G. D. Brewer, R. E. Morris: Study of LH_2 Fueled Subsonic Passenger Transport Aircraft, NASA CR-144935, Lockheed -California Company, January (1976).
[23] G. D. Brewer: Is LH_2 the High Cost Option for Aircraft Fuel? In: Proc. 17th IECEC-Conf., Vol. 3, pp. 1191–1197, Los Angeles (1982).
[24] Anon.: An Exploratory Study to Determine the Integrated Technological Air Transportation System Ground Requirements of Liquid Hydrogen Fueled Subsonic, Long-Haul Civil Air Transports. The Boeing Commercial Airplane Corp., NASA-CR-2699 (1976).
[25] Anon.: IEA—Programme of Research and Development on the Production of Hydrogen from Water: Task III, Assessment of Potential Future Markets for the Production of Hydrogen from Water. Consolidated Final Report, pp. 38, 39. Comm. of the European Communities, Brussels (1980).
[26] S. Weiss: The Use of Hydrogen for Aircraft Propulsion in View of the Fuel Crisis. NASA-TM-X-68242 (1973).

[27] Anon.: Wasserstoff als Sekundärenergieträger, Vorschlag für ein Forschungs- und Entwicklungsprogramm. DFVLR-Mitt. 81-10 (1981).
[28] W. J. D. Escher, R. W. Foster, R. R. Tison, J. A. Hanson: Solar/Hydrogen Systems Assessment. DOE/JPL-955492, U.S. Dept. of Energy, Div. of Energy Storage Systems, Vols. 1, 2 (1980); see also: Int. J. Hydrogen Energy 7, 3–20 (1982).
[29] J. E. Johnson: The Economics of Liquid Hydrogen Supply for Air Transportation. In: Adv. Cryog. Eng., Vol. 19, pp. 12–22. New York, Plenum Press (1973).
[30] L. Prandtl: Führer durch die Strömungslehre, 3. Aufl. Braunschweig: Vieweg (1949).
[31] H. Schlichting: Grenzschichttheorie, 483 S. Karlsruhe: G. Braun (1958).
[32] L. Lees: The Stability of the Laminar Boundary-Layer in a Compressible Fluid. NACA-Rep. 876 (1947).
[33] E. Reshotko: Drag Reduction by Cryo-Fuel. Astronaut. Aeronaut. *10, 1* (1978).
[34] R. C. Mulready: Liquid Hydrogen Engines. In: Technology and Uses of Liquid Hydrogen R. G. Scott (ed.), 149–180. New York, Pergamon Press (1964).
[35] D. T. Pratt, K. J. Allwine, P. C. Malte: Hydrogen as a Turbojet Engine Fuel—Technological, Economical and Environmental Impacts. In: Proc., 2nd Int. Symp. on Air Breathing Engines, Sheffield, England, Royal Aeronaut. Soc. (1974).
[36] J. C. Riple, C. F. Baerst: Preliminary Studies of a Turbofan Engine and fuel System for Use with Liquid Hydrogen. In: Proc. Int. symp. Hydrogen in air Transportation, DFVLR, Stuttgart (1979).
[37] A. A. Dupont: Liquid Hydrogen as a Supersonic Transport Fuel. In: Adv. Cryog. Eng., Vol. 12, pp. 1–20. New York, Plenum Press (1967).
[38] J. B. Whitlow, R. J. Jr., Weber, K. C. Civinkas: Preliminary Appraisal of Hydrogen and Methane Fuels in a Mach 2.7 Supersonic Transport. NASA-TM-X-6822 (1973).
[39] G. D. Brewer, R. E. Morris: Minimum Energy Liquid Hydrogen Supersonic Cruise Vehicle Study. NASA-CR-137776, Lockheed Calif. Comp. (1975).
[40] G. D. Brewer: Advanced Supersonic Technology Concept Study—Hydrogen Fueled Configuration. NASA-CR-114718, Lockheed Calif. Comp. (1974).
[41] R. A. Jones, P. W. Huber: Toward Scramjet Aircraft. Astronaut. Aeronaut., *2*, 38–49 (1978).
[42] R. J. Henery, G. Y. Anderson: Design Considerations for the Airframe-Integrated Scramjet. NASA-TM-X-2895 (1973).
[43] P. J. Waltrup, G. Y. Anderson, F. D. Stull: Supersonic Combustion Ramjet (Scramjet) Engine Development in the United States. In: Proc. 3rd Symp. Fut. Airbreathing Engines, Munich: (1976). DGL—Fachbuch 6.
[44] S. Z. Pinckey: International Performance for Langley Scramjet Engine Module, NASA-TM-X-740378 (1977).
[45] J. V. Becker, F. S. Kirkham: Hypersonic Transports. NASA-SP-292 (1971).
[46] G. D. Brewer, R. E. Morris, R. H. Lange, J. W. Moore: Study of the Application of Hydrogen Fuel to Longe-Range Subsonic Transport Aircraft. NASA-CR-1325559, Lockheed Calif. Comp. and Lockheed Georgia Comp. (1975).
[47] G. D. Brewer, G. Wittlin, E. F. Versaw, et al.: Assessment of Crash Fire Hazard of LH_2-Fueled Aircraft. Final Report, NASA-CR-165525 (1981).
[48] G. D. Brewer, R. E. Morris, D. W. Davis, et al.: Final Report of Fuel Systems for LH_2-Fueled Subsonic Transport Aircraft. NASA-CR-145319, Lockheed Calif. Comp. (1977).
[49] F. M. Anthony, J. Z. Colt, R. G. Helenbrock: Development and Validation of Cryogenic Foam Insulation for LH_2 Subsonic Transports. NASA-CR-3404 (1981).
[50] G. R. Cunnington, Jr.: Analysis and Design of Insulation System for LH_2-Fueled Aircraft. In: Proc. Int. Symp. Hydrogen in Air Transportation, DFVLR, Stuttgart (1979).
[51] G. D. Brewer: Some Environmental and Safety Aspects of Using Hydrogen as a Fuel. Int. J. Hydrogen Energy *3*, 461–474 (1979).

[52] J. Grobmann, C. Norgreen: Turbojet Emission, Hydrogen Versus JP. NASA-TM-X-68258 (1973).
[53] D. E. Broesbeck, W. R. Prince, C. C. Ciepluch: Evaluation of Hydrogen Fuel in a Full Scale Afterburner. NACA-RM-E57H06 (1957).
[54] D. M. Straight, A. L. Smith, H. H. Christenson: Brief Studies of Turbojet Combustor and Fuel System Operation with Hydrogen Fuel at −400 F. NACA-RM-E56K27a (1957).
[55] A. Ferri, A. Agnone: Jet Engine Design that can Drastically Reduce Oxides of Nitrogen. AIAA-Paper, 74–160 (1974).
[56] P. F. Korycinski: Air Terminals and Liquid Hydrogen Commercial Air Transports. Int. J. Hydrogen Energy *3*, 231–250 (1978).
[57] G. D. Brewer (ed.): LH_2—Airport Requirements Study, Lock-heed Calif. Comp., NASA-CR-2700 (1976).
[58] J. E. Johnson: The Economics of Liquid Hydrogen Supply for Air Transportation. In: Adv. Cryog. Eng., Vol. 19. New York, Plenum Press (1974).
[59] J. A. Havens: A Description and Assessment of the SIGMET LNG-Vapor Dispersion Model Rep. CG-M-3-79, NTIS, Springfield, VA 22161 (1979).
[60] D. N. Gideon, A. A. Putnam: Dispersion Hazard from Spills of LNG on Land and on Water. Cryogenics *17*, 9–15 (1977).
[61] N. Gifford, A. Frandkin, Jr.: An Outline of Theories of Diffusion in the Lower Layers of the Atmosphere. In: Meteorology and Atomic Energy (D. H. Slade, ed.) Rep. TID-24190, NTIS, Springfield.
[62] J. A. Raymer: Operation on an Aircraft Engine Using Liquefied Methane Fuel. In: Adv. Cryog. Eng., Vol. 26, pp. 1001–1006. New York, Plenum Press (1981).
[63] R. D. Witkofski: Dispersion of Flammable Vapor Clounds Resulting from Large Spill of Liquid Hydrogen. NASA-TM-83131 (1981).
[64] J. Fay: Unusual Fire Hazard of LNG Tanker Spills. Comb. Sci. Technol. 7, 225–237 (1973).
[65] Y. Tsujikawa, M. Hirano: Effects of Precooling of Suction Air on the Performance of Liquid Hydrogen-Fueled Supersonic Aircraft Engine. Int. J. Hydrogen Energy *13*, 691–700 (1988).
[66] Y. Tsujikawa, H. Asakura: Optimization of the Precooler of Hydrogen Fueled Gas Turbine. J. Heat Recovery Systems *8*, 433–443 (1988).
[67] Y. Tsujikawa, T. Sawada: On the Utilization of Hydrogen as a Fuel for Gas Turbine. Bull—JSME *23*, 1506–1513 (1980).
[68] Y. Tsujikawa, Y. Tsukamoto, S. Fujii: Performance Analysis of Scramjet Engines, Bull. Univ. Osaka Prefecture A *37*, 1–13 (1988).
[69] Y. Tsujikawa, M. Nagaoka: Determination of Cycle Configuration of Gas Turbine and Aircraft Engines by Optimization Procedure. In: Proc., 35th ASME Int. Gas Turbine and Aeroengine Congress, June 11–14, Brussels, Belgium (1990).
[70] A. A. Desoky, A. S. K. Halaf, El-Mahallawy, F. M.: Combustion Process in a Gas Turbine Combustor when using H_2, NH_3 and LPG Fuels. Int. J. Hydrogen Energy, *15*, 203–312 (1990).
[71] S. C. Lee: Turbulent Mixing of Coaxial Jets between Hydrogen and Air. Int. J. Hydrogen Energy *11*, 807–816 (1986).
[72] I. Souche, A. Chatalic, B. G. Bregeon: Kinetics of OH Recombination in the Post Combustion Zone of an Atmospheric Pressure Hydrogen–Oxygen Flame. Int. J. Hydrogen Energy *14*, 677–680 (1989).
[73] H. P. Trinh, K. M. Isaac, Y. S. Chen: Numerical Study of Hydrogen Air Mixing Flowfield in a Typical Combustor Geometry. Int. J. Hydrogen Energy *14*, 737–748 (1989).
[74] A. Sotheran et al.: Some Practical Aspects of Staged Premixed, Low Emission Combustion Trans. ASME *107*, 2–9 (1985).

[75] H. P. Alder (ed.): Hydrogen in Air Transportation, Feasibility Study for Zürich Airport, Switzerland. Int. J. Hydrogen Energy *12*, 571–586 (1987).
[76] N. G. L. Jones: A Schematic Design for a HAZOP Study on a Liquid Hydrogen Filling Station. Int. J. Hydrogen Energy *9*, 115–122 (1984).
[77] G. D. Brewer: Hydrogen Fueled Aircraft—1983 Update. In: Recent Developments in Hydrogen Technology (K. D. Williamson, Jr., F. J. Edeskuty, eds.), Vol. 1, pp. 56–153. Cleveland Ohio, CRC Press (1986).
[78] G. R. Cunnington, R. T. Parmley: Aerodynamic Surface Cooling for Laminar Flow Control for Hydrogen-Fueled Subsonic Aircraft, SAE-paper 801155 (1980).
[79] G. D. Brewer: Hydrogen Aircraft Technology. Boca Raton Fa, CRC (1991).

Ground Transport

[1] L. W. Jones: Liquid Hydrogen as a Fuel for the Future. Science *174*, 367–370 (1971).
[2] M. R. Swain, R. R. Adt: The Hydrogen–Air Fueled Automobiles. In: Proc. 7th IECEC, paper 729217. New York, ACS (1972).
[3] W. J. D. Escher: On the Higher Energy Form of Water (H_2O*) in Automotive Vehicle Advanced Power Systems. In: Proc. 7th IECEC, paper 729119. New York, ACS (1972).
[4] W. E. Winsche, K. C. Hoffmann, F. J. Salzano: Economics of Hydrogen Fuel for Transporation and Other Residental Application. In: Proc. 7th IECEC, paper 729215. New York, ACS (1972).
[5] L. W. Jones: Liquid Hydrogen as a Fuel for Motorvehicles: A Comparison with Other systems. In: Proc. 7th Intersoc. Energy Conv. Eng. Conf. (IECEC), paper 729213. New York, ACS (1972).
[6] W. F. Stewart, F. J. Edeskuty: Alternate Fuels for Transporation, Part 2: Hydrogen for the Automobile. Mech. Engineering, June 1974.
[7] A. Gann: On the Application of Hydrogen as a Fuel for Automotive Vehicles. ESRO TT-132, *14*, S (1973), DLR-Mitt.73–22, Über die Verwendùng von Wasserstoff als Treibstoff für Automobile, DFVLR (1973).
[8] T. Ohta: Liquid Hydrogen as Engine Fuel. Cryog. Eng. (Tokyo) *8*, 52–58 (1973).
[9] L. O. Williams: Hydrogen Powered Automobiles Must Use Liquid Hydrogen. Cryogenics *13*, 693–698 (1973).
[10] Neuen Kraftstoffen auf der Spur—Alternative Kraftstoffe für Kraftfahrzeugen, Wasserstoff. pp. 285–578, Bonn: BMFT, Referat für Öffentlichkeitsarbeit (1974). ISBN 3-8725-3-0992.
[11] W. Peschka: Wasserstoff als Alternativkraftstoff im Kraftfahrzeug. Int. Verkehrswesen *32*, 447–453 (1980).
[12] D. R. Cloyd, W. J. Murphy: Handling Hazardous Materials. NASA-SP-5032 (1965).
[13] K. C. Hoffmann et al.: Metal Hydrides as a Source of Fuel for Vehicular Propulsion. In: Proc., Int. Automotive Eng. Congr., Paper 690232, Detroit, Mich. (1969).
[14] R. H. Wiswall, Jr., J. J. Reilly: Metal Hydrides for Energy Storage. In: Proc., 7th IECEC, Paper 729210. New York, ACS (1972).
[15] J. J. Reilly, R. H. Wiswall, Jr.: The Formation and Properties of Iron–Titanium Hydride. Inorg. Chem. *13*, 218–222 (1974).
[16] J. J. Reilly, R. H. Wiswall, Jr.: The Reaction of Hydrogen with Alloys of Magnesium and Nickel. Inorg. Chem. *7*, 2254 (1968).
[17] H. Buchner: The Hydrogen/Hydride Energy Concept. In: Proc., 2nd World Hydrogen Energy Conf., Vol. 4, pp. 1749–1792 (1978); see also: J. Töpler, O. Bernauer, H. Buchner: The Use of Hydrides in Motor Vehicles. J. Less Common Metals *74*, 385–399 (1980); H. Buchner: Perspectives for Metal Hydride Technology. Prog. Energy Combust. Sci. *6*, 331–346 (1980).

[18] H. Buchner: Energiespeicherung in Metallhydriden (Innovative Energietechnik). Wien New York, Springer (1982).
[19] O. Sultan, M. Shaw: Study of Automotive Storage of Hydrogen Using Recyclable Liquid Chemical Carriers. TEC-75/003, ERDA, Ann Arbor (1975).
[20] M. Taube, P. Taube: Liquid Organic Carrier for Hydrogen as Automobile Fuel. EIR Würenlingen Re. (1979), see also: Int. J. Hydrogen Energy *8*, 213–227 (1983).
[21] W. Justi, A. Winsel: Kalte Verbrennung. Wiesbaden, Franz Steiner (1962).
[22] W. Vielstich: Brennstoffelemente. Weinheim, Verlag Chemie (1965).
[23] K. V. Kordesch: In: Handbook of Fuel Cell Technology, N. J. Englewood Cliffs (C. Berger, ed.). Prentice Hall (1968).
[24] H. Dietz, H. Jahnke, H. Reber: Forschungsarbeiten über elektrochemische Energiequellen. Bosch Tech. Ber. *3*, August 1971.
[25] H. V. Dören, K. J. Euler: Brennstoffelemente. VARTA Fachbuchreihe, Bd. *6*, Düsseldorf: VDI-Verlag (1971).
[26] H. G. Plust: Aktuelle Probleme bei elektrischen Antriebssystemen für Straßenfahrzeuge. ATZ *13*, Hefte 9 und 11 (1971).
[27] H. Weh: Problematik der Energiespeicher und des elektrischen Antriebs. ATZ 76, Heft 6 (1974).
[28] J. G. Finegold, W. D. Van Vorst: Engine Performance with Gasoline and Hydrogen Fuels, A Comparative Study. In: Proc., The Hydrogen Economy (Miami) Energy (THEME)-Conf. 1974.
[29] F. Sturm: Kraftgasverwendung im Luftschiff "Graf Zeppelin". Zeitschr. d. Vereins Deutscher Ingenieure (VDI), September 1929.
[30] K. H. Weil: The Hydrogen I.C. Engine—Its Origins and Future in the Emerging Energy-Transportation-Environment System. In: Proc., 7th Intersoc. Energy Conversion Engineering Conf., pp. 1355–1363. New York, ACS (1972).
[31] Anon.: Use of Hydrogen for Driving Engines, Tests in Holland. Genie Civil *72*, 224 (1918).
[32] H. R. Ricardo: Further Notes on Fuel Research. In: Proc. Inst. Automobile Engrs. *18*, 327–341 (1924).
[33] A. F. Burstall: Experiments on the Behavior of Various Fuels in a High-Speed Internal Combustion Engine. In: Proc., Inst. Automot. Engrs. *22*, 358–371 (1927).
[34] R. A. Erren, W. H. Campbell: Hydrogen from Off-peak Power—a Possible Commercial Fuel. Chem. Trade J. *92*, 328–339 (1933).
[35] R. A. Erren, W. H. Campbell: Hydrogen: A Commercial Fuel for I. C. Engines and other Purposes. J. Inst. of Fuel (Great Britain) *12*, 281 (1933).
[36] R. A. Erren: Der Erren-Wasserstoffmotor. ATZ, 41, Heft 19, 523 (1939).
[37] H. C. Gerrish, H. H. Foster: Hydrogen as an Auxiliary Fuel in Compression-Ignition Engines. NACA Rep. No. 535 (1935).
[38] M. Oehmichen: Wasserstoff als Motortreibmittel. Deutsche Kraftfahrtforschung, Heft 68 (1942).
[39] R. O. King, W. A. Wallace, B. Mahapatra: The Hydrogen Engine and the Nuclear Theory of Ignition. Can. J. Res. F. *26*, 264 (1948).
[40] R. O. King, M. Rand: Oxidation, Decomposition, Ignition and Detonation of Fuel Vapors and Gases. *27*: Hydrogen Engine, Can. J. Technol. *33*, 445–469 (1955).
[41] R. G. Murray, R. J. Schoeppel, C. L. Gray: The Hydrogen Engine in Perspective. In: Proc., 7th IECEC, Paper 729217. New York, ACS (1982).
[42] R. E. Billings, F. E. Lynch: History of Hydrogen Fueled Internal Combustion Engines. Publ. No. 73001, Technology Appl. Center, Univ. New Mexico, Albuquerque, N. M., April 1973.
[43] H. S. Homan, P. T. C. De Boer, W. J. McLean: The Effect of Fuel Injection on NO_x-Emissions and Undesirable Combustion for Hydrogen-Fueled Piston Engines. Int. J. Hydrogen Energy *8*, 131–146 (1983).

[44] P. C. T. De Boer, W. J. McLean, H. S. Homann: The Performance and Emissions of Hydrogen-Fueled Internal Combustion Engines. Int. J. Hydrogen Energy *1*, 153 (1976).
[45] P. N. Blumberg: Nitric Oxide Emissions from Stratified Charge Engines: Prediction and Control. Combust. Sci. Technol. *8*, (1973).
[46] H. K. Newhall: Combustion Process Fundamentals and Combustion Chamber Design for Low Emissions. SAE Trans. paper 75100 (1975).
[47] R. R. Adt, Jr., M. R. Swain, J. M. Pappas: Hydrogen Engine Performance Project. U.S. Dept. of energy (DOE), Second Annual Report, Contr. No. EC-77C03-1212, 1980. See further DOE/CS/31212-1. Washington DC., May 1983.
[48] M. R. Swain, J. M. Pappas, R. R. Adt, Jr., W. J. D. Escher: Hydrogen-Fueled Automotive Engine Experimental Testing to Provide an Initial Design-Data Base. In: Alternate Fuels, SP-480, SAE-paper 810350, 163–180 (1981).
[49] K. Lohner, H. Müller: Gemischbildung und Verbrennung im Ottomotor (Die Verbrennungskraftmaschine Bd. 6). Wien: Springer (1967).
[50] F. A. F. Schmidt: Verbrennungskraftmaschinen. Wien, Springer (1967).
[51] C. A. McCarley, W. D. Van Vorst: Electronic Fuel Injection Techniques for Hydrogen Powered I.C. Engines. Int. J. Hydrogen Energy *5*, 179–204 (1980).
[52] F. Pischinger, M. Schaffrath: Untersuchungen an einem Wasserstoffmotor und Maßnahmen zur Prozeßverbesserung. ATZ *78*, 445 (1976).
[53] K. W. Drexl, H. P. Holzt: Untersuchungen über die Gemischbildung und Verbrennung im Wasserstoffmotor. Entwicklungslinien in der Kraftfahrzeugtechnik. Köln: Verlag TÜV Rheinland (1977).
[54] S. Furuhama: Two Stroke Hydrogen Injection Engine. Tech. Note. Int. J. Hydrogen Energy *4*, 571–576 (1979).
[55] J. G. Finegold: Hydrogen: Primary or Supplementary Fuel for Automotive Engines. Int. J. Hydrogen Energy *3*, 83–104 (1978).
[56] M. Leiker: Die Gasmaschine. Wien, Springer (1953).
[57] D. Downs, A. D. Walsh, R. W. Wheeler: A Study of the Reactions that Lead to Knock in the Spark Ignition Engine. Phil. Trans. Roy. Soc. London A *243*, 463–524 (1951).
[58] R. O. King, S. V. Hayes, A. B. Allan, et al.: The Hydrogen Engine: Combustion Knock and the Related Flame Velocity. Trans. E.I.C. *2*, 143–148 (1958).
[59] K. Binder, G. Withalm: Mixture Formation and Combustion in Interaction with the Hydrogen Storage Technology. In: Proc. 3rd World Hydrogen Energy Conf., Tokyo, Vol. 2, pp. 1103–1117. Oxford, Pergamon Press (1980).
[60] A. H. Gathercole, J. P. Bindon, L. W. Roberts: Detonation Limited Air Fuel Ratios in a Conventional High Speed Spark Ignition Engine Fueled with Hydrogen. In: Proc., South Afr. Conf. in Transportation, Univ. of Cape Town (1979).
[61] M. A. Gammie, J. P. Bindon: The Effect of Mixture Strength and Spark Advance on Detonation Intensity in a Small Squish-Chambered Hydrogen Fueled Spark-Ignition Engine. In: Proc., 3rd World Hydrogen Energy Conf., Tokyo, Vol. 2, pp. 1001–1014 Oxford, Pergamon Press (1980).
[62] S. Furuhama, M. Hiruma, Y. Enemoto: Development of a Liquid Hydrogen Car. Int. J. Hydrogen Energy *3*, 61–81 (1978).
[63] W. Peschka: Liquid Hydrogen as a Vehicular Fuel—A Challenge for Cryogenic Engineering. In: Proc., 4th. World Hydrogen Energy Conf., Vol. 3, pp. 1053–1070 New York, Pergamon Press (1982); see also: Int. J. Hydrogen Energy *9*, 515–525 (1984).
[64] C. Carpetis: Comparison of the Expenses Required for the Onboard Fuel Storage Systems of Hydrogen Powered Vehicles. Int. J. Hydrogen Energy *7*, 61–78 (1982).
[65] W. Peschka: Wasserstoff als Alternativkraftstoff im Kraftfahrzeug. Int. Verkehrswesen *32*, 447–453 (1980).

[66] W. Peschka: Liquid Hydrogen for Automotive Vehicles-Experimental Results. ASME-Paper No. 81-HT-83 (1981).
[67] J. J. Donelly, Jr., W. C. Greayer, R. J. Nichols, W. J. D. Escher, et al.: Hydrogen-Powered Versus Battery-Powered Automobiles. Int. J. Hydrogen Energy *4*, 411–444 (1979).
[68] K. V. Kordesch: City Car with H_2-Air Fuel Cell/Lead Battery (One Year Operating Experience). In: Proc., 6th IECEC, Boston, Mass., Paper No. 719015 (1971).
[69] C. E. Winter, W. L. Morgan: General Motors Electro Van. J. Soc. Autom. Engrs., paper 670182, Detroit (1967); see also: C. Marks, E. A. Rishavy, F. A. Wyczalek, SAE-Paper 670176, Automotive Engineering Congr., Detroit, Mich., January 8–13 (1967).
[70] Anon.: Proceedings of the Fuel Cell in Transporation Applications Workshop, August 15–17 (1977), Los Alamos Sci. Lab., LA-7279-C (1978), see also: Application Scenario for Fuel Cells in Transportation, I.A.-7634-MS (1979).
[71] P. L. Underwood, P. B. Dieges: Hydrogen and Oxygen Combustion for Pollution Free Operation of Existing Standard Automotive Engines. In: Proc., 6th IECEC, SAE-Paper 719046 (1971).
[72] N. E. Morgan, W. D. Morath: Development of a Hydrogen–Oxygen I.C. Engine Space Power System. NASA Rep. by Vickers, July 1965.
[73] R. E. Billings: Hydrogen Storage for Automobiles Using Cryogenics and Metal Hydrids. In: Proc., The Hydrogen Economy Energy (THEME)-Conf., Miami (1974).
[74] W. F. Stewart, F. J. Edeskuty, K. D. Williamson, H. M. Lutgen: Operating Experience with a Liquid Hydrogen Fueled Vehicle. Los Alamos Sci. Lab., LA-UR-74-1637. See also: In: Adv. Cryog. Eng., Vol. 20, pp. 82–89. New York, Plenum Press (1974).
[75] J. G. Finegold, W. D. Van Vorst, et al.: The UCLA Hydrogen Car: Design, Construction and Performance. Trans. SAE 730507 (1973).
[76] A. F. Bush, W. D. Van Vorst: The UCLA Hydrogen Car. In: Adv. Cryog. Eng., Vol. 19, pp. 23–27. New York, Plenum Press (1974).
[77] W. Peschka, C. Carpetis: Cryogenic Hydrogen Storage and Refueling for Automobiles. Int. J. Hydrogen Energy 5, 619–626 (1980).
[78] W. F. Stewart: Liquid Hydrogen as an Automotive Fuel. Adv. Cryog. Eng., Vol. 25, pp. 822–830. New York, Plenum Press (1980).
[79] W. F. Stewart: Liquid Hydrogen Fueled Buick. Los Alamos Sci. Lab. Rep. LA-8605-MS, November 1980.
[80] W. Peschka, F. J. Edeskuty, W. F. Stewart: Liquid Hydrogen Storage and Refueling for Automotive Applications. In: Proc. 3rd Miami Int. Conf. on Alternative Energy Sources. December 15–18, Miami Beach, Florida (1980). See also: Alternative Energy Sources 3 (T. N. Veziroglu, ed.), Vol. 5, pp. 407–418. Berlin Heidelberg New York, Springer (1983).
[81] S. Furuhama, Y. Kobayashi, M. Iida: A LH_2 Engine Fuel System on Board Cold GH_2 Injection Into Two Stroke Engine with LH_2-Pump, ASME-Paper No. 81-HT-81.
[82] W. Peschka: Operating Characteristics of a LH_2-Fueled Automotive Vehicle and of a Semi-Automatic Refueling Station. Int. J. Hydrogen Energy 7, 661–670 (1982).
[83] S. Furuhama, Y. Kobayashi: A Liquid Hydrogen Car with a Two Stroke Direct Injection Engine and LH_2-Pump. Int. J. Hydrogen Energy 7, 809–820 (1982). See also: SAE-Paper 820349 (1982).
[84] S. Furuhama, Y. Kobayashi: Development of Hot-Surface-Ignition Hydrogen Injection Two Stroke Engine. In: Proc., 4th. World Hydrogen Energy Conf., Pasadena, Vol. 3, pp. 1009–1020. New York, Pergamon Press (1982).
[85] W. J. D. Escher: The Hydrogen Fueled Internal Combustion Engine, A Technical Survey of Contemporary U.S. Projects. Escher-Foster Ass. ETA-Rep. PR-51, St. Johns Mich. (1975).

[86] W. J. D. Escher: Survey of Liquid Hydrogen Container Techniques for Highway Vehicle Fuel System Application. Escher-Foster Techn., St. Johns, Mich., DOE-Rep. No. HCP/M2752-01, Novermber (1978).
[87] W. J. D. Escher: Hydrogen as an Automotive Fuel: Worldwide Update. Escher-Foster Technology Inc. Rep., St. Johns, Mich. (1982).
[88] H. May D. Gwinner: Möglichkeiten der Verbesserung von Abgasemissionen und Energieverbrauch bei Wasserstoff-Benzin-Mischbetrieb, MTZ, No. 4 (1981); see also: Possibilities of Improving Exhaust Emissions and Energy Consumption in Mixed Hydrogen–Gasoline Operation. Int. J. Hydrogen Energy *8*, 121–130 (1983).
[89] W. Peschka, C. Carpetis: A System Consideration of the Tank for Liquid Hydrogen Fueled Vehicles and the Resulting Tank Concept for a Passenger Car. In: Proc., 1st World Hydrogen Energy Conf., Miami Beach (1976).
[90] W F. Stewart: Operating Experience with a Liquid Hydrogen Fueled Buick and Refueling System, Hydrogen Energy Process. In: Proc., 4th World Hydrogen Energy Conf., Vol. 3, pp. 1071–1093. New York, Pergamon Press (1982); see also: Int. J. Hydrogen Energy *9*, 525–539 (1984).
[91] W. F. Stewart: Experimental Investigation of Liquid Hydrogen Vehicle Onboard Storage and Refueling System. Los Alamos National Lab. Final Report, US Dept. of Energy (DOE), July 1982.
[92] W. F. Stewart, F. J. Edeskuty: Liquid Hydrogen as a Vehicular Fuel. Mech. Eng. *24*, 24–29 (1981).
[93] W. F. Stewart, W. J. D. Escher: Liquid-Hydrogen Automotive Onboard Storage and Servicing System Project: A Progress Report. In: Alternate Fuels. SP 480, 181–188, SAE-Paper 810351 (1981).
[94] M. Ikegami, K. Miwa, M. Shioji: A Study of Hydrogen Fueled Compression Ignition Engines. Int. J. Hydrogen Energy 7, 341–354 (1982).
[95] K. Enemoto, S. Furuhama, Y. Kobayashi: Ignitability of Hydrogen Air Mixtures by Hot Surfaces and Hot Gases. In: Proc., 3rd World Hydrogen Energy Conf., Vol. 2, pp. 1149–1164. New York, Pergamon Press (1980).
[96] M. Ikegami, K. Miwa, M. Shioji: A Study of Hydrogen Fueled Diesel-Combustion. Prep. of Japan Soc. of Mech. Engrs., Paper No. 750-15 (1975); see also: Proc., 3rd World Hydrogen Energy Conf., Vol. 2, pp. 969–984. New York, Pergamon Press (1980).
[97] H. Tanabe, M. Ohnishi, G. T. Sato, et al.: Experimental Study of the Transient Hydrogen Jet Using a Fast Response Probe. In: Proc., 3rd World Hydrogen Energy Conf., Vol. 2, pp. 985–1000. New York, Pergamon Press (1982).
[98] H. Tanabe, T. Sorihashi, N. Suzuki, et al.: Experimental Study on the Transient Hydrogen Jet. In: Proc. 4th World Hydrogen Energy Conf., Vol. 3, pp. 1115–1128. New York, Pergamon Press (1982).
[99] L. O. Williams, D. E. Spond: A Storage Tank for Vehicular Storage of Liquid Hydrogen. Appl. Energy *6*, 98–112 (1980).
[100] C. A. McCarley, W. D. Van Vorst: Electronic Fuel Injection Technique for Hydrogen Powered I.C. Engines. Int. J. Hydrogen Energy *5*, 178–204 (1980).
[101] W. Drexl, H. Holzt, M. Gutmann: Characteristics of a Single Cylinder Hydrogen Fueled I.C. Engine Using Various Mixture Formation Methods. Daimler-Benz AG, Central Research (1976).
[102] C. A. McCarley: Development of a High Speed Injection Valve for Electronic Hydrogen Fuel Injection. In: Proc., 3rd World Hydrogen Energy Conf., Vol. 2, pp. 1119–1134. New York, Pergamon Press (1980).
[103] K. Suzuki, Y. Uchiyama, J. Hama: Research of Hydrogen Fueled Spark Ignition Engine. In: Proc., 3rd World Hydrogen Energy Conf., Vol. 2, 1027–1040. New York, Pergamon Press (1980).

[104] H. L. Minkin, H. F. Hobart, I. Warshwawski: Performance of Turbine-Type Flowmeters in Liquid Hydrogen. Cryogenic Eng. News *3*, 16–20 (1968).
[105] H. F. Hobart, H. L. Minkin, I. Warshwawski: Life Test of Small Turbine-Type Flowmeters in Liquid Hydrogen. NASA-Lewis Res. Center, NASA-TN-D-7323 and E-7371 (1983).
[106] W. E. Keller: Cryogenic Instrumentation of and above Liquid Hydrogen Temperature-Present and Future. In: Adv. Cryog. Eng. Vol. 18, pp. 189–301. New York, Plenum Press (1973).
[107] Anon.: Liquid Hydrogen Flow by NMR Technique. Instrum. Control Syst. *39*, *87* (1966).
[108] P. M. Ordin: Review of Hydrogen Accidents and Incidents in NASA-Operation. NASA-TM-X-71565 (1974), see also: Proc., 9th IECEC (1974).
[109] E. F. Johnson: Fire Protection Development in LNG-Fueled Vehicle Operation. Fire J. 66, No. 6, 231–245 (1972).
[110] D. Shooter, A. Kalekar: The Benefits and Risks Associated with Gaseous Fueled Vehicles. Rep. to the Massachusetts Turnpike Authority, Arthur D. Little Case 7440-2, May 1972.
[111] J. M. Arvidson, J. Hord, D. B. Mann: Efflux of Gaseous Hydrogen or Methane Fuels from the Interior of an Automobile. NBS-Tech. Note 666 (1975).
[112] J. G. Finegold, W. D. Van Vorst: Crash Test of Liquid Hydrogen Automobile. In: Proc., 1st World Hydrogen Energy Conference, Vol. 3. Miami Beach, Clean Energy Res. Inst. (1976).
[113] W. Peschka: Hydrogen Combustion in Tommorrow's Energy Technology. Int. J. Hydrogen Energy *12*, 481–500 (1987).
[114] W. Peschka: The Status of Handling and Storage Techniques for Liquid Hydrogen in Motor Vehicles. Int. J. Hydrogen Energy *12*, 735–764 (1987).
[115] T. Krepec, T. Giannacopoulos, D. Miele: New Electronically Controlled Hydrogen-Gas Injection Development and Testing. Int. J. Hydrogen Energy *12*, 855–862 (1987).
[116] S. Furuhama: Hydrogen Engine Systems for Land Vehicles. Int. J. Hydrogen Energy *14*, 907–914 (1989).
[117] W. Peschka: Liquid Hydrogen-Cryofuel in Ground Transportation. In: Adv. Cryog. Eng., Vol. 31, pp. 1035–1046. New York, Plenum Press (1986).
[118] S. Furuhama, T. Fukuma: Liquid Hydrogen Fueled Diesel Automobile with Liquid Hydrogen Pumps. In: Adv. Cryog. Eng. *31*, 1047–1056. New York, Plenum Press (1986).
[119] R. E. Knowlton: Safety in New Uses of Hydrogen Energy. In: Adv. Cryog. Eng. *31*, 1057–1062. New York, Plenum Press (1986).
[120] W. Peschka: Liquid Hydrogen for Automotive Vehicles—Status and Development in Germany. In: Cryogenic Processes and Equipment, ASME, pp. 97–104 (1984).
[121] S. Furuhama, T. Fukuma, T. Kashima: Liquid Hydrogen Fuel Supply System for Hot Surface Ignition Turbocharged Engine. In: Cryogenic Processes and Equipment, ASME, pp. 105–114 (1984).
[122] W. F. Stewart: Refueling Considerations for Liquid–Hydrogen Fueled Vehicles. In: Cryogenic Processes and Equipment, ASME, pp. 115–122 (1984).
[123] R. E. Knowlton: The Safety of Hydrogen as a Ground Transportation Fuel. In: Cryogenic Processes and Equipment, ASME, pp. 123–130 (1984).
[124] A. L. Bain, R. R. Tison, R. J. Sprafka: Hydrogen Fueled–Hydrogen Transport Rail System-NASA Proposal. In: Cryogenic Processes and Equipment, ASME, pp. 131–136 (1984).
[125] W. Peschka: Flüssiger Wasserstoff als Motorenkraftstoff der Zukunft. Maschinenwelt und Elektrotechnik *43*, No. 8/9, 1–10 (1988).
[126] R. D. Mc. Carty, J. Hord, H. M. Roder: Selected Properties of Hydrogen. NBS Monograph 168 (1981).

[127] S. J. Collocott: Cryogenic Liquid Level Monitor Using a Frequency to Voltage Converter. Cryogenics *23*, 327–328 (1983).
[128] D. Hagedorn, D. Leroy, P. Dullenkopf: Monitor for the Quality Factor in Two-Phase Helium Flow Using a Low Temperature Oscillator. In: Adv. Cryog. Eng. *31*, 1295–1307. New York, Plenum Press (1986).
[129] J. Hord (ed.): Selected Topics on Hydrogen Fuel, NBSIR 75-803, Chapter 5: Instrumentation for Cryogenic Hydrogen Fuel, Chapter 1: H_2-Fueled Automobiles (1975).
[130] Y. Rotenberg: Numerical Simulation of Self Pressurization in a Small Cryogenic Tank. In: Adv. Cryog. Eng., Vol. 31, pp. 963–971. New York, Plenum Press (1986).
[131] T. Krepec, T. Tebelis, C. Kwok: Fuel Control Systems for Hydrogen Fueled Automotive Combustion Engines—A Prognosis. Int. J. Hydrogen energy *9*, 109–114 (1984).
[132] S. Furuhama, Y. Kobayashi: Development of a Hot-surface-ignition Hydrogen Injection Two-stroke Engine. Int. J. Hydrogen Energy *9*, 205–214 (1984).
[133] K. S. Varde, G. M. Frame: A Study of Combustion and Engine Performance Using Electronic Hydrogen Fuel Injection. Int. J. Hydrogen Energy *9*, 327–332 (1984).
[134] W. Peschka: Liquid Hydrogen as a Vehicular Fuel—A Challenge for Cryogenic Engineering. Int. J. Hydrogen Energy *9*, 515–523 (1984).
[135] W. F. Stewart: Operating Fueled Buick and Refueling System. Int. J. Hydrogen Energy *9*, 525–538 (1984).
[136] B. Bogdanovic: Magnesium Hydride: A Homogeneous-catalysed Synthesis and Its Use in Hydrogen Storage. Int. J. Hydrogen Energy *9*, 937–943 (1984).
[137] M. Taube, D. Rippin, W. Knecht, et al.: A Prototype Truck Powered by Hydrogen from Organic Liquid Hydrides. Int. J. Hydrogen Energy *10*, 595–600 (1985).
[138] K. S. Varde, G. A. Frame: Development of a High-Pressure Hydrogen Injection for SI Engine and Results of Engine Behaviour. Int. J. Hydrogen Energy *10*, 743–748 (1985).
[139] W. Peschka, W. Nieratschker: Experience and Special Aspects on Mixture Formation of an Otto-Engine Converted for Hydrogen Operation. Int. J. Hydrogen Energy *11*, 653–660 (1986).
[140] W. Peschka: Liquid Hydrogen Fueled Automotive Vehicles in Germany—Status and Development. Int. J. Hydrogen Energy *11*, 721–728 (1986).
[141] H. Braes, N. Regar, W. Strobl: Drive Systems to Protect the Environment. Automotive Technology International, pp. 249–255, MCI CP 66/8/88 (1988).
[142] N. Regar, W. Strobl, R. Heuser: Pkw-Antriebe mit Elektro-und Wasserstoffspeicher (Stand der Entwicklungen und Perspektiven). In: Proc. 2. Aachener Kolloqium, Fahrzeug-und Motorentechnik, pp. 463–483 (1989).
[143] J. E. Blackford, P. Halford, D. H. Tantam: Expanders and Pumps. In: Cryogenic Fundamentals (G. G. Haselden, ed.), pp. 403–489. London, Academic Press (1971).
[144] K. P. Martin, R. B. Jacobs, R. J. Hardy: Performance of Pumps with Liquefied Gases. In: Adv. Cryog. Eng. *2*, 295–302. New York, Plenum Press (1960).
[145] G. H. Caine, L. Schäfer, D. Burgeson: Pumping of Liquid Hydrogen. In: Adv. Cryog. Eng. *4*, 241–254. New York, Plenum Press (1960).
[146] C. F. Goltzmann: High Pressure Liquid Hydrogen and Helium Pumps. In: Adv. Cryog. Eng. *5*, 289–298. New York, Plenum Press (1960).
[147] S. C. Collins: Hydrogen–Helium Liquefier. Paper A2, pp. 8–11. In: Adv. Cryog. Eng., Proc., 1956. CEC Boulder, Co. New York, Plenum Press (1960).
[148] S. C. Collins: Helium Liquefier. Science, *116*, 282–299 (1952).
[149] A. G. Cryomec: Supercharger, Europ. Patent 0174269 A3, (1987), US Patent No. 4, 639, 197.
[150] S. A. Socsil: Tiefsttemperaturpumpe für Flüssiggas. Ecublens, Lausanne, P 28 44 719.0 (1977).

[151] R. H. Hinds: The Musashi 7. Int. J. Hydrogen Energy *12*, 427–429 (1987).
[152] M. A. Deluchi: Hydrogen Vehicles: An Evaluation of Fuel Storage, Performance, Safety, Environmental Impacts, and Cost. Int. J. Hydrogen Energy *14*, 81–130 (1989).
[153] A. H. Seilly: Colenoid Actuators—A New Concept in Extremely Fast Acting Solenoids. SAE-paper 810462 (1981).
[154] M. Takiguchi, S. Furuhama: Combustion Improvement of Liquid Hydrogen Engine for Medium Duty Trucks. SAE-Techn. paper 870535 (1987).
[155] M. Takiguchi, T. Pichainarong, T. Matsushita, S. Furuhama, Characteristics of Combustion Pressure Vibration in Hydrogen Fuel Injection Hot Surface Ignition Engine. SAE Techn. Paper 871611 (1987).
[156] W. Peschka, G. Schneider: Kapazitive Füllstandanzeigen für LH_2-Kraftfahrzeugtanks. In: Proc. 14. DKV-Jahrestagung, Deutscher Kälte- und Klimatechnischer Verein, pp. 107–118 (1988).
[157] Y. Rotenberg, M. Burrows, R. Mc Neil: Vibration Effects on Boil-off Rate from a small Liquid Hydrogen Tank. Int. J. Hydrogen Energy *11*, 729–736 (1986).
[158] T. Krepec, T. Giannacopoulos, D. Miele: New Electronically Controlled Hydrogen-Gas Injector Development and Testing. Int. J. Hydrogen Energy *12*, 855–862 (1987).
[159] B. Bogdanovic, B. Spliethoff: Active Mg_2-Mg-Systems for Hydrogen Storage. Int. J. Hydrogen Energy *12*, 863–874 (1987).
[160] S. Furuhama, T. Fukuma: High Output Power Hydrogen Engine with High Pressure Fuel Injection, Hot Surface Ignition and Turbocharging. Int. J. Hydrogen Energy *11*, 399–407 (1986).
[161] W. Peschka: Liquid Hydrogen Reciprocating Pumps for Automotive Application. In: Adv. Cryog. Eng. *35*, 1783–1790. New York, Plenum Press (1990).
[162] W. F. Stewart: Hydrogen as a Vehicular Fuel. In: Recent Developments in Hydrogen Technology (K. D. Williamson, Jr., ed.), Vol. 2, pp. 69–146. Cleveland Ohio, CRC Press (1986).
[163] G. Withalm, W. Gelse: The Mercedes Benz Hydrogen Engine for Application in a Fleet Vehicle. In: Proc., 6th World Hydrogen Energy Conf., Vol. 3, pp. 1185–1198. Oxford, Pergamon Press (1986).

Outlook to Future Applications 7

The use of liquid hydrogen as an energy carrier in the future strongly depends on the role which hydrogen can take in general in future energy systems. Within the scope of this book it is not possible to go in detail into the numerous publications and studies concerning hydrogen's role in future energy systems and the arguments in this area. It is intended to provide an overview about the current state of the art of liquid hydrogen and the possible developments that can be derived from it.

7.1 Hydrogen as an Energy Carrier in the Future

The potential for development and application in aviation and space flight as well as in earth-bound traffic that is described in previous chapters lead to the conclusion that, with the exception of space flight, a quick, overall introduction of hydrogen is not necessary. Also, apart from special applications and countries with a suitable energy structure, it cannot be assumed that the development of production plants and the infrastructure for a new energy carrier will take place exclusively to meet the needs of air and earth bound traffic.

Clearly a result of oil policies, all plans for the future are essentially based on the conservation of oil for the transportation sector and the production of synthetic fuels from coal and other fossil residues like oil shale and tar sands for example. Furthermore, the commercial production of synthetic fuels directly from coal and hydrogen has been considered since oil reserves are becoming more and more scarce. However, here fuel is obtained by the thermal decomposition of water via coal with the emission of carbon dioxide and, depending upon the kind of coal, a more or less strong emission of sulfur dioxide.

Hydrocarbons have been used successfully as energy carriers for transportation means for almost a century. Their use represents a good compromise with respect to handling, storage and distribution as well as to mass and volume related energy density (see Table 23). Also with respect to economy their use seemed to be the easiest solution. This can only be changed by overcoming this established inertia.

In particular the term "economy", however, has to be subjected to a critical examination because pollution and health defects of the population due to "economic technology" are difficult to quantify and usually are not taken into consideration. Hydrogen though, does not solve the global energy problem but can considerably reduce environmental problems under certain conditions if it is produced as an energy carrier from water and in an environmentally safe manner. Although it has to be considered that in contrast to many opinions the energy from hydrogen is not cheaper than energy from any other energy carrier, like coal, oil, natural gas or other sources, otherwise it already would have been used for energy on a large scale. The energy from hydrogen is and will remain about 2 to 3 times more expensive than that from fossil energy carriers. The same is true for liquid hydrogen which corresponds to energy costs of synthetic gasoline or methanol from coal.

Looking to other energy carriers, hydrogen will only be acceptable for economical reasons when, as can be expected, the costs for the environment are included in future criteria regarding the economy. If this would be applied to fossil energy carriers, hydrogen would currently compare well. However, quantifying environmental damage, is still a difficult problem to solve so that this question is often not asked, especially not when it has to do with the different taxation of fossil energy carriers.

It is very probable that hydrogen, produced by processes which do not harm the environment like for instance from electrolysis via electricity from nonpolluting primary energy, i.e. hydro-power, as well as from nuclear energy, could be more economical than its fossil alternatives despite higher production costs.

Works and experience with Hydrogen show that it will be used as an additional energy carrier to the degree to which the people all over the world become conscious of the problems caused by atmospheric carbon dioxide and methane to include the resulting greenhouse effect and its negative effects on the global climate [1–6]. There is currently much speculation on the existence of this danger but it probably will not be known for certain after 10–15 years—but then with absolute certainty. The question remains, however, if it will not be too late for counter-measures because a process has been initiated in the atmosphere which can no longer be stopped.

A Brief Description of Current Facts

100 years of increased industrialization together with the explosive increase in the earth's population in the last two decades caused an immense increase in the consumption of fossil energy carriers and this consumption is still on the rise. The reversal of the consequences of the increase in the CO_2

content in the atmosphere is no longer possible within a time foreseeable by mankind, fact that is recognized by only too few.
In the last hundred years the CO_2 content in the atmosphere increased about 25% (270 ppm to 350 ppm) and currently increases more than 0.3% (1 ppm) every year and that annual increase will have doubled in about 10–15 years.
About half of the currently predictable increase in CO_2 corresponds to the amount produced by mankind whereas the rest is removed from the cycle through absorption and sedimention in the ocean or is subjected to the photosynthesis and respiration cycle of algae and plants. In this process, CO_2 is exchanged between the atmosphere and the oceans and continents which serve as storage reservoirs. Human activities like the devastation of forests and the increased combustion of fossil energy carriers currently cause an increase of CO_2 in the atmosphere of about 3 billion metric tons per year [7, 8].
If all available fossil energy reserves were exhausted their carbon would initially be in the atmosphere in form of CO_2 from where it once was removed by nature to the currently recognized value through geological epochs by the carbonate–silicate cycle with the help of biological processes [8–11]. A relapse into this chapter of the earth's history would correspond, with strong restrictions, to the condition on Venus, which, with a very large percentage of CO_2 in the atmosphere of 98% and a surface temperature of about 450 °C, is and in all probability will remain a dead planet. There is currently a negative feedback mechanism on the Earth which stabilizes the global temperature within certain limits. If, for example, the global temperature of the oceans decreases less water vapor enters the atmosphere which causes less rain and thus less weathering processes partially caused by HCO_3^- ions and erosion of the continents. Thus less CO_2 is removed from the atmosphere whereas there are no changes in the carbonate–silicate cycle which, with an appropriate time delay, releases about half the originally bound CO_2 in the subduction zones of the tectonic plates. Therefore, the CO_2 content of the atmosphere increases which counteracts the original decrease in temperature due to an increasing greenhouse effect. In the same way, the rise of the temperature of the ocean in the long term causes more rain, weathering and erosion which thus causes a reduction of the CO_2 content in the atmosphere and therefore a decrease in temperature. A perceptible decrease of the CO_2 content can—apart from the storage effect of the oceans—only be achieved by removing the carbon from the cycle, i.e., through sedimentation (carbonate–silicate cycle, formation of coal deposits, etc.) as was the case during the earth's evolution and which still takes place today. A perceptible effect of these processes can be expected only after a period of millions of years [10, 11].
Some of these mechanisms which serve to remove the carbon from the cycle could begin to show signs of saturation as a result of nonlinear effects.

In addition, there is the worldwide devastation of forests, primarily the rain forests, as well as the infringement of healthy growth of algae and plankton due to an increased pollution of the oceans.
All that can be done—in contrast to the currently exaggerated ozone problem—is to develop appropriate techniques to slow down the increase in CO_2 and methane and to apply them. Thus, at least the future maximum value in the atmosphere could level off at a value lower than at with further unrestrained use of fossil energy.
Still, an increase of the greenhouse gases CO_2 and CH_4 has to be expected which is directly related to the increase of the earth's population which can increase to 10–12 billions within the next 2–3 decades.
But one thing is certain: there will be perceptible changes in the climate. Though—with respect to specific regions it is still uncertain what effect this will have.
In addition to a reasonable solution to the energy problem and measures to stabilize the earth's population one of many necessary long-term steps to slow down the increase of CO_2 and methane would be the application of hydrogen, produced in an environmentally safe manner, in as many combustion processes as possible, or in the ideal case in all combustion processes.
Furthermore, the fact that in the combustion of fossil energy carriers the price of the fuel has an immense impact on the economy and from a worldwide perspective is of practically explosive political importance, has not sufficiently been taken into consideration. But it goes practically unnoticed that the oxygen which is needed for the combustion did not cost anything, i.e., it was drawn from the atmosphere and thus was a burden to everyone[1]. Instead of imposing a tax on the fossil energy carrier, a more effective way to prevent the misuse of the atmosphere as a CO_2 dump site would be to impose a tax on the oxygen which is drawn from the atmosphere and which is technically irreversible.
Here, hydrogen plays a special role because its production from water by means of non-fossil primary energy frees exactly the amount of oxygen which is later needed for its combustion, so that there is no disturbance of the oxygen balance in the atmosphere.
In general, hydrogen is well known in the energy technology of the past. Town gas, used for a long time, contained hydrogen at least up to 60 per cent volume.
The handling of large amounts of hydrogen in industry is basically state of the art. However, it is mostly used as raw material and not as an energy carrier.

---

[1] Example: 2,000 MW power station, thermal efficiency 40%, 10,000 t coal per day, 2–3 freight trains; 30,000 t oxygen, 40,000 CO_2 per day. The atmosphere is used as a CO_2 dump site!

Its compatibility with the different energy systems and energy forms is advantageous in its use as an energy carrier. Thus, hydrogen can be produced by different forms of primary energy and be changed easily into other energy forms like heat, through combustion, or electricity by means of electrochemical processes (fuel cells).

That does not necessarily mean that a global application of hydrogen as a secondary energy carrier is able to solve the problems of storage and distribution of energy better than electricity. This also applies to the coupling of solar electricity with the production of hydrogen. Although this process is suitable to reduce the disadvantages of solar energy regarding availability, storage and distribution it cannot bring out the specific advantages of hydrogen like clean combustion or its general suitability for applications in transportation technology. At least initially electricity, a very high-quality energy form, is produced by means of solar energy. Through the follow-on transformation into the chemical energy of hydrogen and the expenditure for transporation, storage and distribution its value is reduced. This situation is even worse with the generally senseless retransformation of hydrogen energy into electrical energy which can only be justified in special cases like transportation applications. Proven disadvantages of electrical energy like the problems associated with storage can basically and effectively be eliminated through a global electrical network. This type of global energy network could balance out the daily and seasonal variations of supply and demand in electrical energy without storage requirements. Location problems regarding electrical energy production would be pushed into the background since there are currently unused reserves in hydropower as well as undeveloped locations suitable for nuclear energy and solar energy which could be identified through such a global energy network. The high-voltage direct--current conversion technology can already be used for these applications. Further developments in the field of superconductivity could also have an immense impact. All the continents with the exception of the Antarctic and Australia could be connected by power lines run above the ground or above the continental shelf. In the case of Australia, relatively short distances between the Arafura shelf and the Sunda shelf could be partially bridged by deep-sea cables.

Therefore, from this point of view, the coupling of energy technology and hydrogen in general is not necessary but can be appropriate for special application areas. Transportation in general could be such an area if the motivation is to do without fossil energy carriers. The advantages of hydrogen with respect to clean combustion, storage in liquid state in the vehicle and the infrastructure required by the traffic are definitely evident here.

The problem of carbon dioxide in the atmosphere can no longer be neglected in connection with the fossil energy carriers. Rather, the solution will soon place strong limitations on the future use of fossil energy carriers [1–6, 12]. With respect to several proposals to remove the carbon dioxide from the exhaust gases of fossile heating systems and factories [13] and to direct it into the deep sea or with regard to the production of hydrocarbons from atmospheric carbon dioxide by means of hydrogen, it is doubtful whether the costs for the installation of this type of gigantic plants in the scope of a "planetary technology" would not far exceed those which would introduce the hydrogen-energy technology[2].

This clearly shows that the motivation to use hydrogen and the question regarding its economy under the conditions outlined above will be more a result of its characteristics as an unequalled nonpolluting energy carrier and "clean energy technology" which it provides rather than a "breakthrough" in the reduction of the current production costs which is often considered to be an essential prerequisite for the introduction of hydrogen-energy technology.

Furthermore, it can be seen that the consistent use of environmentally detrimental technology will also become more intolerable from an economical point of view. From a more general point of view it is widely accepted, to prevent environmentally detrimental technologies from the start instead of spending significant effort to mitigate the damage caused from the application of this technology. Thus the question arises, is it not strongly indicated to initiate energy technologies which cause principally only a minimum of envirnomental damages. It must be understood that this type of development requires considerably long periods of time. Decisions which affect the public cannot exclusively be left to the producers of consumer goods, the operators of power plants and companies which develop fossil energy carriers[3].

The liquid hydrogen technology as it is described within the scope of this book through individual examples reveals applications in transportation as well as general energy storage as a result of the potential of mobile and stationary hydrogen storage. In further applications, developments can take place which affect large fields of energy technology. They can also provide a transition phase for a hydrogen-energy technology which remains to be established to a reasonable extent. Earthbound traffic in general is representative of an area which can currently be developed through special applications.

Applications of cryogenic technology in energy will become more and more important since gas liquefaction in the future will be carried out in the

---

[2] See: J. O'M. Bockris, E. W. Justi Wasserstoff, die Energie für alle Zeiten, S. 28. München: Udo Pfriemer Verlag 1980.

[3] To compare it with the immense investments for tapping new oils reservoirs the latter would most be able to be a decisive factor in hydrogen-energy technology with only slightly reduced investments.

same magnitude as natural gas today in the interest of reducing storage and transporation costs.

7.2 Motor Vehicle Applications

Initial applications are especially appropriate for vehicle fleets, i.e., vehicles which are used within the same general area, have low infrastructure requirements and receive routine repair work and maintenance centrally. The same could basically be applicable for public institution vehicle fleets as well as for taxis.

For example, public service buses, which are maintained and refueled in central bus depots could take advantage of alternate fuels in urban traffic inter alia liquid hydrogen. There are already several serious proposals for just such a project. Especially two-stroke Diesel engines could be converted to environmentally safe hydrogen operation with very little effort. A prerequisite is the development of a high-pressure injection pump for liquid hydrogen which is suitable for automotive application with corresponding reliability and lifetime. The general goal should be dual-fuel operation, i.e., operation with hydrogen or conventional fuel as, to a certain extent, has been successfully demonstrated, so that the use of this type of vehicle is not hindered even with an initially weak infrastructure for liquid hydrogen.

Additional early application areas are agricultural vehicles in countries with a weak foreign exchange, like for example Africa or South America where hydrogen could be produced by means of electrolysis using the existing surplus of hydroelectric power.

The operation of passenger vehicles with hydrogen in general, with the exception of fleet vehicles and special applications, will surely not be one of the first steps taken to introduce hydrogen-energy technology. The technology's strict requirements for the automobile drive system with regard to reliability; lifetime; behaviour in accidents; light, compact construction[4]; driving quality; etc., should not be underestimated. Hydrogen technology must be further developed with respect to the needs of the automotive technology. In the end effect this supports the use of hydrogen in general.

Furthermore, it is considerably important to examine hydrogen internal combustion engines in order to improve the characteristics of the engine operation with conventional fuel. While hydrocarbons are considered to be "convenient" fuels with regard to the tolerances which have to be met in mixture formation and combustion, hydrogen requires the absolute optimal mixture composition as well as the most favorable ignition timing over the entire operating range. Deviations which only cause increased fuel

---

[4] Automotive technology approaches aviation technology in this regard.

consumption and increased harmful substances in the exhaust gas, using conventional fuel, cause unsatisfactory engine operation under hydrogen operation. The mixture formation and control processes which function under extreme conditions in hydrogen internal combustion engine operation can contribute to better fuel economy and reduced emission of harmful substances when using conventional fuels.

With regard to the exhaust pollutants, which can consist only of nitric oxides[5] with the hydrogen engine, the formation mechanism is easier to comprehend than with hydrocarbons where carbon monoxide, soot particles as well as carcinogenic substances in form of poly-aromatic hydrocarbons can occur. In contrast to hydrocarbons the emission of nitrogen oxides can be reduced well below the levels established by several regulations—the "California Test" for example—through the operation in the lean mixture range (wide flammability range of hydrogen–air mixtures) attainable with hydrogen engines. The ability to operate with a lean mixture is also basically used in the conventional Diesel engine as a result of controlled combustion and has earned it a reputation as an environmentally safe engine. This is refuted by current experience since currently Diesel engines are not driven in the lean mixture range in the interest of high power and high acceleration. The environmental pollution caused by increased nitrogen oxide emission, accompanied by soot emission and carcinogenic substances like benzo[*a*]pyrene [70] has evidently not been made clear to the public up to now for several reasons[6].

Soot formation and the accompanying emission of carcinogenic poly-aromatic hydrocarbons like benzo[*a*]pyrene[7] for example represents an inherent problem with Diesel operation with hydrocarbons which is not easily solved. The strong nitrogen oxide emission in the range of the stoichiometric mixture can basically be reduced in a manner similar to hydrogen engines through water injection into the engine intake air. This is basically also applicable for conventional Otto-cycle engines and is being considered as additional internal cooling for race car engines. However, it does not represent a generally practical solution because it is necessary to use demineralized water and antifreeze agents in case of low outside temperatures. On the other hand, the method of cryogenic mixture formation used with hydrogen and methane (see page 156) drastically

---

[5] The hydrogen peroxide which forms during operation with an extreme lean mixture is already decomposed on the metal walls in the exhaust system as a result of a catalytic reaction.

[6] Especially in Europe, institutions and vehicle passengers obviously accept without question the fact that large-caliber exhaust pipes of trucks and buses are located at a height of the windows and inlet ports of the car ventilation systems where their damage can have the most effect.

[7] The name is just as much a derivation from "Benz" as benzene (see page 117) or "Benzin", the german word for gasoline.

reduces the nitrogen oxide emission with external and internal mixture formation even at operation in the stoichiometric range. Thus it can replace the water injection which essentially only represents a temporary solution for test and demonstration vehicles.
According to the current level of research the consistent application of digital electronics for the control of the mixture formation and course of combustion is the technically proven method of reducing the emission of exhaust pollutants from internal-combustion engines. While the formation of pollutants in the engine are already reduced[8] here, the exhaust gas catalyst is only able to convert the carbon monoxide with the nitric oxide in the exhaust gas more or less into carbon dioxide and nitrogen under certain operating conditions. With the Otto-cycle engine the necessary conditions are best satisfied at medium to upper part load range[9]. They cannot be satisfied at all with the Diesel engine.
Finally it remains to be seen if research and development in the field of hydrogen internal combustion engines not only provides means of propulsion for transportation which are unsurpassed in their environmental safety but also considerably contribute to the reduction of the emission of exhaust pollutants from conventional fuels.

7.3 Rail Vehicle Applications

Currently in railbound vehicles Diesel–electrical propulsion processes or pure electrical propulsion processes are used. For the electrification, the latter requires an appropriate network to supply the electricity which is only sensible from an economic point of view in densely populated areas. Diesel–electrical propulsion processes are especially suitable for sparsely populated areas and for long distances like in Canada or the U.S. because of its onboard energy supply. The use of liquid hydrogen, produced via electrolysis and electricity from hydropower and nuclear power as a fuel would correspond to "indirect electrification"[10] and currently is of particular interest in Canada. Beside Diesel engines and gas turbines, electrical propulsion via hydrogen–air fuel cells is taken into consideration as a means of propulsion (see [14–18, 71, 73, 75]). There are also similar considerations regarding future applications in high-speed railway systems [19].

---

[8] From the technical point of view, it is more sensible to avoid directly the formation of pollutants instead of removing them by appropriate additional measures.

[9] At ground level this corresponds to a speed of about 80–100 km/h which corresponds to the speed limit in the U.S., Canada and Japan.

[10] According to the author's knowledge, this term was first coined in Canada in 1983 by D. Scott, at that time at the Institute for Hydrogen Systems, Mississauga, Ontario.

7.4 Applications in Water Transportation

As a result of several studies, ship propulsion via hydrogen basically seems to be possible but many technical questions remain to be examined [20–24]. The storage of hydrogen in its liquid state is preferable to other storage methods although several problems occur with regard to the low density of this fuel and the accompanying higher storage volumes [25–27]. Merchant ships which are loaded with freight of high average density—ore or steel for example—appear to be extremely suitable for hydrogen because there is a surplus of volume in the hull for the storage of larger tanks compared to the tanks of conventional fuels. Ships with a load of low average density and thus have only limited available volume—container ships for example—may have problems which can only be solved by the enlargement of the ship's hull. Furthermore, ships with limited available volume have problems with respect to stability since the freight often has to be piled on the deck and therefore the center of mass is relatively high. Compensation by the current customary deep location of the fuel tank does not result in considerable improvement with hydrogen as a result of the low mass of the fuel (only about one third compared to hydrocarbons) [27]. Significant applications could result for tankers for the transportation of liquid hydrogen which basically can be developed from current LNG tankers [28–30]. With regard to the large hydropower reserves[11] not being used to their full extent for example in South America, Africa and also in Greenland[12] it would be possible to transport locally produced electrolysis hydrogen after liquefaction in tankers to appropriate harbors in consumer intensive areas as is currently done with liquid natural gas. To continue the possibilities this could also be carried out with underwater tankers as might be the case with liquid natural gas.
The application of hydrogen improves the efficiency, capacity, amount of weight and construction volume of the ship's propulsion system. While Diesel engines in ships, especially two-stroke engines, can be converted basically with rather minimum effort to hydrogen operation with its advantages with regard to fuel economy and emission of pollutants, and if necessary enable dual-fuel operation, gas turbines enable improvement of efficiency and construction volume with hydrogen operation in a manner similar to jet engines (see Chap. 6).

---

[11] The worldwide hydropower potential acc. to [33] is $4.4 \cdot 10^4$ TWh/a, 37% of that in Asia, 23% Africa, 13% each North and South America. About 20% of that are economically usable. Further use of 20% via "indirect electrification" would result in a yearly equivalent of about 100.10^6 metric tonnes of liquid hydrogen!

[12] The usable potential of the glacier runoff in South Greenland (South of the polar circle) [31] corresponds according to [32] to about $1.1 \cdot 10^5$ GWh per year or to a yearly equivalent of at least $2 \cdot 10^6$ metric tonnes of liquid hydrogen.

Electrical propulsion by hydrogen–air fuel cells appears promising because a considerable reduction of the fuel consumption or higher speed can be achieved due to the high overall efficiency. Encouraging possibilities are provided by the application of superconducting electrical motors which even for high power can be small and light as a result of their strong magnetic fields and enable the usual low speed of revolutions for ship propellers (about 100–200 r.p.m).
This enables the installation of the propulsion engines in the tail of the hull without using the long actuation shaft which is usually necessary [34, 35].
Although the superconductors which are currently technically usuable, have to be cooled below the temperature range which can be obtained with liquid hydrogen the application of hydrogen as a heat sink provides substantial more effective cooling if necessary with magnetocaloric refrigerators (see Chap. 4)[13].
No statement can currently be made about the technical applicability of the ceramic superconductors, discovered in 1986, based on yttrium-barium copperoxide. Since these materials have very high transition temperatures in the range of up to about 120 K, they basically would be of utmost interest for technical application.
These developments could be concluded with submarine vessels like submarine tankers for liquid hydrogen for example which use a hydrogen–oxygen propulsion via gas turbines or electric propulsion via fuel cells.

7.5 Applications in Aviation and Space Flight

Most of the essential applications of liquid hydrogen in aviation in the future have been mentioned in Chap. 6. In addition to special applications including the military field, extensive use in civil aviation can be expected at the earliest in the generation of airplanes after the next. While technical problems which still have to be solved, especially with respect to the flying apparatus, are not insurmountable obstacles, the required high capital investments for ground installations at airports is the primary hindrance to the political–economic decisions which have to be made. Beside supersonic airplanes, early areas of application are foreseeable in the military field for special airplanes with extreme operating ranges as well as for helicopters.

[13] It is not necessary to emphasize that there are further possible applications for liquid hydrogen as a heat sink for example for cooling freight or to extract nitrogen out of the air which can be used on board as a protective gas for safety reasons (inerting).

Hydrogen seems to be made for the supersonic range—use in intercontinental routes will follow. Due to the good combustion properties of hydrogen higher flight altitudes are attainable without reducing the power of the propulsive unit so that the effects of the sonic boom can practically be eliminated [78]. This causes the further development of propulsion concepts which have been under research for a long time. They are partly based on shock induced combustion and supersonic combustion [79]. The goal of this development is the external combustion where the thrust caused by the external shock induced combustion is absorbed by the appropriately designed underside of the airplane. All these propulsion unit concepts can only be carried out practically with hydrogen.
In space flight, liquid hydrogen will also be the most important rocket fuel in the future for chemical rocket propulsion. Due to foreseeable developments in the field of air-breathing hybrid propulsion units an extension of the current use in the upper stages to the lower stages of rocket carriers will take place and thus hydrogen will be a universal fuel [36][14].
Further development will finally lead to air and space vehicles which can use conventional runways. Hybrid propulsion will be used in these air/space vehicles (trans-atmospheric vehicles) which combust H_2 with air in the lower layers of the atmosphere and use H_2/O_2 rocket engines only at altitudes of more than 30 km. Thus earth orbits can be approached with considerably less transported oxygen and larger payload [82–84].
Sometimes the emission of water vapor into the stratosphere by the jet engines of high flying airplanes or by rocket engines is criticized with respect to the ozone layer as well as with respect to changes of the global albedo. Since the stratosphere does not have strong local vertical exchange mechanisms it is assumed that the mean duration of stay of water vapor is very long there. In any case, the air of the stratosphere near the equator tends to flow from layers of high altitudes to those of low altitude near the polar regions. These air flows carry along ozone and other gases, sometimes also water vapor [80] and concentrate with them at the polar regions. Nevertheless, the formation of condensation trails have to be considered. They can occur in the upper troposphere at an altitude of about 8–12 km as a result of the low temperature, though only in relatively humid air and can contribute to the greenhouse effect to a small degree. These conditions usually do not exist in the stratosphere as a result of an increasing temperature and very low humidity. Rather in the stratosphere, the efficient mechanism of photodissociation takes place to a large degree where water is decomposed into hydrogen and oxygen by very high energetic ultraviolet

[14] Also see ISTRA: Ein luftatmender balistischer Raumtransporter für Europa. P. A. Kramer, R. D. Bühler Z. F. Flugwissenschaften und Weltraumforschung 7, 237–241 (1983).

radiation via intermediate reactions. The hydrogen passes off into interplanetary space because the gravitational field of the earth cannot effectively prevent hydrogen (and also helium) from escaping. In this manner, the planet Venus has probably lost the main portion of its original water (runaway greenhouse) [81]. On the earth on the other hand, due to the pressure and temperature conditions very little water vapor passes off into the stratosphere because almost all the steam condenses between the the troposphere and stratosphere and thus is not exposed to the very effective photodissociation of the stratosphere. However, if water vapor is somehow introduced into the stratosphere, as for example through the exhaust jet of a propulsion system, it is possibly removed relatively quickly. The influence of the protons which exist in the solar wind particularly during periods of increased solar activity (sun spot maxima, solar flares) is not taken into consideration here. The questions regarding individual reactions of the photodissociation, primarily in connection with negative effects on the ozone balance, have not been completely settled. It has to be realized though that apart from the technical application of halogenated hydrocarbons, several natural "ozone killers" exist in the atmosphere which is often not clearly mentioned. Inter alia, these are methane (CH_4), nitrous oxide (N_2O), nitric oxides (NO_x), methyl iodide, methyl bromide and methyl chloride. In particular they occur in the continental shelf regions as a result of anerobic processes from certain bacteria which consume nitrogen or from biological metabolic processes which decompose organisms. Their influence on the ozone balance compared to the effect caused by human activities in terms of magnitude is not yet known.

Furthermore, applications regarding the energy supply of lunar or planetary bases or industrial plants are possible for future space missions. For example it is possible to satisfy the initial energy and water needs of future lunar bases by delivering hydrogen while the oxygen which is required for the production of energy and water can be obtained from lunar minerals.

Despite the current situation, the use of hydrogen as a jet mass in nuclear–thermal propulsion systems will gain importance in the future because "fast travel" in the inner and outer planetary belt, which is a basic prerequisite for manned missions, is attainable in this manner.

With regard to transfer time, electrical propulsion systems are even more suitable, particularly in the outer planetary belt, but there they need a nuclear–electrical power system with a costly radiator for the release of waste heat. On the other hand, in the inner planetary belt, electrical propulsion systems can be supplied by solar generators and in particular enable freight traffic between orbital stations.

Hydrogen will definitely be used as jet mass in the further development of nuclear–thermal propulsion systems. Based on the present solid-state nuclear reactors and reactors with gaseous nuclear fuel, this development will lead to reactor temperatures where the fissionable material is in the plasma state. These propulsion systems have the problem of transmitting the energy released at nuclear fission to the jet-mass hydrogen in common. This can be achieved by means of convective heat transfer in solid-state reactors and through direct thermal collision processes and optical radiation in reactors in a gaseous or plasma state. While these propulsion systems do not correspond to an energetic use of hydrogen this is the situation in future propulsion systems which are based on controlled thermonuclear fusion—the fusion propulsion system—where hydrogen simultaneously serves as an energy carrier. Although the fusion propulsion system would be the most powerful propulsion unit for missions within our planetary system there are currently no known practical means of applying these concepts. It can definitely be said that the magnetic plasma confinement procedures currently being researched for energy production (see [37]) can hardly be considered for space propulsion due to the immense mass, which is also required for stationary energy production. The "inertial confinement" processes could be more promising. In this process, a solid hydrogen target is heated up to a temperature of about 10^8 K in a fraction of a nanosecond by the appropriate input of energy and afterwards energy is set free for 10 to 100 nanoseconds through nuclear fusion. After being directed by a magnetic field this fusion plasma can be used as a pulsed jet to produce thrust where the goal is a pulse frequency between 10 to 100 cycles per second. Key problems which still have to be solved are the heating process as well as the recovery of the energy required for heating from the fusion plasma. Although the application of nanosecond pulse lasers were successfully demonstrated in experiments [38] they will hardly be suitable for application in space flight as a result of the low efficiency, complexity and large size of this type of equipment.

Conceptually, the compression and heating of the target by pulsed particle jets, like protons for instance, offer considerably better opportunities for development [39]. Not much research has been made in this area. In contrast to the production of pulsed laser radiation the energy losses involved with the production of the particle jet are one to two orders of magnitude smaller. Problems lie in the development of particle accelerators pulsed in nanoseconds which have to provide a particle energy in the range of 10 MeV with currents of about 10^6 A for light ions (see [40–48]).

As mentioned before, another problem is the recovery of the energy which is needed for the function of the propulsion from the expanding fusion plasma which provides the thrust. In particular magnetoplasmadynamic energy conversion could provide a solution where electrical energy which is necessary for maintaining the function of the propulsion system is recovered from the jet through interaction with magnetic fields. The recovery of the energy which is released during controlled nuclear fusion, via magnetoplasmadynamic converters—analogous to the induction generator in electrical power technology—will generally receive increased attention in future attempts to produce energy by controlled thermonuclear reactions if nuclear fusion is established as a "clean source of energy" [49–51]. A prerequisite for this is the use of thermonuclear reactions where the released energy is primarily available as kinetic energy of charged particles, because it is only possible to recover energy via electromagnetic fields and to avoid detrimental radioactive side effects in this manner.

Table 37 shows this in detail with the energy balances of different nuclear fusion reactions. Of the reactions mentioned here, research has concentrated on reaction 3 because these thermonuclear reactions can take place at lower temperatures than the other reactions with corresponding higher energy profit. However, the large amount of energy (14.1 MeV) is neutron energy, i.e., energy contained in uncharged particles which in addition to the problems caused by the neutrons excludes direct interaction with electromagnetic fields and thus direct energy conversion. The energy released by this thermonuclear reactions is emitted in the form of kinetic energy of the neutrons from the reactor. This energy has to be converted into heat in an appropriate shielding jacket through neutron absorption. The concepts which are the basis for stationary energy production through nuclear fusion intend to use a jacket of liquid lithium to absorb the neutron energy and release it via heat exchangers to a conventional steam power plant. The problems of radiation load and the handling of accidents are basically associated with the liberation of highly energetic neutrons. In this respect a comparison to nuclear fission is very informative where only about 6% of the energy released is in the uncharged particles (neutrons, gamma particles) while this percentage can be about 80% in reaction 3 and still ranges between 40 and 60% in the combination of reactions 1 and 3. From this point of view it could be concluded that, apart from the radioactive fission products in nuclear fission reactors the problems of radiation load are small compared to those of future first generation nuclear fusion reactions which use a combination of reactions 2 and 3 (Table 37).

Only reactions 4, 5 and 6 will provide the justification for nuclear fusion to be considered as a clean energy source as well as to permit applications

Table 37. Energy balance of nuclear fusion reactions

1. $D + D \rightarrow {}^3\text{He}$	$+ n$
(0.82 MeV)	(2.45 MeV)
2. $D + D \rightarrow T$	$+ H$
(1.01 MeV)	(3.02 MeV)
3. $D + T \rightarrow {}^4\text{He}$	$+ n$
(3.5 MeV)	(14.1 MeV)
4. $D + {}^3\text{He} \rightarrow {}^4\text{He}$	$+ H$
(3.6 MeV)	(14.7 MeV)
5. $H + {}^6\text{Li} \rightarrow {}^3\text{He} + {}^4\text{He} + 4.0\,\text{MeV}$	
6. $H + {}^{11}\text{B} \rightarrow 3\,{}^4\text{He} + 8.9\,\text{MeV}$	

in stationary energy production and space flight. However, they probably cannot be carried out via quasi-stationary plasma confinement in a magnetic field but probably only through inertial confinement and pulsed heating via charged particle radiation.
Finally, it can be concluded that in general the significance of controlled nuclear fusion as an energy source should not only be viewed with regard to the production of electrical energy. This is the ideal method of hydrogen production from water via direct high temperature processes. In this process "indirect electrification" of the controlled nuclear fusion is made possible by hydrogen which is used here as a primary and secondary energy carrier.

7.6 Liquid Hydrogen in Stationary Energy Technology

Several investigations showed that in the future hydrogen-energy technology the storage of energy in the form of hydrogen will play a definite role [50–61]. Its use in transportation enables seasonal storage[15] of energy in the form of liquid hydrogen which is especially economical compared to alternative storage processes with pressure vessels or metal hydrides [52–57, 62–64, 76, 77].
Furthermore, the production of steam via a so-called hydrogen–oxygen steam generator should be mentioned. It is basically the modified combustion chamber of a hydrogen-oxygen rocket engine and supplies high-pressure, high temperatures steam which can serve as an instantaneous standby for the turbines to cover peak power requirements and to balance control actions (see [65–69, 74]).
The advantages are high volume related energy exchanges which enable very compact, economical plants where constant operating conditions can be maintained as well as extremely short required start-up times with typical values in the range of about 1 second.
The economical production and storage of the required hydrogen and oxygen are problematic as well as the production of the necessary combustion chamber pressure in the range of up to 200 bar during the steam feed in the high-pressure stage and 80 bar in the intermediate-pressure stage of the power plant. In order to produce a high pressure in the combustion chamber the storage of hydrogen in its liquid state is preferred. On one hand the work required from the pumps to supply the combustion chambers can be held within bounds based on experience with rocket engines, on the other hand questions arise regarding the economy of the process.

---

[15] This can also refer to underground storage which is economical as long as appropriate locations like caverns, depleted oil and natural gas fields etc. with corresponding low leakage are available for the gaseous hydrogen which has to be stored.

One disadvantage is the necessary cooling of the steam delivered by the high-pressure combustor to the maximum allowable steam temperature of the plant of about 560 °C by means of water injection because in this manner the high temperature (3,000–4,000 °C), which was provided by the hydrogen–oxygen combustion, is decreased thermodynamically. Thus the range of high temperature where the relatively expensive hydrogen–oxygen technology offers economical advantages cannot be used. These problems could be reduced but not eliminated by the development of high temperature materials and blade cooling (see Chap. 6, [93]) for steam and gas turbines. Here, the application of a real high-temperature process for the production of electrical energy, for example the use of an MPD[16] hydrogen–oxygen generator as a topping stage for a combined gas–steam turbine power station, would be an alternative which could be examined more closely [72].

Finally if the assumption is made that future power plants not only provide electrical energy and possibly district heating but also hydrogen and oxygen then opportunities will also necessarily arise for the technical use of liquid hydrogen as an energy carrier for economical reasons.

7.7 Future Prospects

In addition to the consistent and increasing use of non-fossil primary energy, the use of hydrogen as a clean energy carrier, free of CO_2 and methane, requires in particular, a public acceptance of a global energy policy which must have the goal of an energy technology which is much more in equilibrium with the environment than the energy technology of today. Only then it is possible to overcome current hindrances like classic economic thinking or certain forms of lobbyism that are not sufficiently oriented to real environmental problems.

Although it has to be taken into consideration that primarily the specific application of hydrogen in transportation mentioned here can serve as an example to initiate an energy policy which consistently leads away from the use of fossil energy carriers. However, it must be emphasized, that with respect to CO_2 emissions, transportation worldwide occupies only the third place behind household and industry.

But the problem of sufficient global availability of primary energy still remains. If fossil energy carriers must be abandoned then in the long-term only nuclear energy and solar energy including its indirect energy forms such as wind and hydropower are available.

The energy problems of a world population of 10–12 billion people cannot

[16] Magnetoplasmadynamic generator also called magnetohydrodynamic (MHD) generator.

be solved by forced direct or indirect use of solar energy and large efforts to conserve energy. Increased development and improvement of nuclear energy with simultaneous reduction in the demand for fossil energy carriers would certainly be opportune. But can this be accomplished according to a long-term energy policy which is not oriented to election periods if, from a global point of view, the increased use of coal, oil and natural gas has priority? Are not the Third World countries more dependent than ever upon cheap fossil energy carriers which are not in balance with the environment?

In order to overcome the problems of survival mankind has created the state which functions best if give and take is balanced for the individual, i.e., when there is equilibrium!

Since the initiation of the technical industrial revolution, at the latest, this no longer applies to the global economy to the same degree. The successful development of an economy which is in equilibrium with the earth's system will be a matter of life and death for mankind in the future. In certain fields of future energy technology hydrogen would be a first step in the right direction!

References

[1] H. Flohn: Klimaänderung als Folge der CO_2-Zunahme? Phys. Bl. *37*, 184–190 (1981).

[2] W. Bach, (ed.): The Carbon Dioxide Problem. Experentia *36*, 767–812, 1017–1025 (1980).

[3] J. Williams, (ed.): Carbon Dioxide. Climate and Society. New York, Pergamon Press (1978).

[4] G. M. Woodwell, E. V. Pecan: Carbon and the Biosphere. Brookhaven Symp. in Biology No. 24, U.S. Atomic Energy Com. Div. of Technical Inf., Oak Ridge, Tenn. (1973).

[5] J. S. Olson, H. A. Pfuderer, Y. H. Chan: Changes in the Global Carbon Cycle and the Biosphere. RNL-IEIS-109. Oak Ridge, Tenn. (1978).

[6] J. A. Laurmann: Impacts of CO_2-Induced Climate Change, Strategic Issues and their Treatment. In: Proc. 4th World Hydrogen Energy Conf., Vol. 4, pp. 1785–1987 New York, Pergamon Press (1982).

[7] B. Bolin, Ed.: The Greenhouse Effect, Climatic Change, and Ecosystems. Wiley (1986).

[8] L. Margulis: Symbiosis in Cell Evolution: Life and Its Environment on the Early Earth. New York, Freeman (1981).

[9] Berner et al.: The Carbon—Silicate Geochemical Cycle and Its Effect on Atmospheric Carbon Dioxide over the Past 100 Million Years. Am. J. of Sci., 283, No. 7, 641–683 (1983).

[10] W. S. Broeker, G. H. Denton: The Role of Ocean—Atmosphere Reorganizations in Glacial Cycles. Geochim. Cosmochim. Acta, 53, No. 10, 2465–2501 (1989).

[11] J. F. Kasting: Theoretical Constrainsts on Oxygen and Carbon Dioxide Concentrations in the Precambrian Atmosphere. Precambrian Res., 34, Ns. 3, 4, 205–226 (1987).

[12] R. M. Zweig: Hydrogen—Prime Candidate for Solving Air Pollution Problems. In: Proc. 4th World Hydrogen Energy Conf., Vol. 4, pp. 1789–1805 New York, Pergamon Press (1982).

[13] W. Häfele, H. Barnert, W. Sassin: Künftige fossile Brennstoffe-Ihre Nutzung und Einbettung in moderne Energiesysteme. DFVLR-Nachrichten *35*, 8–19 (1982).
[14] H. Loeken: Hydrogen as Motor Fuel for Locomotives, Cars and Ships. Teknisk Ukeblad *82*, 555–558, 563–565 (1935).
[15] R. W. Foster, E. J. D. Escher: Hydrogen Fueled Railroad Motive Power Systems, A Feasibilit Study. Escher-Foster Techn. Ass., St. Johns, Mich. (1976).
[16] R. T. Alpaugh, W. J. Escher, R. W. Foster, M. Novil: Hydrogen-Fueled Railroad Motive Power Systems. A North American View. In: Proc., 2nd World Hydrogen Energy Conf., Vol. 4, pp. 1793–1827 New York, Pergamon Press (1978).
[17] B. A. Steinberg, D. S. Scott: A Systems Analysis Comparing Conventional and Hydrogen Powered Rail Locomotives. In: Proc. 17th IECEC, Paper 829372, pp. 1178–1183, Los Angeles (1982).
[18] W. D. Van Vorst, R. L. Wolley: Hydrogen Fueled Surface Transportation. In: Hydrogen: Its Technology and Implications Vol. 4: (K. E. Cox., K. D. Williamson, eds.). Boca Raton, Florida, CRC-Press (1979).
[19] T. Ishigohka: A Conceptual Design of a Hydrogen Fueled Magnetically Levitated Superconducting High-Speed Train. In: Proc., 4th World Hydrogen Energy Conf., Vol. 3, pp. 1095–1104 New York, Pergamon Press (1982).
[20] H. W. Carhart, W. A. Affens, B. D. Boss, et al.: Hydrogen as a Navy Fuel, Special Study. Naval Res. Lab., Washington, D. C., NRL-7754, NTIS (1974).
[21] C. F. Sindt: Transmission of Hydrogen. In: Selected Topics of Hydrogen Fuel (J. Hard, ed.). NBS-Spec. Publ. 419, (1975).
[22] E. Quandt: Investigation of Hydrogen Fueled Naval Vehicles. In: Hydrogen Energy, Part B (T. M. Veziroglu, ed.). New York, Plenum Press (1974).
[23] A. E. Ford: Hydrogen Fueled Turbine Boat Demonstration. SAE-paper 770797, September (1977).
[24] E. N. Cart, Jr., et al.: Alternative Energy Sources for Nonhighway Transportation. U.S. Dept. of Energy (DOE), Transportation Energy Conservation Div., Washington, D. C. (1977).
[25] H. Silla: Possible Future Maritime Fuels. Webb Inst. of Naval Architecture, Glen Glove, N. A. (1977).
[26] Anon.: Alternate Fuels for Maritime Use. National Academy of Sci., Maritime Res. Board, Washington, D. C. (1980).
[27] J. P. Archibald: Hydrogen Fueled Ships. In: Proc., Int. Conf. on Alternate Energy Sources, Miami Beach (1980); see also: Alternative Energy Sources 3 T. N. Veziroglu, ed.). Vol. 5, Berlin Heidelberg. New York, Springer (1983).
[28] R. C. Fooks: Some Important Factors in LNG-Tanker Design Selection. In: Adv. Cryog. Eng., Vol. 19, pp. 269–275 New York, Plenum Press (1973).
[29] A. Patuhov: Status Report on LNG Tanker Design, Adv. Cryog. Eng., Vol. 19, pp. 282–291. New York, Plenum Press (1973).
[30] J. L. Howard: Near-Team Trends in LNG-Tankships Design. In: Adv. Cryog. Eng., Vol. 19, pp. 276–281. New York, Plenum Press (1973).
[31] C. F. Kollbrunner, H. Stauber: Unerschöpfliche saubere Wasser- und Energiequellen in Grönland. Inst. für bauwissenschaftliche Forschung, Heft 25. Zürich, Leemann (1973).
[32] V. Bundschuh: Nutzung der Gletschereisenergie. In: Energiequellen für morgen? Teil 6: Nutzung der Wasserenergie. Programmstudie im Auftrag des BMFT, Arbeitsgemeinschaft der Großforschungseinrichtungen (AGF/ASA), Nr. ASA-ZE/03/75 (1975).
[33] World Energy Conference 1980: In: Proc., World Energy Conference Munich (1980).
[34] A. J. Stewart, J. H. Springer, T. J. Doyle: Effectiveness of Superconducting Electric Drives. Naval Eng. J. 91, No. 2 April (1979).

[35] B. Gamble, T. A. Keim: Superconducting Generator Design for Airborne Application. In: Adv. Cryog. Eng., Vol. 25, pp. 127–136. New York, Plenum Press (1979).
[36] H. Künkler H.: Air Precooling before Compression Effect on the Air Breathing Engine of a Space-Craft Launch Vehicle. 5th Annual DGLR Meeting, DGLR, Köln, NTIS (1972).
[37] R. Hancox: Fusion Reactors: 1969–80. Nucl. Fusion *20*, 1064–1968 (1980).
[38] R. E. Kidder: Laser Driven Isentropic Hollow-Shell Implosion: The Problem of Ignition. Nucl. Fusion *19*, 223–234 (1979).
[39] F. Winterberg: The Possibilit of Precessing a Dense Thermonuclear Plasma by an Intense field Emission Discharge, Phys. Rev. *174*, 212–220 (1968).
[40] G. Yonas, J. W. Poukey, K. R. Prestwick, et al.: Electron Beam Focussing and Application to Pulsed Fusion. Nucl. Fusion *14*, 731–740 (1974).
[41] Sh. A. Goldstein, R. Lee: Focussed Intense Ion Beams Using Self-Pinched Relativistic Electron Beams. Phys. Lev. Letts. *35*, 1079–1082 (1973).
[42] P. A. Miller, R. I. Butler, M. Conan, et al.: Propagation of Pinched Electron Beams for Pellet Fusion. Phys. Rev. Letts. *39*, 92–98 (1977).
[43] J. A. Nation, R. N. Sudan (eds.): In: Proc., 2nd Int. Conf. High Power Electron and Ion Beam Research and Technology, Cornell Univ., Ithaka, Lab of Plasma Studies (1977).
[44] J. D. Johnson, R. N. Sudan: High Power Ion Diodes for Inertial Confinement Fusion Experiments. Sandia Labs., Albuquerque, N. M., Rep. RS-5244/1003 (1978).
[45] G. A. Moses, R. Spencer: Compact Electron Beam or Light-Ion Beam Fusion Reactor Cavity Design using Non-Spherical Blast Waves. Nucl. Fusion *19*, 1386–1389 (1979).
[46] G. Yonas: Developments in Sandia Laboratories Particle Beam Fusion Programme, Plasma Physics and Contr. Nucl. Fusion Res. Proc. 7th Int. Conf., Innsbruck (1978).
[47] J. S. Humphries, Jr.: Intense Pulses Ions Beams for Fusion Applications. Nucl. Fusion *20*, 154 (1989), see also: Physics Today, *21*, 231, December (1980).
[48] Anon.: Untersuchungen zur Eignung von Schwerionenstrahlen für den Trägheitseinschluß. Ges. Für Schwerionenforschung, GSI-Bericht, Darmstadt, Juni (1980).
[49] R. F. Post: Direct Conversion of Fusion Energy to Electricity, paper 709004. In: Proc. 5th Intersoc. Energy Conv. Eng. Conf. (IECEC), Las Vegas, Nev. (1970).
[50] W. Peschka: Hochtemperatur—Energiesysteme unter Verwendung von Plasmareaktoren und induktiven magnetoplasmadynamischen Wandlern. DLR-FB-67-59, DFVLR-Ber. (1967); see also: W. Peschka, Some Notes on Future Inductive MPD-Converters with Controlled Nuclear Fusion. Proc., 5th IECEC (1970).
[51] W. Peschka: Neue Energiesysteme für die Raumfahrt. München: Goldmann (1972).
[52] C. Carpetis, W. Peschka: Untersuchung der Wasserstoffspeicherung mit Kryoadsorbern. Abschlußbericht Projekt FA-057-76 EHC. EG-Kommission, Brüssel (1978).
[53] C. Carpetis: A System Consideration of Alternative Hydrogen Storage Facilities for Estimation of Storage Costs. Int. J. Hydrogen Energy *5*, 423–439 (1980).
[54] C. Carpetis: Estimation of Storage Costs for Large Hydrogen Storage Facilities. Int. J. Hydrogen Energy 7, 191–205 (1982).
[55] C. Carpetis: Storage, Transport and Distribution of Hydrogen. In: Hydrogen as an Energy Carrier (C. J. Winter, J. Nitsch eds.). pp. 249–289, Berlin Heidelberg, New York Tokyo, Springer (1988).
[56] F. Lawaczeck: Storage of Surplus Electrical Energy as Hydrogen. Tek. Tidskr. *31*, 395–401; *32*, 407–412 (1929).
[57] J. Hord, (ed.): Selected Topics on Hydrogen Fuel. NBS Spec. Publ. *419*, 208 (1975).
[58] L. A. Booth, J. D. Balcomb, F. J. Edeskuty: Combined Nuclear and Hydrogen Economy: A Long-Term Solution to the World's Energy Problems. In: Proc., 8th IECEC-Conf., A. H. Smith, (ed.): pp. 396–403. Amer. New York, Amer. Inst. of Aeron. and Astron. (1973).

[59] T. S. Jayadevaiah, S. C. Chin: Economics of a Hydrogen Storage Peaking Power Plant. ASME paper No. 74-WA/PW R-6 (1974).
[60] F. J. Salzano, E. A. Cherniavski, et al.: Role of Hydrogen in Electric Energy Storage. In: Hydrogen Energy, Part B (T. N. Veziroglu, ed.). New York, Plenum Press (1975).
[61] C. J. Kippenhan, R. C. Corlett: Hydrogen-Energy Storage for Electrical Utility Systems. In: Hydrogen Energy, Part B. (T. N. Veziroglu, ed.). New York, Plenum Press (1975).
[62] J. P. Ackerman, J. J. Barghausen, L. E. Link: Assessment Study of Devices for the Generation of Electricity from Stored Hydrogen. Argonne Nat. Lab. Rep., ANL-75-71, December (1975).
[63] W. R. Parrish: Economic Study of Electrical Peaking Alternatives. In: Hydrogen Energy, Part 5, (T. N. Veziroglu, ed.). New York, Plenum Press (1975).
[64] W. R. Parrish: Hydrogen in the Electrical Utility Industry. In: Selected Topics on Hydrogen Fuel, NBS Spec. Publ. 419 (1975).
[65] W. J. Escher, et al.: A Non-Polluting Noiseless Engine for Powerplant Applications with Specific Orientation to a High Speed Ground Transportation Systems. Rocketdyne, RIP-13, Canoga Park, Calif. (1970).
[66] R. M. Reese, A. D. Carmichael: Cycle for the Propulsion of Deep Submersibles. 6th IECEC, pp. 563–576. New York, SAE (1971).
[67] H. Wojkowski, W. Schnurnberger, H. J. Sternfeld: Abschätzung erzielbarer Wirkungsgrade und Kosten bei der Verstromung von Wasserstoff. EG-Abschlußbericht zu FA 404–78–7, EHD, Brüssel (1981).
[68] D. E. Wright, A. D. Lucci, J. Campbell, J. C. Lee: Hydrogen Turbine Power Conversion System Assessment. NASA-CR-135298, RI/RD 77-252, April (1978).
[69] R. S. Colladay: Thermal Feasibility of Using Methane or Hydrogen Fuel for Direct Cooling of a First Stage Turbine Stator. NASA-TN-D-6042, NTIS (1970).
[70] G. Stärk, W. Beßlein, et al.: Untersuchungen der Partikelemission von Dieselmotoren auf direkt wirkende Mutagenität. MTZ *44*, 263–267 (1983).
[71] R. W. Foster, W. J. D. Escher: A Project Plan for Implementing a Hydrogen Fueled-Hydrogen Transport Rail System. E: F Technology Inc. St. Johns, Mich. (1983).
[72] J. M. Smith, L. D. Nichols, G. R. Seikel: NASA-Lewis H_2–O_2 MHD-Program. In: Proc., 14th Symp. on Engineering Aspects of Magnetohydrodynamics, paper III, 7 (1974).
[73] B. A. Steinberg, D. S. Scott: Hydrogen vs. Diesel Fueled Locomotives: A Techno-Economic Appraisal. Int. J. Hydrogen Energy, 9, No. 1/2, 101–108 (1984).
[74] H. J. Sternfeld, P. Heinrich: A Demonstration plant for the Hydrogen/Oxygen Spinning Reserve. Int. J. Hydrogen Energy, 14, No. 10, 703–716 (1989).
[75] K. Kordesch, Ch. Gruber et al.: Fuel Cell Research and Development Projects in Austria. Int. J. Hydrogen Energ, 14, No. *12*, 915–925 (1989).
[76] F. N. Lin, W. I. Moore, S. W. Walker: Economics of Liquid Hydrogen from Water Electrolysis. Int. J. Hydrogen Energy, 10, No. *12*, 811–816 (1985).
[77] M. A. K. Lodhi: Power Potential from Ocean Currents for Hydrogen Production. Int. J. Hydrogen Energy, 13, No. *3*, 151–172 (1988).
[78] G. D. Brewer: Advanced Supersonic Technology Concept Study—Hydrogen Fueled Configuration, NASA CR-14 49 35, Lockheed California Company (1976).
[79] F. Suttrop: Überschallverbrennung; Zweck und eigene Versuchseinrichtungen, WGLR-Jahrbuch, pp. 366–376 (1963).
[80] R. S. Stolarski: The Antarctic Ozone Hole. Sci. Am. pp. 20–26 (1988).
[81] J. F. Kasting et al.: How Climate Evolved on the Terrestrial Planets. Sci. Am., pp. 46–53 (1988).
[82] S. Walmsley, J. Wilson: HOTOL—A Cryogenic Materials Challenge, Adv. in Cryogenic Engineering, Vol. 34. New York, Plenum Press (1988).

[83] R. R. Barthelemy: The National Aero-Space Plane Program. In: Proc. Symp. on National Aero-Space Plane and Space Applications, pp. 1–7, Hawaii: Hawaii Natural Energy Institute, Univ. of Hawaii 1990

[84] Th. Wierzbanowski, J. G. Amstrong.: The Flight Test Program for the Hydrogen Powered NASP/X-30 Research Aircraft. In: Proc. Symp. on National Aero-Space Plane and Space Applications, pp. 21–33, Hawaii: Natural Energy Institute, Univ. of Hawaii 1990

Safe Handling of Liquid Hydrogen 8

The safe handling of hydrogen is currently state of the art in the industrial and commercial field [1–6]. Furthermore, within the scope of the U.S. space programs, liquid hydrogen is produced, transported, stored and expended in large amounts [7–10]. As a result of positive experience, liquid hydrogen can be compared with the alternative liquid energy carriers methane (LNG) and other low boiling hydrocarbons like gasoline for example, with respect to its behavior regarding handling and accidents [11]. According to current experience all three energy carriers mentioned have certain technical safety characteristics which in this regard does not lead to a preference of any one over the other (Fig. 125). Energy carriers have different risks for different areas of application and require detailed safety inspections for each. It must be considered that new application areas could also create new technical safety problems. In contrast to the use of hydrogen in the industrial and commercial fields, in the fields of transportation, distribution and commerce it is necessary to carry out technical safety inspections in a manner similar to the other energy carriers. The positive experience with town gas, which contains a high percentage of hydrogen, as well as with liquid hydrogen within the scope of the space programs and also with automotive vehicles used to demonstrate the applicability of liquid hydrogen in transportation, does not reveal any insurmountable technical safety problems.

There is extensive literature about the safety of handling gaseous hydrogen in general which cannot be examined in detail here. It should only be mentioned that gaseous hydrogen has various characteristics in common with the energy carriers mentioned above, for example it is easily ignited by the weak spark caused by electrostatic discharge of the human body. In this regard, hydrogen has the lowest minimum ignition energy. Hydrogen-air explosions occur in the open similar to other energy carriers, i.e., in general with a relatively low degree of destruction, whereas in enclosed areas the destruction is much larger. In this process, characteristic features of hydrogen are the high flame speed in air, the wide range of flammability and detonability as well as the high pressures in detonations of air–hydrogen mixtures. In contrast to gasoline and methane, due to its

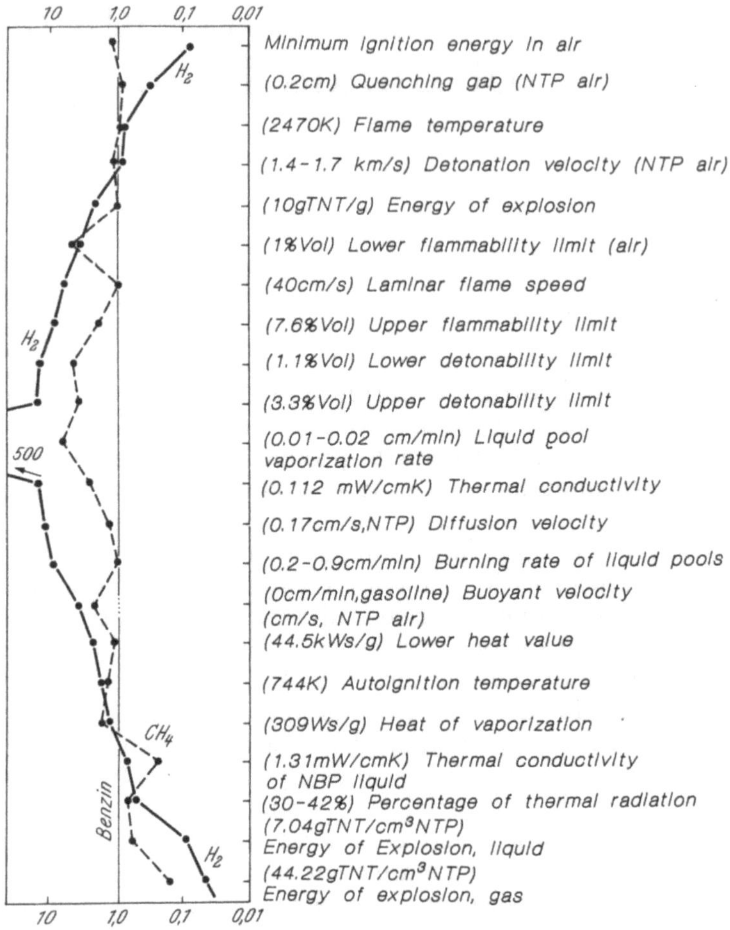

Fig. 125. Safety comparison of characteristic physical properties of hydrogen and methane related to gasoline (data in brackets are valid for gasoline)

strong positive buoyancy in the air gaseous hydrogen disperses very quickly by turbulent convection. It only has a fraction of the thermal flame radiation of hydrocarbons which is absorbed quickly in humid air and has the lowest mixture specific heat in a stoichiometric mixture with air.

When using liquid hydrogen and liquid methane, cryogenic properties associated with the required low temperatures are also exhibited. When handling these cryogenic liquids, large amounts of gaseous hydrogen and methane form due to evaporation in case of breakdown or accidents so that the liquid and gaseous ranges cannot be considered separately.

The following discussion deals primarily with safety questions which are basically related to the characteristics of the cryogenic liquid.

8.1 Materials for the Use of Liquid Hydrogen

While the handling of pure hydrogen at ambient temperature and above as well at high pressures, especially with cyclic stress, causes considerable material problems (see [12–15]), there are several reliable materials for cryogenic hydrogen such as austenitic chrome-nickel steel with a high nickel content for example as well as aluminum alloys with very good weldability. Although the strength of cold-drawn materials generally increases at low temperatures it must be considered that they must also function safely at ambient temperature and during cooldown and warmup phases which are associated with strong material stress as a result of thermal contraction and expansion. Furthermore, cold deformation of the materials should be prevented because with austenitic steel this can lead to a partial conversion to the more brittle martensite structure. Since it is necessary to use cold-drawn materials with a basically cubic, face-centered lattice structure, splinter and shrapnel damages which always occur when using brittle materials like glass and cold brittle synthetics, can simultaneously be avoided in accident situations due to material failure.
Of the nonmetallic materials, fiberglass and carbon fiber synthetics in particular have proven themselves in suspensions or spacers. As ultra-lightweight container material they are also receiving an increased interest as cryocontainers (see [16–28]).
Special attention must be given to the instrumentation, in particular, valves and manometers which as a result of incorrect selection of materials, fail due to hydrogen embrittlement and can cause uncontrolled discharge of hydrogen [29].
Insulating materials, powder and multilayer insulation—superinsulation—should be made of noncombustible materials because there is a danger of explosion with leaks in the outer vessel due to condensation of air in the interstitial space of the vacuum. Multiple layer insulation is an exception which according to current regulations in various countries like Germany for example has to be made of incombustible material whereas based on U.S. experience aluminized mylar foil for example can be used for the insulation of liquid hydrogen containers.

8.2 Handling of Hydrogen as a Cryogenic Liquid

From the physical point of view the handling of liquid hydrogen basically does not differ from that of other cryogenic liquids. Due to its flammability though several additional facts have to be taken into consideration.
With all cryogenic liquids it is possible that gases in the air—primarily water vapor—condense into the liquid and accumulate which can cause pipes to become clogged and valves to be blocked. Air can condense with

liquid hydrogen as well as with liquid helium, liquid neon and liquid nitrogen. At equilibrium this condensation product has about 52% oxygen at ambient pressure [30–32]. In the same manner pipes which carry liquid hydrogen, which do not have thermal insulation, or vacuum-insulated pipes that lose their vacuum, can cause the condensation of the air and dripping which can cause fire and explosion in the presence of inflammable materials [34–39] due to the higher oxygen concentration. The blockage of storage vessel vent lines of large vessels by ice plugs is prevented by a sufficiently large ventilation system [33] which contains diffusion baffles in the form of molecular sieves in order to supress the back-diffusion of air. Acting as "air traps" they absorb the gases in the air. In smaller vessels, especially for mobile applications, where it is not possible to install large ventilation systems, it is recommended to keep the cross sections of the vent lines as small as possible in order to keep the back-diffusion of air in the vent line as small as possible through a corresponding high flow velocity of the hydrogen.

Therefore, special attention has to be given to the cross sections of the vent lines and valves because different requirements exist for different cryogenic liquids in order to safely carry off the amount of gas which results from evaporation of the liquid in accidents. A characteristic parameter for the gas volume is the ratio of the temperature difference T from ambient temperature to the boiling point of the liquid to the evaporation enthalpy per unit of volume h_v (see Table 38) (see [30]).

Therefore the vent line cross section is especially critical with helium whereas hydrogen behaves more like nitrogen and methane. Since small hydrogen storage containers are often designed from helium containers or helium containers are used for the storage of hydrogen, hydrogen storage containers are generally grossly oversized with respect to the vent line cross section. Therefore, while helium containers generally can be used for the storage of hydrogen the reverse is not necessarily true.

Table 38. Data of cryogenic liquids

	Critical pressure (bar)	Critical Temp. (K)	Boiling Temp. (K)	Evaporation enthalpy h_v (J/g)	(J/cm³)	T/h_v (K·cm³/J)
Helium	2.3	5.2	4.2	20.2	2.5	118
Hydrogen	12.8	33.2	20.4	447	31.8	8.8
Nitrogen	33.5	126.1	77.3	199	160	1.39
Methane	45.8	190.7	111.7	508	215	0.88

Table 39. Volumetric efficiency of LH_2 containers based on pressure

Pressure p (bar)	Temp. T (K)	$\frac{\Delta p}{\Delta T}$	Density of the liquid	Coefficient of space g/L	Volumetric efficiency %
1.0	20.4	0.30	71.2	67.6	95.0
1.5	21.8	0.36	69.4	65.9	92.5
2.0	22.9	0.45	67.9	64.5	90.6
2.5	23.9	0.5	66.5	63.2	88.8
3.0	24.7	0.62	65.3	62.0	87.0
4.0	26.1	0.71	63.0	59.8	84.0
5.0	27.3	0.83	61.0	57.9	81.3
6.0	28.2	1.11	59.0	56.0	78.7
8.0	29.9	1.25	55.0	52.2	73.3
10.0	31.4	1.42	50.4	47.9	67.2
12.0	33.0	1.66	38.2	36.2	50.8
13.2	33.2	5.0	29.9	28.4	39.9

In order to avoid undue stress on the container as a result of thermal expansion of the liquid a 5% void must still exist when the maximum operating pressure is reached. According to Table 39 the following pressure dependent volumetric efficiencies are attained for liquid hydrogen:
This shows that the maximum operating pressure should not be chosen too high in the interest of good volumetric efficiency. Since at temperatures near the critical point the pressure, which is related to the temperature increase of the liquid, increases considerably, operating pressures above about 6–8 bar should be avoided by all means. Furthermore, mobile storage containers with high operating pressures present an increased accident risk.
While stationary and mobile containers for liquid hydrogen are currently sufficiently sturdy based on experience in dealing with accidents, the external tubing to include instrumentation cause certain problems. Here a compromise between sturdy design and good deformability in accidents without causing leakage must be found. Due to lack of experience especially with smaller containers further experimental operational test are required.
Injuries Caused by Cryogenic Liquids. All cryogenic liquids can cause injuries similar to burns when they come in direct or indirect contact with the skin or other parts of the body. This is also true for structural elements of cryogenic plants which are at low temperature as well as for cold gas which escapes during the evaporation of cryogenic liquids. Due to the heat capacity of the tissue which is supplied with blood and the formation of a gas film, which is a poor conductor of heat, between the cryogenic liquid and the tissue at body temperature—Leidenfrost's phenomenon—the skin and parts of the body do not cool down directly with the contact with cryogenic liquids. Thus, in many cases severe injuries can be avoided because there is often enough time to leave the dangerous area and to

eliminate the cause. When handling cryogenic liquids in open vessels and at accidents there is also the danger that clothing will soak up the cryogenic liquid which is avoided in laboratories and in industry by wearing appropriately repellent protective clothing. The eyes are also vulnerable to splashes of cryogenic fluid when handling open cryogenic liquids. This danger cannot be excluded when handling cryogenic liquids which boil or show a tendency toward to boiling delay. Therefore, when handling open cryogenic liquids it is necessary to wear safety glasses or face shields [30, 34–36].

The only direct response to cryogenic injuries is to warm the injured body tissue with water to normal body temperature as fast as possible. The use of warm water which should not be warmer than approximately 40 °C is very effective for this purpose [30, 40].

According to previous experience, cryogenic liquids with a large cooling capacity per unit of mass, i.e., high density and large heat of evaporation like liquid nitrogen, liquid oxygen and liquid methane for example are more dangerous than liquid hydrogen or liquid helium with regard to cold burns. However, the increased cooling effect of the cold gas due to the larger heat conductivity and increased heat transfer should not be underestimated. The increased heat transfer is caused by the relatively high flow velocity which is obtained at a low overpressure due to the low density of the helium and hydrogen gases when issuing from leaks or other openings.

While there exists sufficient documented experience from industry concerning cold injures in connection with cryogenic liquids of high density, like liquid nitrogen and especially liquid oxygen, this is not the case for liquid helium and liquid hydrogen. On one hand this could be due to the statistically larger amounts of liquid nitrogen and liquid oxygen compared to liquid hydrogen, and on the other hand to the stricter safety regulations for handling liquid hydrogen. Liquid helium is an exception in this regard since it is only dealt with in relatively small amounts in industry compared to liquid hydrogen and moreover it is primarily used in laboratories.

Danger of Asphyxiation Due to Oxygen Deficiency in Closed Areas. While the toxicity of liquefied gases like fluorine or carbon monoxide and in some respects also for carbon dioxide for example must be taken into consideration there is also the danger of asphyxiation due to oxygen deficiency when nonpoisonous cryogenic liquids like liquid nitrogen, liquid helium, or liquid hydrogen escape in closed or poorly ventilated rooms. Death occurs after about 8 minutes in room with an oxygen content of about 8 Vol.%[1], at 4 Vol.% after about 40 second [9, 30, 31]. This is basically also valid for hydrogen–air mixtures, however, it has to be considered that,

[1] Normal breathing air contains about 21 Vol.% of oxygen.

in contrast to nitrogen, hydrogen has a positive buoyancy in air even as a cold gas and the formation of a homogenous mixture with air is very difficult. Furthermore, for safety reasons, closed rooms require good ventilation because of the flammability of hydrogen and the small lower flammability limit of 4 Vol.% in air. This also guarantees an acceptable breathing air quality if purging is carried out, or where entering the room is prevented by appropriate safety measures.

8.3 Hydrogen as Flammable Liquid

Liquid hydrogen is always a source for considerable amounts of gaseous hydrogen. From this point of view, technical safety questions regarding the handling of liquid hydrogen are mainly determined by the properties of gaseous hydrogen for which extensive experience has long been available. When handling liquid hydrogen, like with other combustible materials, coincidence of flammable materials, air or oxygen, and source of ignition must be prevented, for example, elimination of ignition sources and monitoring the accumulation of hydrogen gas in the air. In the end effect, an accident occurs as an uncontrolled coincidence of these three conditions. The following describes two characteristic accident situations for liquid hydrogen, i.e., large spills of liquid hydrogen as well as the accumulation of solidified air or oxygen in liquid hydrogen.

Uncontrolled Large Spills of Liquid Hydrogen. This type of situation occurs when containers in tankers or airplane tanks for example leak or burst (see Chaps. 4 and 6) or when liquid hydrogen lines leak. The relatively low heat of evaporation and the very low boiling temperature of liquid hydrogen result in completely different behavior compared to gasoline, liquid methane (LNG) and liquid gas. When liquid hydrogen escapes from containers or lines into the environment, hydrogen evaporates very quickly and has a positive buoyancy in air after it has been warmed up to about 30 K. Thus, hydrogen rises very quickly, mixes intensively with the air as a result of violent turbulent convection and disperses within a very short time.
Spill experiments with amounts up to 20 m^3 of liquid hydrogen [41–43] also show that in the open and with unlimited volume (less than two walls and the ground) combustion always occurs by deflagration, and detonation will not take place as long as the ignition was not induced by means of explosive igniters. The increase in pressure above the boiling liquid hydrogen when the hydrogen-air mixture is ignited was relatively low in the open but increased considerably when it was partially enclosed. The increase in pressure of a 5 liter spill after ignition in a U-shaped enclosure ($60 \times 60 \times 60$ cm) for example corresponded to a 600 liter spill without enclosure. Furthermore, reproducible detonation could be demonstrated

in this type of enclosures constructed with four surfaces (three sides and the ground).

Spills in closed rooms require special attention. According to [44], a H_2-air mixture, which could detonate formed after spilling 65 liter LH_2 in a closed room of approximately 100 m^3 volume (log cabin) after a maximum of 2 seconds, deflagration on the other hand was attained with amounts below 30 liter with an overpressure of less than 0.5 bar. Therefore, detonation does not necessarily occur in an enclosure as long as the enclosed volume is large enough compared to the volume of the spilled liquid.

Moreover, in order for a detonation to occur from deflagration, a certain distance is required, the so-called detonation induction distance. In a complicated way, it is dependent upon the mixture composition, pressure, temperature, the geometry of the enclosure and the tpe of ignition source [45, 46]. The influence of the size of the "fireball"[2] which is smaller in direct ignition of the flammable hydrogen–air mixture than in time delay before ignition, is important in the open. The flame burns vertically when there is no wind, but the flame shifts downwind, strongly dependent upon the wind velocity. The amount of time the spill burns is only about 2–5% of the time for an energetically equivalent amount of hydrocarbons like gasoline for example. Tests also showed that the area of the ignitable hydrogen–air mixture cannot be determined from the dispersion of the cloud formed during the mixture of cold hydrogen gas with air by condensed water drops. A considerable portion of the ignitable hydrogen–air mixture is above the cloud.

The tests also showed that in case of a line rupture and continuous escape of liquid hydrogen the area of ignitable hydrogen–air mixture near the ground is considerably larger. At a leakage rate of about 1,000 L/min it can extend over a distance of about 100 m downwind [41–43].

On one hand, these tests provided substantial information on the behavior of liquid hydrogen in spills and following ignition, and contributed to a more realistic and positive assessment of the behavior of liquid hydrogen as a rocket propellant as well as a fuel. On the other hand, they identified the necessity of gathering more information about the dispersion of hydrogen clouds in the air, i.e., their dispersion in space and dependent upon the time, in order to examine the reliability of existing physical models for the formation and dispersion of the combustible hydrogen-air mixture in spills of liquid hydrogen. The following shortly described test program was carried out by NASA between 1979 and 1980 at White Sands, New

---

[2] The fireball is defined by the burning area directly after the ignition. From experience, according to [11, 32] the value for the diameter D is $7.93 \cdot M_f^{1/3}$ and the duration of the fireball is $t \sim 0.47 \cdot M_f^{1/3}$ (D in meters, t in seconds, M_f mass of fuel in kg).

Fig. 126. View of a NASA liquid hydrogen spill facility (Source: NASA-Langley Research Center)

Mexico for this reason. Special attention was given to the instrumentation used to make measurements [47][3].

In the experimental facility shown in Fig. 126, liquid hydrogen is released with the help of gaseous helium (6.9 bar pressure) from a pressurized 5.7 m^3 spill-dewar (about 400 kg LH_2) via a 30 m line with 152 mm inner diameter by means of a diffuser (see Fig. 127). The composition of the hydrogen clouds as well as their temperature and propagation behavior as a function of the prevailing wind velocity was also examined.

Sensors were installed at various height levels on nine appropriately arranged 19.5 m instrument towers (see Fig. 128) in order to record temperature measurements, degrees of turbulence and hydrogen contents in the air. Furthermore, evacuated (ball-shaped) sample bottle clusters (see Fig. 129) were used in order to take gas samples which were later tested to determine their composition by means of a gas chromatograph. There were a total of seven spill experiments which were carried out under the

[3] These tests were financed by the U.S. government with public funds and are not subjected to Copyright restrictions. The author wishes to express his gratitude to the NASA-Langley Research Center especially to Dr. R. D. Witkofski for providing these documents, in particular Figs. 126 to 132.

Fig. 127. End section of the liquid hydrogen spill line, valve, spill pond and diffuser (Source: NASA-Langley Research Center)

conditions described in Table 40. The course of the experiments was documented by photographs and on film. The experiments confirm the strong dependency of the distribution of the hydrogen cloud near the ground upon the wind velocity (Fig. 130). They also showed that with a fast spill rate (Tests 1–6), a flammable hydrogen–air cloud forms near the ground for a relatively short time (typically: 20 seconds) which however quickly rarefies below the lower flammability limit (4 Vol.%) (see Figs. 131 and 132) due to the evaporation of liquid hydrogen, the strong buoyancy of the gas and the associated vaporization induced turbulent convection. The propagation of the cloud on the ground ranged from 50 to 100 m downwind under prevailing test conditions. The cloud's rate of climb was about 0.5 to 1.0 m/s.

From correlating the measurements of the hydrogen sensors and the sample bottles, a method for the calculation of hydrogen concentration in the air as a function of the local temperature measured in the hydrogen cloud was derived. This so-called adiabatic hydrogen–air mixing model proved to be very effective for the determination of the hydrogen concentration inside the cloud and will also be important for future applications.

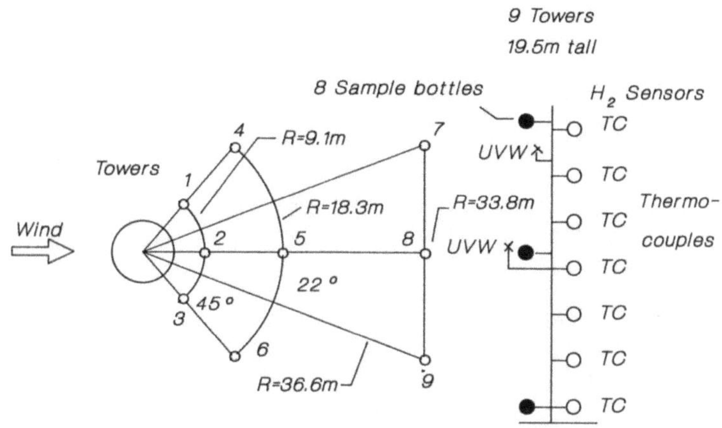

Test No. and date (1980)	Spill time (sec)	Windspeed (m/s)	Ambient temp. C	Relative humidity,%	Data obtained: Motion picture	Sample bottles	Thermo couples	Turbu-lence	H_2 sensors	Still photos
(1) Aug. 1	60	2.7 3.1	30	18	x					x
(2) Sept. 25	40	1.3 1.8	24	49	x	x	x	x	x	x
(3) Okt. 10	85	4.5	26	27	x					
(4) Okt. 22	33	3.1 3.6	15	43	x	x	x	x	x	x
(5) Nov. 24	24	6.3	12	43	x	x	x	x	x	x
(6) Dez. 18	35	2.2	15	29	x	x	x	x	x	x
*(7) Dez. 18	240	3.1	17	29	x		x	x	x	x

*2.8 m³

Fig. 128. Location of instrument towers and typical instrumentation array. *TC* thermocouple arrays, *UVW* turbulence sensors (Source: NASA-Langley Research Center)

Table 40. Test conditions for the 5.7 m³ LH_2 spill experiments from [47]

Test	Date (1980)	Duration of spill (sec)	Wind velocity (m/s)	Air Temp. (°C)	Relative humidity of the air (%)
1[a]	01.08.	60	2.7–3.1	30	18
2[b]	25.09.	40	1.3–1.8	24	49
3	10.10.	85	4.5	26	27
4	22.10.	33	3.1–3.6	15	43
5	24.11.	24	6.3	12	43
6[c]	18.12.	35	2.2	15	29
7	18.12.	240	3.1	17	29

[a,b] were not considered in the evaluation.
[c] 2.8 m³.

Fig. 129. Location of typical hydrogen sample bottle cluster to extract gas from the H_2 cloud (Source: NASA-Langley Research Center)

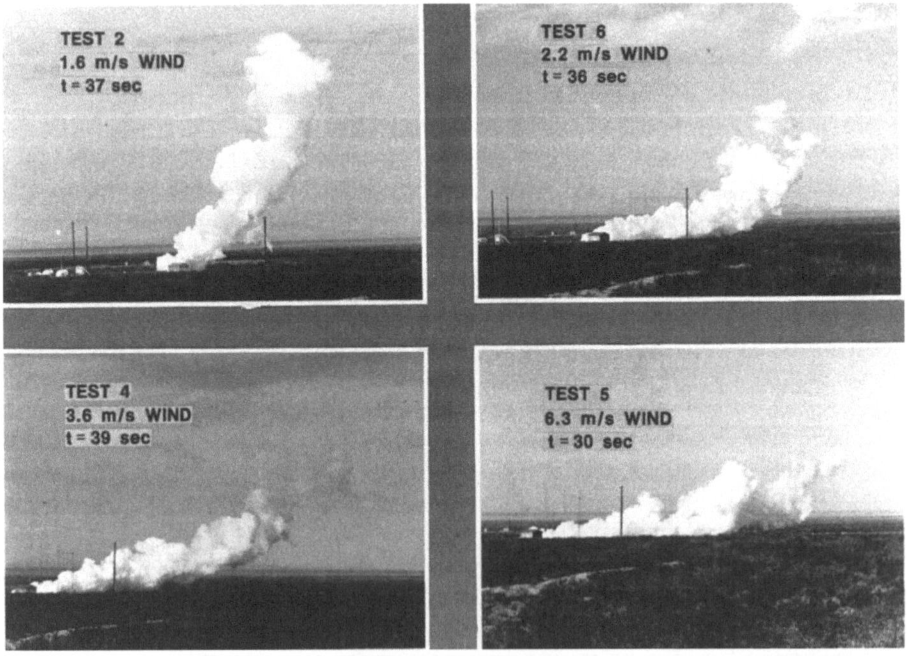

Fig. 130. Dispersion of the visible H_2 clouds depicting wind effects (Source: NASA-Langley Research Center)

While short-term spills (Tests 1 to 6) mainly correspond to the formation of a leak in liquid hydrogen containers, prolonged spills (Test 7), for example a rupture of a liquid hydrogen pipeline, behave differently with regard to the dispersion of the hydrogen cloud. In particular the vaporization rate of the liquid could decrease due to the intensive cooling of the ground. Furthermore, the conclusions drawn in [47] from the spill tests compared to the behavior of liquid natural gas (LNG) in accidents [48–52] are informative. While, for instance, based on current experience LNG storage tanks are surrounded by dikes in order to retard the spread of the liquid, according to [47] it seems to be more appropriate to do without dikes in the case of liquid hydrogen. Because of its rapid evaporation and mixture with the air due to excessive turbulent convection in the open the spilled liquid is dispersed over as large an area as possible. In the interest of rapid evaporation the ground material should be coarse gravel instead of a smooth surface [11, 30, 31]. According to this concept, which still has to be examined very closely, the accumulation of liquid at undesired locations should be prevented by a canalization via open ditches.

An important question concerning the storage of large amounts of liquid hydrogen is the distance which has to be maintained between stationary storage vessels and buildings for safety reasons. An overview of these distances can be found in [11]. According to this overview, liquid hydrogen also proves to be comparable with liquid natural gas and gasoline in this respect.

Accumulation of Solidified Air and Oxygen in Liquid Hydrogen. If the liquid hydrogen storage has improper ventilation [41, 42] solidified air and oxygen can accumulate in liquid hydrogen so that the resulting mixture could be explosive. As a result of the low solubility of oxygen and nitrogen in liquid hydrogen, the solidified gases always accumulate on the bottom of the inner vessel. Beside the usual ignition sources it is also possible to ignite this mixture through sharp blows and shocks or violent movements of the storage vessel because the required ignition energy can be activated through friction of the oxygen crystals on the inner wall of the vessel. Tests revealed that the accumulation of pure oxygen results in detonation whereas solidified air only results in local deflagration [42]. Air enriched by 40 Vol.% oxygen always results in detonation after ignition by shock or fuse wire.

Since the condensation of air in liquid hydrogen can be connected with oxygen enrichment this should be avoided by all means in liqid hydrogen containers. This can be achieved through an appropriate ventilation system and overpressure in the container. Moreover, during a periodic routine check it is appropriate to flush out larger containers with inert gas, to warm it up to ambient temperature and to evacuate the inner vessel.

To conclude this chapter concerning safety aspects of handling liquid hydrogen it should be mentioned that it was only possible to provide a

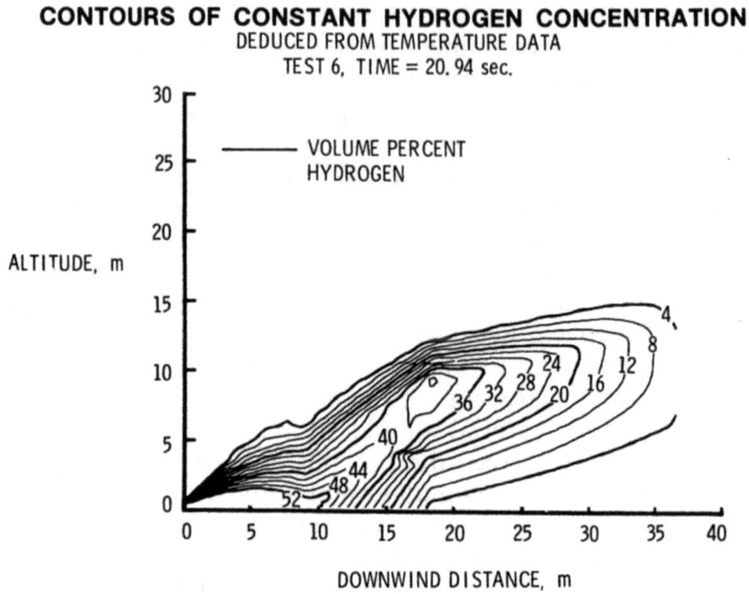

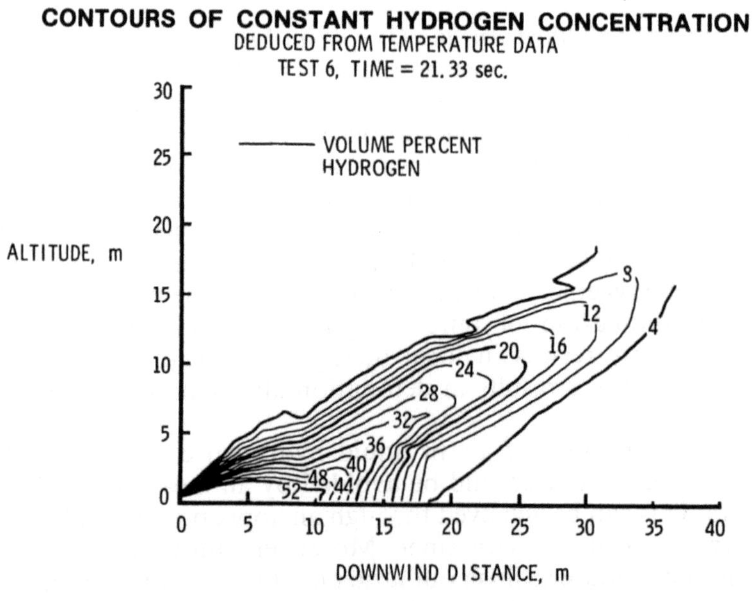

Fig. 131. Hydrogen concentration of the "hydrogen cloud" in Test 6 (see Table 32 at two consecutive times; source: NASA-Langley Research Center)

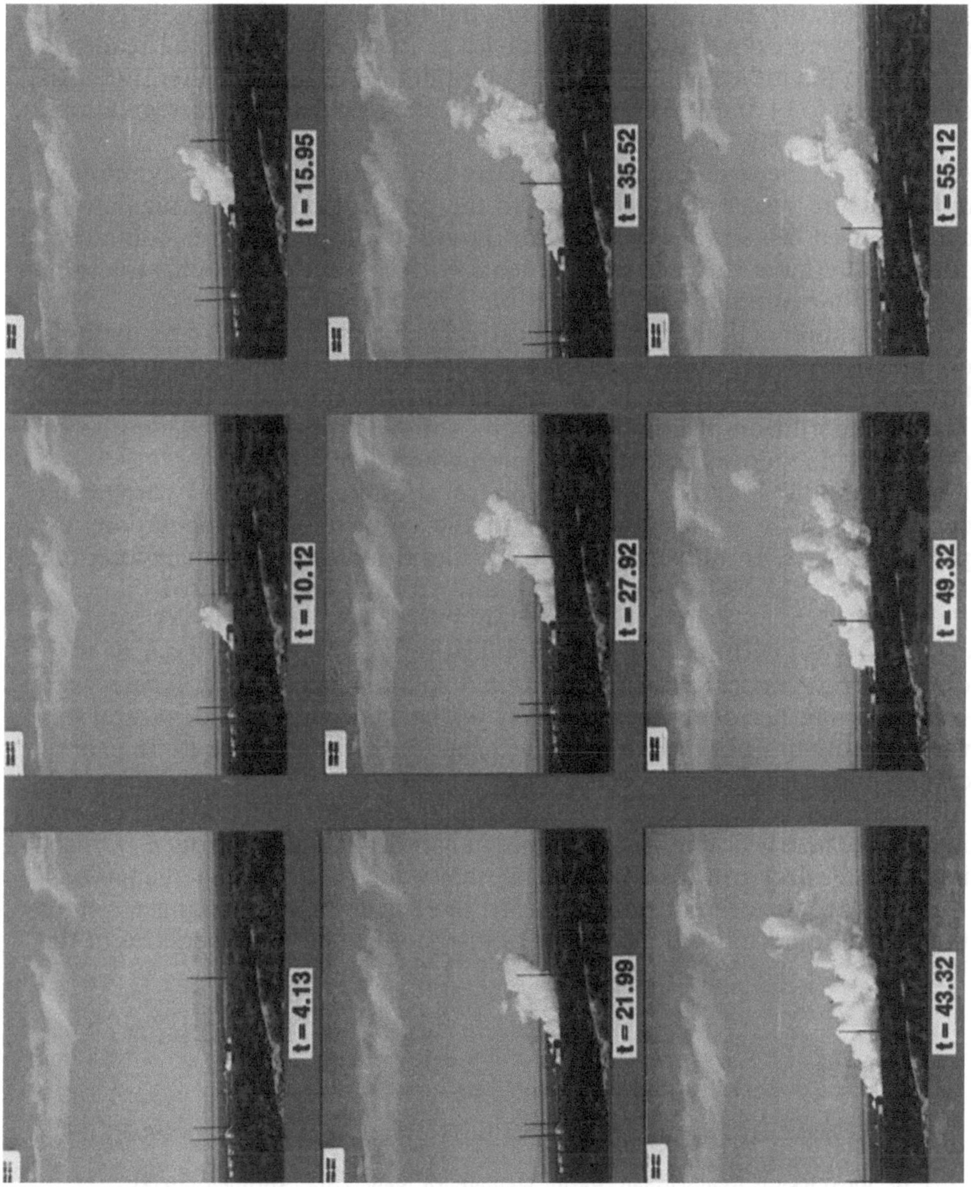

Fig. 132. Photos of clouds from Test 6 (see Table 32) (Source: NASA-Langley Research Center)

short insight into the typical safety questions regarding handling liquid hydrogen including a comparable description of alternatives, i.e., liquid natural gas and gasoline or low boiling hydrocarbons. In order to provide an overview, the extensive regulations and instructions for handling these materials were only described in individual cases. The interested reader is referred to the indicated literature. A compilation of the U.S. literature can be found in [11] and [53] and deals with safety guidelines and regulations (see also [54–57]).

According to the regulations in Germany and in the European Community only pressure vessel which conform to the accident prevention regulations can be used as storage vessels for liquid hydrogen. These containers additionally have to meet requirements with respect to the construction, production, equipping and installation.

Transportation of liquid hydrogen in street vehicles is subject to regulations on pressurized gas and street safety regulations. Traffic which crosses boundaries is also subject to the regulations of international traffic laws. The transportation of liquid hydrogen of more than 100 kg is considered to be very dangerous and requires permission.

Regarding filling and transportation of containers, liquid hydrogen is considered to be a pressurized gas in the context of the regulations on pressurized gas. In railway regulations, liquid hydrogen is not among the gases allowed to be transported by train without special controls.

Finally, it can be concluded that liquid hydrogen has not yet attained customary use with respect to its application as an energy carrier as with other energy carriers which contributed to their acceptance. There were several severe accidents in the forties in the U.S. with liquid natural gas which finally did not have a negative effect on its acceptance as an energy carrier under longterm aspects.

Although, safety aspects are different with liquid hydrogen and particularly in Europe there is not much experience in handling liquid hydrogen—which in the end effect can also be seen in safety instructions and regulations. According to the current state of the art there is no potential danger greater than with liquid natural gas which is also confirmed by experience in the U.S.A.

References

[1] J. P. Schödel: Hydrogen—a Safety Risk. In: Hydrogen as an Energy Vector. EC-EUR 6085, Brussels, October 1978.
[2] W. Bartknecht: Explosionen. Berlin Heidelberg New York, Springer (1978).
[3] B. Lewis, G. von Elbe: Combustion, Flames and Explosion of Gases, 2nd ed. New York, Academic Press (1961).
[4] Anon.: Standard for Gaseous Hydrogen at Consumer Sites. Compr. Gas Ass. (CGA)—Pamphlet G-5. 1 (1965).

[5] R. G. Zalosh, T. P. Short: Compilation and Analysis of Hydrogen Accidents. Rep. Dep. of Energy (DOE), Contract No. EE-77-C-02-4442 (1977).
[6] R. G. Zalosh, T. P. Short, Comparative Analysis of Hydrogen Fire and Explosion Incidents. Rep. Factory Mutual Res. Corp. March 1978.
[7] G. J. Caras: Prevention, Detection and Suppression of Hydrogen Explosions in Aerospace Vehicles. NASA-CR-78268 (1966).
[8] W. E. Baker, J. J. Kulesz: Workbook for Predicting Pressure Wave and Fragment Effects of Exploding Propellant Tanks and Gas Storage Vessels. NASA-CR-134906 (1975).
[9] Anon.: Hydrogen Safety Manual. NASA-N-75-72909 (1958).
[10] P. M. Ordin: Review of Hydrogen Accidents and Incidents in NASA Operations. NASA-TM-X-71565 (1974).
[11] J. Hord: Is Hydrogen Safe? NBS-Technical Note 690, 34 p. (1976). See also: NBS-Monograph 168, Selected Properties of Hydrogen, 292 p. (1981).
[12] Anon.: Wasserstoffversprödung. Ergebnisse des Forschungs- u. Entwicklungsprogramms "Korrosion und Korrosionsschutz". DECHEMA 1 (1974–1977).
[13] R. L. Mills, F. J. Edeskuty: Hydrogen Embrittlement of Cold-Worked Metals. Chem. Eng. Progr. *52*, 477–480 (1956).
[14] M. E. Smith: Hydrogen Embrittlement of Metals—a Bibliography with Abstracts. Rep. FCR-1964, NTIS/PS-75/049, January 1975.
[15] R. L. Mills, F. J. Edeskuty: Tests for Hydrogen Embrittlement of Steels Used in the Tank Farm Cylinder. Los Alamos Sci. Lab., Note: LA-3602 (1966).
[16] C. W. Keller: Fiberglass Supports for Cryogenic Tanks. NASA-Lewis Res. C., NASA-CR-120937, Lockheed Missiles and Space Co., Sunnyvale, Calif., Rep. No. LMSC-D281476, October 1972.
[17] C. A. Hall, D. E. Spond: Low Thermal Flux Glassfiber/Metall Vessels for LH_2 Storage Systems. In: (T. N. Veziroglu, ed.) Hydrogen Energy, Part A. New York, Plenum Press (1974).
[18] G. Hartwig: Low-Temperature Properties of Epoxy Resins and Composites. In: Adv. Cryog. Eng., Vol. 24, pp. 63–75. New York, Plenum Press (1978).
[19] R. E. Schramm, M. B. Kasen: Static Tensile Properties of Boron-Aluminium and Boron-Epoxy Composites at Cryogenic Temperatures. In: Adv. Cryog. Eng., Vol. 22, 205–213 New York, Plenum Press (1978).
[21] E. I. Augsburger, W. Dietsche, H. Kinder, J. Becker: Thermal Conductivity of Several Fibre-Reinforced Composites between 2 K and 300 K. Cryogenics *20*, 666 (1980).
[22] Z. G. Khim: Testing of Fiberglass-Reinforced Polyester Composites. In: Adv. Cryog. Eng., Vol. 26, pp. 280–285. New York, Plenum Press (1980).
[23] M. B. Kasen, R. E. Schramm: Current Status of Standardized Nonmetallic Cryogenic Laminates. In: Adv. Cryog. Eng., Vol. 28, pp. 271–278. New York, Plenum Press (1982).
[24] D. J. Radcliffe, H. M. Rosenberg: The Thermal Conductivity of Glass-Fibre and Carbon-Fibre/Epoxy Composites from 2 K to 80 K. Cryogenics *22*, 245–249 (1982).
[25] J. V. Gauchel, J. L. Olinger, D. C. Lupton: Characterization of Glass-Reinforced Composites for Cryogenic Applications. In: Adv. Cryog. Eng., Vol. 28, pp. 211–222. New York, Plenum Press (1982).
[26] S. S. Wang, E. S. M. Chim: Degradation of Fiber-Reinforced Composite Materials at Cryogenic Temperatures. In: Adv. Cryog. Eng., Vol. 28, pp. 191–210. New York, Plenum press (1982).
[27] G. Hartwig: Reinforced Polymers at low Temperatures. In: Adv. Cryog. Eng., Vol. 28, pp. 179–190. New York, Plenum Press (1982).
[28] A. Khalil, K. S. Han: Mechanical and Thermal Properties of Glass–Fiber Reinforced Composites at Cryogenic Temperatures. In: Adv. Cryog. Eng., Vol. 28, pp. 143–252. New York, Plenum Press (1982).

[29] R. O. Voth: Safety at Hydrogen Pressure Gauges. In: Adv. Cryog. Eng., Vol. 17, pp. 182–188. New York, Plenum Press (1972).
[30] F. J. Edeskuty, R. Reider, K. D. Williamson, Jr.: Safety. In: (G. Haselden, ed.) Cryogenic Fundamentals. London New York, Academic press (1971).
[31] D. B. Chelton: Safety in the Use of Liquid Hydrogen. In: (R. B. Scott, ed.) Technology and Uses of Liquid Hydrogen. New York, Pergamon Press (1964).
[32] J. Hord: Explosion Criteria for Liquid Hydrogen Test Facilities. NBS-Rep. 10734, 1972.
[33] J. C. Aydelott, C. M. Spruckler: Venting of Liquid Hydrogen Tankage. NASA-TN-D-5263, 1969.
[34] R. M. Neary: Handling Cryogenic Fluids. Nat. Fire Prot. Ass. Quart. 54, 63–70 (1970).
[35] F. J. Edeskuty, K. D. Williamson, Jr.: Storage and Handling of Cryogens. In: Adv. Cryog. Eng., Vol. 17, p. 56–68 New York, Plenum Press (1972).
[36] F. J. Edeskuty, R. Reider: Liquefied Hydrogen Safety. Los Alamos Sci. Lab., Rep. LA-DC-9569 (1968).
[37] W. W. Connolly: Practical Safety Standard for Commercial Handling of Liquefied Hydrogen. In: Adv. Cryog. Eng., Vol. 12, pp. 192–197. New York, Plenum Press (1967).
[38] D. S. Allan: Safety Aspects of Liquid Hydrogen. SAE-Paper 994B (1965).
[39] L. H. Cassutt, F. E. Maddocks, W. A. Sawyer: Study of Hazards in Storage and Handling of Liquid Hydrogen. In: Adv. Cryog. Eng., Vol. 5, pp. 55–61. New York, Plenum Press (1960).
[40] J. Asse: Liquefied Hydrogen Safety. Review. J. Amer. Soc. Saf. Eng. *14*, 18–23 (1969).
[41] Anon.: On an Investigation of Hazards Associated with the Storage and Handling of Liquid Hydrogen. Final Rep. C-61002, Contract No. AF18 (600). 1678, Arthur D. Little Inc., DDC Access. No. AD 324194, March 1960.
[42] M. G. Zabetakis, D. S. Burgess: Research on the Hazards Associated with the Production and Handling of Liquid Hydrogen. WADC Tech. Rep. 60–141, December 1961, See also: U.S. Dept. of Interior, Bureau of Mines Rep. RI 5707 (1961).
[43] M. G. Zabetakis, A. L. Furno, G. J. Martindill: Explosion Hazards of Liquid Hydrogen. In: Adv. Cryog. Eng., Vol. 6, pp. 185–194. New York, Plenum Press (1961).
[44] M. G. Zabetakis, A. L. Furno, H. G. L. Perlee: Hazards in Using Liquid Hydrogen in Bubble Chambers. Bureau of Mines Rep. No. 6309 (1963).
[45] M. G. Zabetakis: Flammability Characteristics of Combustible Gases and Vapors. Bureau of Mines Bul. (1965).
[46] L. E. Bollinger, M. C. Fong, J. A. Laughrey, R. Edse: Experimental and Theoretical Studies on the Formation of Detonation Waves in Variable Geometric Tubes. NASA-TN-D1983 (1963).
[47] R. D. Witkofski, J. E. Chirivella. Experimental and Analytical Analyses of the Mechanisms Governing the Dispersion of Flammable Flouds Formed by Liquid Hydrogen Spills. In: Proc. 4th World Hydrogen Energy Conf., Vol. 4, 1659–1674. New York, Pergamon Press (1982); see also: Int. J. Hydrogen Energy 9, 425–436 (1984).
[48] D. S. Burgess, M. G. Zabetakis: Fire and Explosion Hazards Associated with Liquefied Natural Gas. N 63–18682, available from NASA Sci. and Tech. Inf. Facility (1972).
[49] E. M. Drake: Vapor Dispersion from Spills on LNG on Land. In: Adv. Cryog. Eng., Vol. 20, pp. 134–142. New York Plenum Press (1974).
[50] J. Hord (ed.): Selected Topics on Hydrogen Fuel. NBS-Spec. Publ. SP 419 (1975).
[51] T. L. Bowen: Investigation of Hazards Associated with Using Hydrogen as a Military Fuel. Naval Ship Res. and Dev. Center, rep. 4541, Bethesda, MD (1975).
[52] Anon.: An Approach to Liquefied Natural Gas (LNG). Safety and Environmental Control Research. U.S. Dept. of Energy (DOE), DOE/EV-0002 (1978).
[53] J. Hord: Hydrogen Safety: An Annotated Bibliography of Regulations, Standards and Guidelines. Int. J. Hydrogen Energy 5, 579–584 (1980).
[54] J. R. Bartlitt: Hydrogen Isotope Processing in Fusion Power Application. In: Recent

Developments in Hydrogen Technology, Vol. 1. pp. 19–54, Cleveland, CRC Press (1986).

[55] M. Berman: Hydrogen Behaviour and Nuclear Safety. In: K. D., Williamson Jr., F. J. Edeskuty, eds.): Recent Developments in Hydrogen Technology. Vol. 2, pp. 35–68. Cleveland, CRC Press (1986).

[56] F. J. Edeskuty, J. J. Haugh, R. T. Thompson: Safety Aspects of Large Scale Combustion of Hydrogen. In: Proc. 6th World Hydrogen Energy Conf., Vol. 1, pp. 147–158. Pergamon Press (1986).

[57] F. J. Edeskuty, W. F. Stewart: Slush Hydrogen Safety. 81p, Los Alamos National Lab. Rep. LA-UR-91-672, 1991 to be published also as chapt. 12 in P. M. Ordin: Hydrogen Safety Handbook, NASA Report, in preparation, 1991

Physical and Technical Data of Liquid Hydrogen 9

Very detailed and extensive literature exists about the physical and technical data for hydrogen. The following data in the form of tables and diagrams represent a small selection of values which are often used and were chosen with respect to the handling of liquid hydrogen including safety aspects. The data is not limited to liquid hydrogen alone because liquid hydrogen can always be a potential source of large amounts of gaseous hydrogen due to evaporation. The diagrams (in particular figs. 133 to 147) only take para-hydrogen into consideration in the interest of a concise description. In many cases, the originals which were used as a model and which contained a maximum of information, were simplified and schematized for the sake of clarity. For more detailed information refer to the sources as well as appropriate computer programs[1], i.e., mainly to R. D. McCarty, J. Hord, H. M. Röder; National Bureau of Standards Monograph 168, Selected Properties of Hydrogen (Engineering Design Data), 523 p. (1981), in which there is also a detailed compilation of relative literature. For the permission to publish this data as well as the support provided, the author wishes to express his gratitude to the National Bureau of Standards and especially to J. Hord of the NBS Boulder Division.

[1] B. A. Younglove: Interactive Fortran Program to Calculate Thermophysical Properties of Six Fluids, NBS-Technical Note 17048, 1982.

Physical Properties of Hydrogen, Methane and Gasoline[g,h]

		Hydrogen	Methane	Gasoline
Molecular weight	—	2.016	16.043	107
Triple point pressure	[At]	0.0695	0.1169	—
Triple point temperature	[K]	13.803	90.680	180–220
Normal boiling point (NBP) temperature	[K]	20.268	111.632	310–478
Critical pressure	[At]	12.759	45.387	24.5–27
Critical temperature	[K]	32.976	190.56	540–569
Density at critical point	[g/cm³]	0.0314	0.1604	0.23
Density of liquid at triple point	[g/cm³]	0.0770	0.4516	—
Density of solid at triple point	[g/cm³]	0.0865	0.4872	—
Density of vapor at triple point	[g/m³]	125.597	251.53	—
Density of liquid at NBP (ρ_L)	[g/cm³]	0.0708	0.04226	0.70
Density of vapor at NBP	[g/cm³]	0.00134	0.00182	0.0045
Density of gas at NPT (ρ_G)	[g/m³]	83.764	651.19	4400
Density ratio ρ_L/ρ_G	—	845	649	156
Heat of fusion	$[Wsg^{-1}]^i$	58.23	58.47	161
Heat of vaporization	$[Wsg^{-1}]^i$	445.59	509.88	309
Heat of sublimation	$[Wsg^{-1}]^i$	507.39	602.44	—
Heat value (low)	$[kWsg^{-1}]^i$	119.93	50.02	44.5
Heat value (high)	$[kWsg^{-1}]^i$	141.86	55.53	48
Specific heat C_p of NPT gas	$[Wsg^{-1}K^{-1}]^i$	14.89	2.22	1.62
Specific heat C_p of NBP liquid	$[Wsg^{-1}K^{-1}]^i$	9.69	3.50	2.20
Specific heat ratio C_p/C_v of NPT gas	—	1.383	1.308	1.05
Specific heat ratio C_p/C_v of NBP liquid	—	1.688	1.676	—
Viscosity of NPT gas	$[gcm^{-1}s^{-1}]$	0.0000875	0.000110	0.000052
Viscosity of NBP liquid	$[gcm^{-1}s^{-1}]$	0.000133	0.001130	0.002

Thermal conductivity of NPT gas	[$mWcm^{-1}K^{-1}$]	1.897	0.330	0.112
Thermal conductivity of NBP liquid	[$mWcm^{-1}K^{-1}$]	1.00	1.86	1.31
Surface tension of NBP liquid	[Nm^{-1}]	0.00193	0.01294	0.0122
Dielectric constant of NPT gas	—	1.00026	1.00079	1.0035
Dielectric constant of NBP liquid	—	1.233	1.6227	1.93
Index of refraction of NPT gas	—	1.00012	1.0004	1.0017
Index of refraction of NBP liquid	—	1.110	1.2739	1.39
Adiabatic sound velocity in NPT gas	[ms^{-1}]	1294	448	154
Adiabatic sound velocity in NBT liquid	[ms^{-1}]	1093	1331	1155
Compressibility factor[a] Z in NTP gas	—	1.0006	1.0243	1.0069
Compressibility factor[a] Z in NBP liquid	—	0.01712	0.004145	0.00643
Gas constant (R)	[$Atcm^3g^{-1}K^{-1}$]	40.7037	5.11477	0.77
Isothermal bulk modulus (α) of NBP liquid	[MNm^{-2}]	50.13	456.16	763
Volume expansivity (β) of NBP liquid	[K^{-1}]	0.01658	0.00346	0.0012
Limits of flammability in air	[Vol%]	4.0–75.0	5.3–15.0	1.0–7.6
Limits of detonability in air	[Vol%]	18.3–59.0	6.3–13.5	1.1–3.3
Stoichometric composition in air	[Vol%]	29.53	9.48	1.76
Minimum energy for ignition in air	[mWs]	0.02	0.29	0.24
Autoignition temperature[b]	[K]	858	813	501–744
Hot air-jet ignition temperature[c]	[K]	943	1493	1313
Flame temperature in air	[K]	2318	2148	2470
Fraction of thermal energy radiatet from flame to surroundings[d]	[%]	17–25	23–33	30–42
Flame propagation in NTP air[f]	[cms^{-1}]	265–325	37–45	37–43
Detonation velocity in NTP air[f]	[kms^{-1}]	1.48–2.15	1.39–1.64	1.4–1.7
Diffusion coefficient in NTP[f]	[cm^2s^{-1}]	0.61	0.16	0.05
Diffusion velocity in NTP air[f]	[cms^{-1}]	< 2.00	< 0.51	< 0.17
Buoyant velocity in NTP air[f]	[ms^{-1}]	1.2–9	0.8–6	non buoyant
Maximum experimental safe gap, NTP air[f]	[cm]	0.008	0.12	0.07

Physical Properties of Hydrogen, Methane and Gasoline[g,h] (*Cont.*)

		Hydrogen	Methane	Gasoline
Quenching gap in NTP air	[cm]	0.064	0.203	0.2
Detonation induction distance, NTP air[e,f]	[Z/D]	100	—	—
Steady state vaporization rates of liquid pools without burning	[cm/min]	2.5–5.0	0.05–0.5	0.005–0.02
Burning rates of spilled liquid pools	[cm/min^{-1}]	3.0–6.0	0.3–1.2	0.2–0.9
Flash point	[K]	gaseous	gaseous	230
Energy ratio of explosion				
(g TNT)/(g fuel)		24	11	10
(g TNT)/(cm^3 NBP liquid fuel)		1.71	4.56	7.04
(kg TNT)/(m^3 NTP gaseous fuel)		2.02	7.03	44.22

[a] $Z = pV/RT$.
[b] Ignition at hot surface.
[c] Temperature of hot air jet as it enters fuel vapors or a combustible fuel–air mixture at NTP and causes ignition to occur.
[d] Percentage of heat of combustion (high) that is radiated from the combustion zone. Flames fueled by vaporization of pools of liquid fuels in an air enviroment. Atmospheric moisture absorbs thermal energy radiated from a fire. Hydrogen fires benefit most from this absorption effect. It is estimated that 45% of radiant hydrogen flame energy is absorbed within a distance of 8 m in 25 °C air containing water vapor at 15 Torr partial pressure.
[e] Determined in long zylindrical tube of Diameter *D*.
[f] NTP 1 At, 293.15 K.
[g] Thermophysical properties listed are those of para-hydrogen.
[h] See also: R. D. Mc Carthy, K. E. Cox, K. D. Williamson: Hydrogen: Its Technology and Implications, Vol. 3, Hydrogen Properties, CRC-Press Inc.: Cleveland, Ohio 1975. ISBN: 0-8493-5123-5.
[i] Ws: Watt second, 1WS = 1J

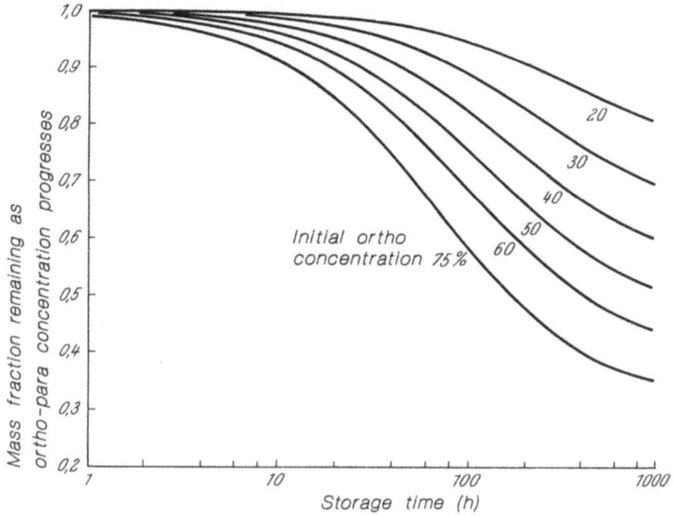

Fig. 133. Mass fraction of liquid hydrogen remaining from boiloff due to ortho-para conversion as a function of time (Source: NBS-Monograph 168)

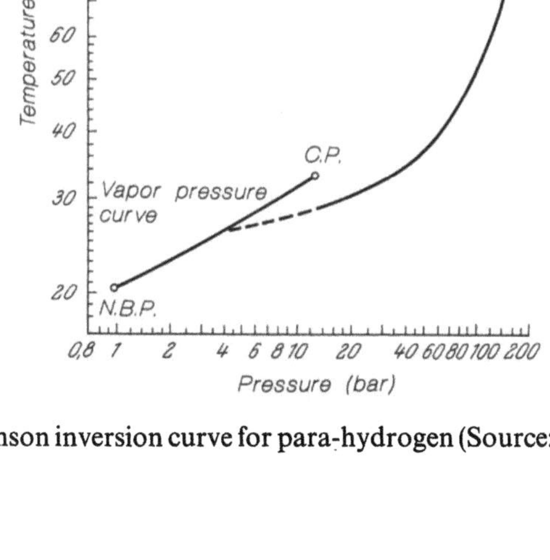

Fig. 134. Joule-Thomson inversion curve for para-hydrogen (Source: NBS-Monograph 168)

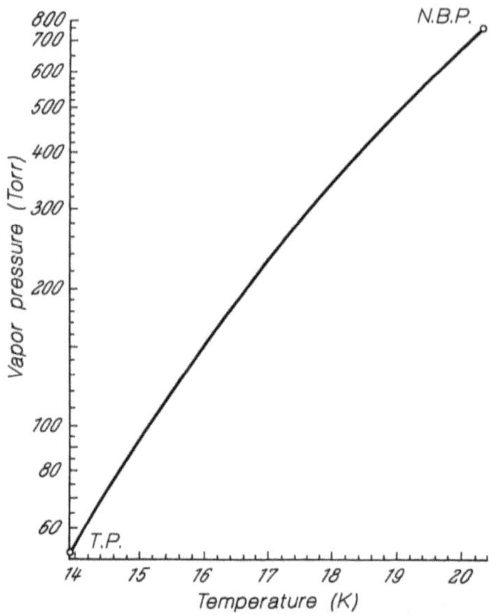

Fig. 135. Vapour pressure of para-hydrogen below 0.101325 MPa (Source: NBS-Monograph 168)

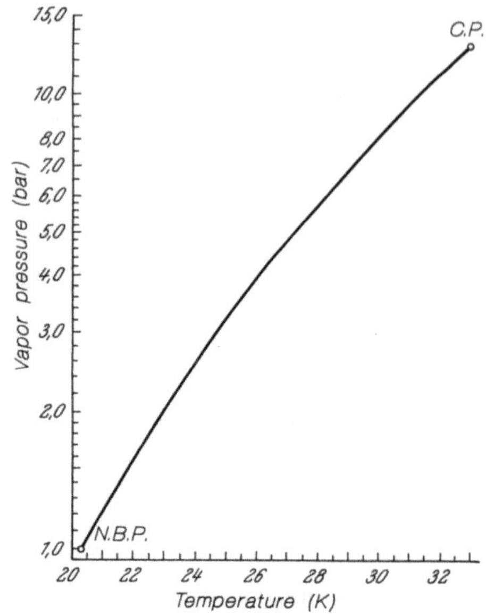

Fig. 136. Vapor pressure of para-hydrogen above 0.101325 MPa (Source: NBS-Monograph 168)

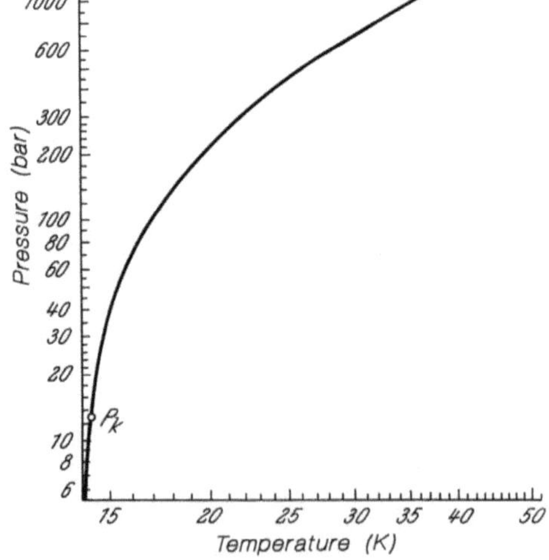

Fig. 137. Melting line for para-hydrogen. p_k critical point pressure (Source: NBS-Monograph 168)

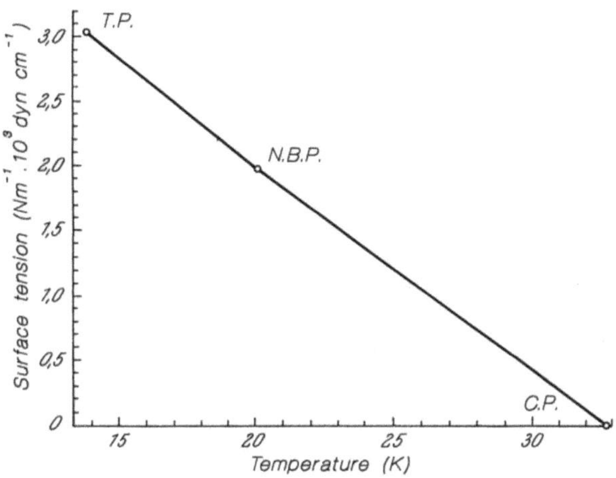

Fig. 138. Surface tension of liquid para-hydrogen (Source: NBS-Monograph 168)

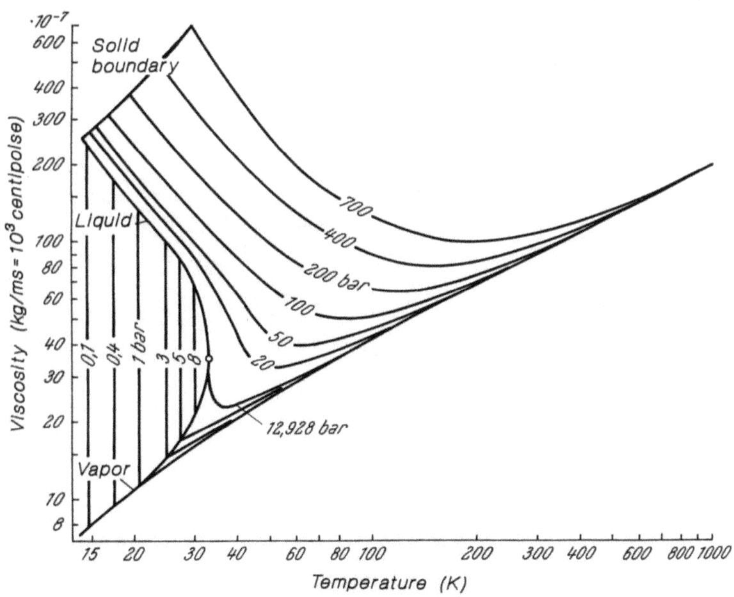

Fig. 139. Viscosity of normal and para-hydrogen (Source: NBS-Monograph 168, diagram simplified)

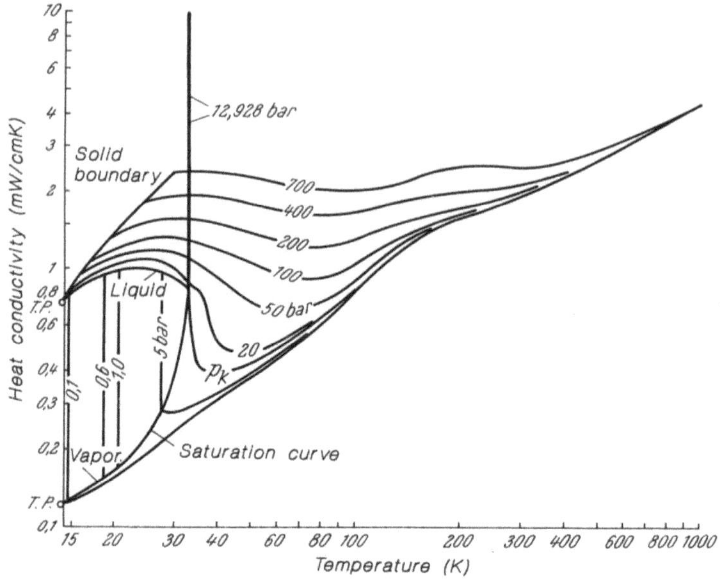

Fig. 140. Heat conductivity of para-hydrogen (Source: NBS-Monograph 168, diagram simplified)

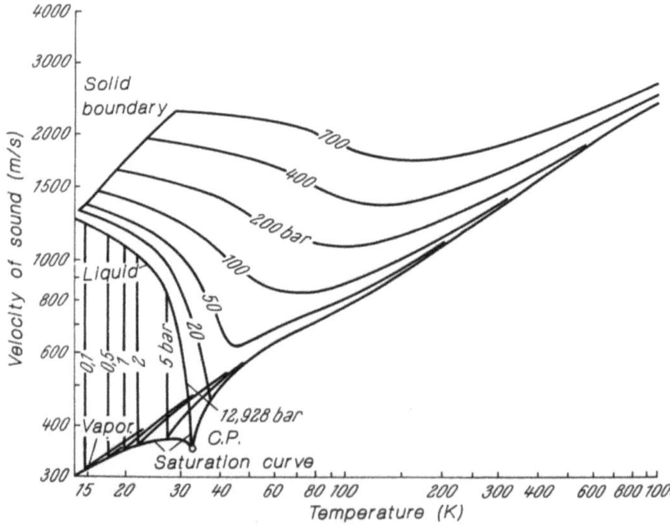

Fig. 141. Velocity of sound for para-hydrogen (Source: NBS-Monograph 168, diagram simplified)

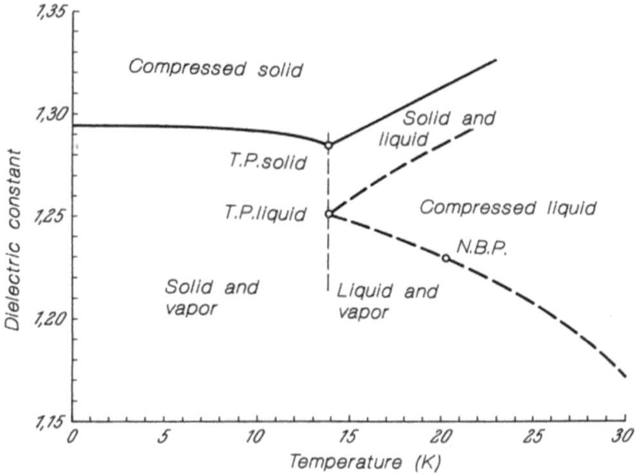

Fig. 142. Dielectric constant of para-hydrogen (Source: NBS-Monograph 168)

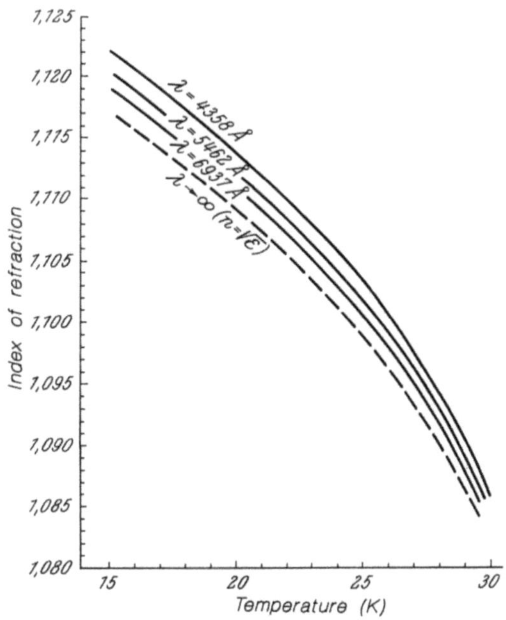

Fig. 143. Index of refraction of saturated liquid para-hydrogen (Source: NBS-monograph 168)

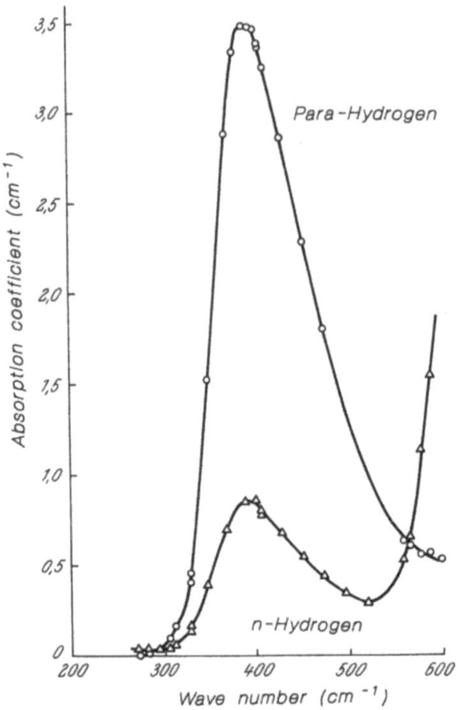

Fig. 144. Far infrared (rotational band) absorption in liquid hydrogen (Source: NBS-Monograph 168)

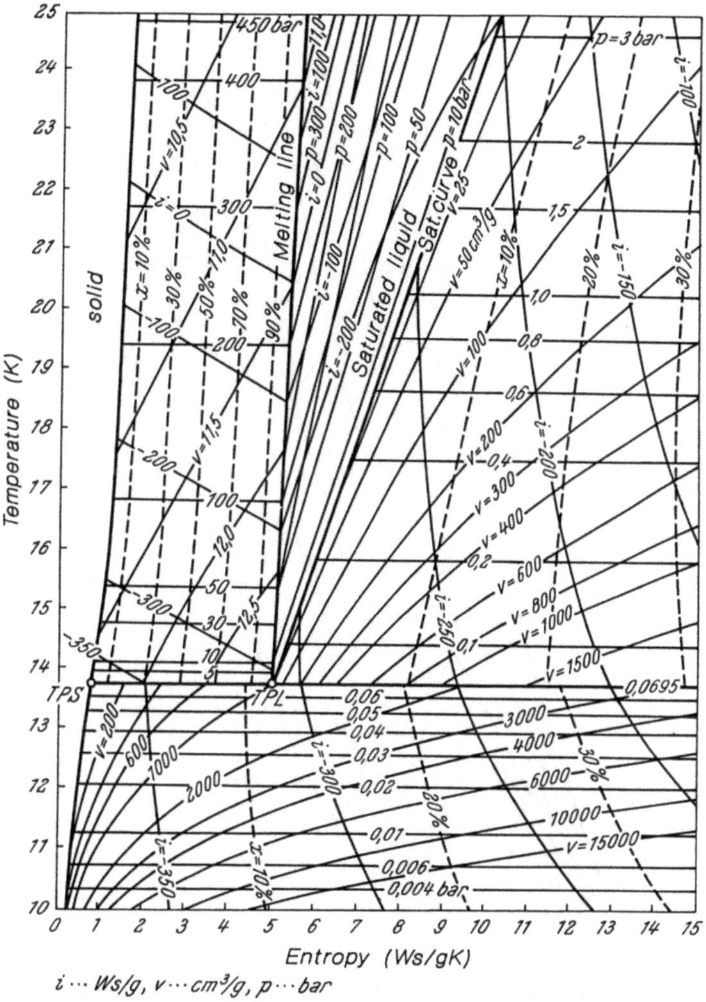

Fig. 145. Para-hydrogen temperature-entropy diagram around the triple point (*TPS*, *TPL*) pressure in bar, enthalpy in Ws/g, (Ws = Joule) specific volume in cm^3/g (Source: NBS-Monograph 168, diagram in extracts)

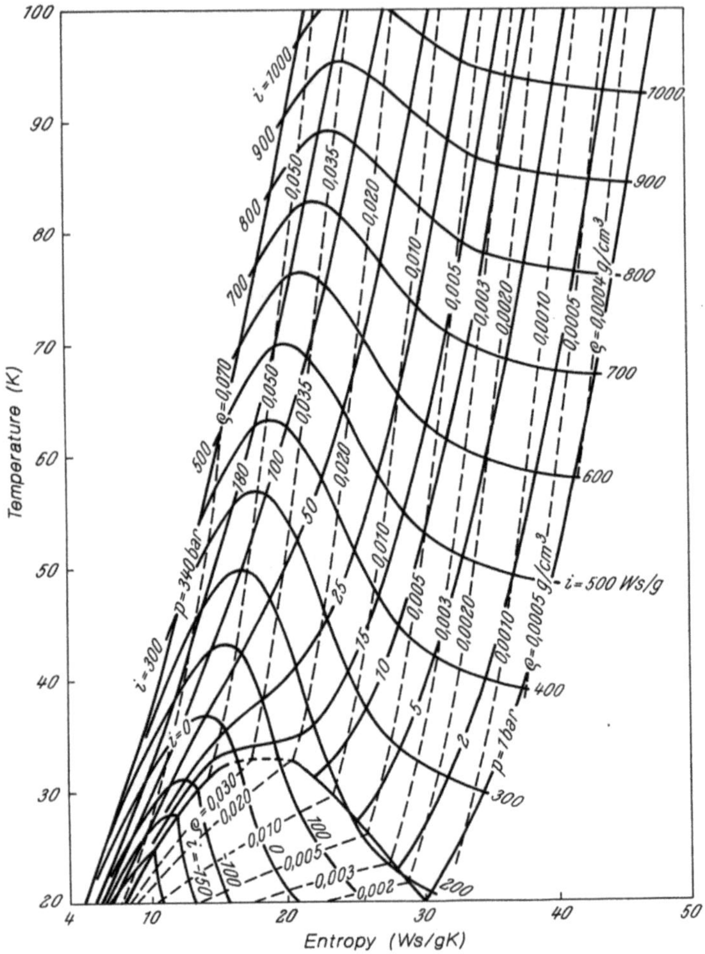

Fig. 146. Para-hydrogen temperature-entropy diagram below 100 K. Pressure in bar, density in g/cm³, enthalpy in Ws/g (Ws = Joule) (Source: NBS-Monograph 168, diagram in extracts)

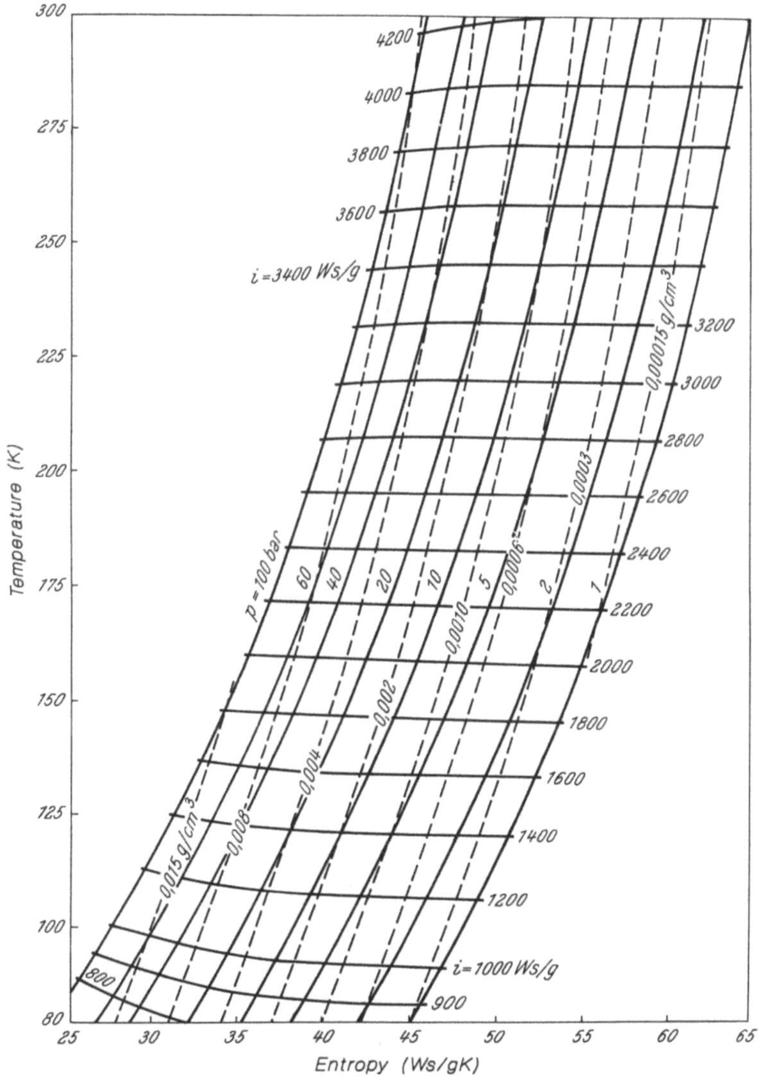

Fig. 147. Para-hydrogen temperature-entropy diagram between 100 K to 300 K. Pressure in bar, density in g/cm^3, enthalpy in Ws/g (Ws = Joule) (Source: NBS-Monograph 168, diagram in extracts)

Subject Index

MIX
Papier aus verantwortungsvollen Quellen
Paper from responsible sources
FSC® C105338

If you have any concerns about our products,
you can contact us on
ProductSafety@springernature.com

In case Publisher is established outside the EU,
the EU authorized representative is:
Springer Nature Customer Service Center GmbH
Europaplatz 3, 69115 Heidelberg, Germany

Printed by Libri Plureos GmbH
in Hamburg, Germany